OLIVE WHICHER
Projective Geometry

OLIVE WHICHER

PROJECTIVE GEOMETRY

CREATIVE POLARITIES
IN SPACE AND TIME

Sophia Books

Sophia Books
Rudolf Steiner Press
Hillside House, The Square
Forest Row, RH18 5ES

www.rudolfsteinerpress.com

Published by Sophia Books 2013
An imprint of Rudolf Steiner Press

First edition 1971; reprinted 1985

© Rudolf Steiner Press 1971

The moral right of the author has been asserted under the Copyright, Designs and Patents Act, 1988

All rights reserved. No part of this publication may be reproduced, stored in a retrieval system, or transmitted, in any form or by any means, electronic, mechanical, photocopying or otherwise, without the prior permission of the publishers

A catalogue record for this book is available from the British Library

ISBN 978 1 85584 379 0

Cover by Morgan Creative using drawings by Olive Whicher
Printed and bound in Malta by Gutenberg Press

Contents

	Foreword	9
I	*Concerning Changes in Classroom Mathematics* . .	13
II	*Geometry and the Development of Thinking* . . .	27
	From Moving Form to the Fixity of Measure .	27
	Experience with Measured Shapes	29
	The Question of the Infinite	38
	The Birth of Modern Geometries; Two Pathways	46
	"Projective Geometry is All Geometry" . .	49
III	*First Step in Projective Geometry; Movement* . .	52
	The Line-woven Net in Step-measure . . .	54
	The Infinitely Distant Point of a Line . . .	58
	The Infinitely Distant Line of a Plane . . .	62
	The Infinitely Distant Plane of Space . . .	64
	Real or Unreal?	70
	Point, Line and Plane	73
	Point, Line and Plane in Mutual Interplay . .	75
	The Axioms of Community of Point, Line and Plane	77
	Two Basic Theorems; Desargues and Pascal .	78
	Desargues, Two-triangle Theorem	79
	Pascal's Theorem	84
IV	*Further Discoveries; Duality and Projection* . . .	89
	The Principle of Duality	94
	Perspective and Projective Relationships . .	95
	The Projection of Three into Three . . .	99
	The Fundamental Theorem	102
	The Theorem of Pappos	103
	Projective Creation of Curves; the Rainbow .	108
	The Harmonic Forms	113
	Harmonic Fours and the Anharmonic Ratio .	115
	The Invariance of the Harmonic Property . .	119

	The Thirteen Configuration and the Diagonal Triangle	120
V	*Projective Laws of Curves*	124
	The Line-woven Net in Growth Measure . .	124
	Projective Concentric Circles	127
	The Theorem of Brianchon	127
	Curves Through Five Points and Touching Five Lines; Pascal and Brianchon	130
	The Theorem of Jacob Steiner	133
	The Tangent and its Point of Contact . . .	134
	Identity of Pointwise and Linewise Circle-curve	136
VI	*Projective Transformations; Collineations* . . .	140
	Projectivity and Involution within the Line .	140
	Projection with One Double Point	142
	Projection with Two Double Points . . .	142
	Involution	144
	Projection and Involution on a Circle-curve	146
	Projection without Fixed Double Points; Potentizing Process	150
	Breathing Involution	151
	Cyclic Projectivity	153
	Circling Involution	153
	Indications concerning the Imaginary Double Elements	155
	Plane Path-curves in Breathing and in Circling Involution; One-dimensional Transformations .	158
	Two-dimensional Projective Transformation of Curves; Homology and Elation	167
	Curve Families and the Harmonic Net . . .	177
	Spiral Matrix	177
	Projective Transformations in Space; Plastic Perspective	180
VII	*Polar Transformations on Circle-curves; Correlations*	185
	The Fundamental Polarity of Space	186
	Pole and Polar with respect to Circle-curves	188

	Self-polar Triangles; Polar Conjugate Pairs	193
	Conjugate Diameters	196
	Polar Reciprocation; Correlations	200

VIII *Polar Forms in Space* 218

    Pole and Polar with respect to the Sphere . . 221
    The Line-line Polarity of Space; Line Congruence. 233

IX *Geometry of the Twentieth Century* 238

    Rudolf Steiner's Indications concerning Space and Counterspace 239
    The Mathematical Concept of Space and Counterspace 243
    Physical and Ethereal Spaces 247
    Three Positive and Three Negative "Dimensions" 255
    Gravitational and Anti-gravitational Forces . 258
    Sun and Earth 262

    Notes and Bibliography 275

    Index 287

Die Tätigkeit des *Raums* und der *Zeit*
ist die Schöpfungskraft, und ihre Verhältnisse
sind die Angel der Welt.

NOVALIS

I saw Eternity the other night
Like a great ring of pure and endless light,
All calm, as it was bright;
And round beneath it, Time in hours, days, years,
Driven by the spheres
Like a vast shadow moved; in which the world
And all her train were hurled.

VAUGHAN

Foreword

The task attempted by this book is to bring to the non-mathematician fundamental truths of Projective Geometry in such a way that he may be inspired in thought and imagination to use the key they provide for the understanding of aspects of nature and of human experience to which scientific thought otherwise remains blind. Like the sleeping beauty surrounded by thorns, this most creative field of human endeavour has been difficult of access to any but the trained mathematician, and then only in the mode of thought of a beautiful, though remote and almost unapplied branch of pure mathematics.

The book is a result of collaboration since 1935 with the late George Adams (1) whose endeavour it was to bring the princess to life for those who may wish to see her face. He would, no doubt, have undertaken the present task differently. I have repeatedly referred to his writings, published and unpublished, and also to notes made in study with him, with the result that his own formulations occur very often throughout the book. I have used the terminology he strove to introduce to bring out the pictorial quality of the new geometrical concepts; many of the illustrations are also from his hand.

It is in the nature of my presentation that I have left aside all algebraic formulae, and in the course of writing I have come to realise that it also accords with my method not to overload the book with proofs. These are easily to be found elsewhere, and if the reader will take pencil and paper and work his way consistently from step to step, he will find that he is being led on a journey of exploration during which his own experience in making drawings will give him as much certainty as the mathematician gains from a formal proof.

The trained mathematician may well look for a more systematic treatment, and wonder at the variety of fields which have been included, only to be outlined, while others have been dealt with at greater length. He will however no doubt appreciate that in so doing I have attempted to bring about a measure of organic wholeness, to show the width and depth and the extremely developable quality of the subject, and above all to awaken the enthusiasm of scientist and artist alike.

I have provided the mathematician with references to lead him further in his study of the relevant literature. The general reader will find that though demands are made on his inner vitality of pictorial thinking, no obstacles of a technical nature will impede his progress.

The sequence of illustrations from great works of art runs like a *leitmotiv* through the whole, without many words of explanation. Such pictures have had their influence throughout the centuries; the truths which come to man through art will one day light up clearly in his thinking.

The notion of Polarity, in the subtle meaning of the term which this geometry gives to it, has come to birth surprisingly recently in the history of man's development. It penetrates further than the familiar concept of *opposites*, such as the extremes in the swing of a pendulum, where rhythmic processes arise through the to and fro movement of a mechanical system. We are concerned with the primary, earthly-celestial polarity, in whose rhythms life itself comes to expression.

The notion of polarity in this deeper sense is potent and it is modern; it underlies the spiritual-scientific teaching of Rudolf Steiner (2), in the context of whose work this book is set.

I am greatly indebted to Heidi Keller-von Asten (29) and to Walter Keller (Dornach, Switzerland) for their indispensable help and to Dr. Peter Gmeindl, mathematician of the Rudolf Steiner School, Munich, for his scholarly aid and for his support in the translation of the English text into German. Frau Keller's energetic enthusiasm enabled her to translate as the English pages were being written; to Walter Keller is due the final preparation for print of very many of the black-and-white drawings. Many drawings had already been prepared by George Adams in collaboration with the draughtsman Louis Loynes of London.

It is in the mood of a servant, who undertakes a task which

he knows he will be unable to finish to perfection, but of the urgency of which he is deeply convinced, that I venture to go to print. May the very imperfections which a reader may well meet with in my presentation, as well as the grandeur of the task itself, be an incentive to him to put his hand also to the plough.

Olive Whicher

Michaelmas 1970

Foreword to the Second Impression

Since this book was first published in English and almost simultaneously in German, I have brought out a number of related publications, notably a revised and enlarged edition of our book *Adams and Whicher: THE PLANT BETWEEN SUN AND EARTH and the Science of Physical and Ethereal Spaces,* which is also available in French and German. This book contains many illustrations, diagrams and coloured plates, with the aim of helping the student of organic morphology to reach beyond the mechanistic view of natural phenomena. An essential nature of mathematical thinking has achieved wonders in science today, so too, an equally essential and moreover all-embracing aspect of mathematics demonstrates the idea of the interweaving synthesis of polarities, which work formatively in all living processes. Science moves on today, towards the overcoming of one-sidedly atomistic and mechanistic concepts.

Olive Whicher

Midsummer 1985
Emerson College,
Forest Row, Sussex.

Reference is made to the *Notes and Bibliography* by the numbers in brackets throughout the text.

I Concerning Changes in Classroom Mathematics

In quite recent years a revolution has been taking place in classroom mathematics which is symptomatic of our time. The abstraction associated with the subject is widely challenged and methods of teaching of a practical and very realistic nature are being tried, with the use of all manner of visual aids. The aim is to encourage initiative and to achieve greater flexibility and critical activity of thought, while placing less accent on the early mastery of set rules. Among the general public also the significance of mathematics for our whole life and culture is becoming more recognised, even by the 'non-mathematician'. We are reminded that mathematics is the most powerful language there is. "Mathematics," said Galileo, "is the language in which God has written the universe."

It is hoped that the new curricula now being devised all over the world will help to overcome the widely prevalent fear and dislike of mathematics. The professional mathematician's love for the strictly logical method of Euclid, with its proofs, which has turned to boredom in the classroom, is being replaced by methods of learning about form and number in the geometrical shapes, through practical discovery and experiment. The approach is more through the will and in accordance with the nature of the young child.

This trend, in line with attempts in other fields of education, to do away with old-fashioned methods and bring more life into the classroom, is welcome. However, the deeper our understanding of the great language of mathematics itself, the easier it will be to allow it to speak creatively to young minds. The real task at hand is not merely that of finding methods more suited to the development of mathematical ability, with a view to its technical and utilitarian uses. There is a far more profound aspect to mathematics in education than this, as any true educator will agree. Just how far-reaching this is, and how deeply connected with the educational unrest of our times, is perhaps not yet widely recognised.

Not only has mathematics developed in the mind of man through the ages, but *the very kind and quality of the thought it embraces is, in turn, a potent force in the development of thinking itself*. In modern time, when old methods are being questioned and new ones sought, it is crucial to bear this fact in mind and to understand its implications clearly. What are the important aspects of mathematical training, over and above the utilitarian, and in what different ways do the various fields of the great landscape of mathematics have a formative influence on the development of thinking?

These are eminently practical questions, because in fact the evolution of thinking, its reflection in the history of mathematics and the development of the individual, are bound up in one great whole. It is the task of education not only to produce good scholars, but to see to it that the development of the individual takes its course in the right way. This touches a problem which is being considered today, not only by the teachers, but by the students themselves. Much of the heart-searching among teachers is born of the general anxiety caused by the growing number of dropouts and misfits which the present system produces. How can the strong and active will of the young child be prevented from becoming blunted and soured?

The place in education of art and practical activities has become undisputed since, among the pioneers of this century, Rudolf Steiner brought new ideas to bear on the problem. Rudolf Steiner's methods are, however, still in advance of other progressive systems in regard to *the way the actual subject-matter is used* in the classroom. The imparting of information, its mere intellectual assimilation, is subordinated to the far more important task of calling forth the unfoldment of the powers latent in the child. This comes about in the wise use of the subject matter at various stages of child development, as well as in the way it is taught. Rudolf Steiner showed in all detail how the great epochs of civilisation in the cultural life of mankind are reflected in the development of the individual, and how this can be an invaluable key to the right use of the various subjects taught (2). To teach the right thing at the right time can have astonishing results.

This key to the art of education should apply above all in teaching the great art of mathematics. (How often do we forget that mathematics is, in fact, an *art!*) The history of the development of mathematics itself, if rightly understood, can be a clear guide in forming a mathematics curriculum more suited to modern children and in tune with the spirit of our time.

It is the aim of the present work to show that an all-embracing field of modern mathematics, until now largely disregarded in education generally and therefore comparatively little known, has in fact an invaluable contribution to make, precisely in regard to present-day educational problems. We refer to what is called modern Projective Geometry.

Hidden away in the general textbooks and regarded as a rather obscure and abstract branch of higher mathematics, projective geometry waits for the appreciation it will receive in the future. To a technically-minded age, it has not presented many immediate aspects for practical application, and this, as well as its abstract guise, has kept it in the background for centuries. The important thing about it is the quality of its forms of thought, and this is modern. This quality of thought should begin to penetrate the teaching of geometry as a whole and play its required part in the development of the individual through the entire period of education. We do not of course mean that projective geometry as such should be taught in the lower school, but it should be possible for every teacher to develop his approach to teaching about form generally and then more specifically about geometry in the spirit of projective geometry, for this geometry embraces the whole of geometry. A new dimension in the study of form is opened up by a knowledge of projective geometry; it underlies the theory of metamorphosis in living forms and also in artistic creation.

Seeing that projective geometry has its roots in the sixteenth century and was fully developed as a discipline in the first third of the nineteenth century, one may well ask why it deserves to be called modern. We shall come to see that this is a true statement, for projective geometry is the outcome of a fundamental change in mathematical and geometrical thought transcending the classical conceptions of ancient Euclidean geometry. First steps in this change resulted in the discovery of the so-called "non-Euclidean" metrical geometries, which have had a profound effect on scientific conceptions of space and of physics. The discovery of the non-Euclidean geometries culminated however in projective geometry, itself not based at all on measure, but embracing the others, including the Euclidean. The resources of projective geometry reach into the far future of mankind and have hardly as yet been touched by modern culture.

The basic concepts of projective geometry are not difficult, but they are new and require a kind of thinking to which we are not yet accustomed, nurtured as we have been for so long on the one-sidedly Euclidean way of thinking and the analytical method. The new geometry requires a qualitative grasp of the mathematical forms; we learn to experience them less bodily, and to cultivate an activity of thought which compasses more than the momentary manifestation of a form in some particular shape, with certain fixed measurements. Thinking must rise beyond the mere contemplation of a finished form or pattern and reach to the experience of geometrical metamorphosis, where one form changes into another without losing its identity. In so doing, we proceed towards the understanding of creative processes which take place in *time*; we penetrate to something which lies behind the fixed form and see that it arises as the outcome of *relationships* between geometrical entities. Thinking thus apprehends the form long before it takes on a fixed shape in space in the quantitative field of measure. This is precisely the kind of step towards which people are groping today in many fields of enquiry and experience.

The kind of thinking which has been derived from and cultivated by classical and analytical geometry alone is no longer adequate. The change in the quality of thinking which is being sought will come about when the spirit of modern projective geometry permeates education, not merely intellectually, but in the far more profound way which belongs to the *art* of mathematics.

A significant comment on the influence of Euclidean geometry has been made by Struik, Professor of Mathematics in the Massachusetts Institute of Technology (3). Struik writes:

"The 'Elements' form, next to the Bible, probably the book most reproduced and studied in the history of the Western World. More than a thousand editions have appeared since the invention of printing and before that time manuscript copies dominated much of the teaching of geometry. Most of our school geometry is taken, often literally, from six of the thirteen books; and the Euclidean tradition still weighs heavily on our elementary instruction. For the professional mathematician these books have always had an inescapable fascination, and their logical structure has influenced scientific thinking perhaps more than any other text in the world."

Indeed, it is particularly in the English tradition that the actual books of Euclid have played such a prominent part in the classroom, yet their influence—possibly for precisely this reason—has been far-reaching. This does not mean that we need have remembered or even learned any geometry at school. The prevailing mathematical discipline has a profound influence on all spheres of life. The task of ancient geometry has been well fulfilled in the life of mankind, and there is still an important place for it in the education of the individual. But the time has come to go forward to wider perspectives, and to add new and freer ways of thinking geometrically to the old, more fixed and rigid thought-forms.

What, then, is projective geometry and what are its tasks in the life and development of mankind?

Born in the cradle of art in Renaissance times, projective geometry, while embracing the metrical (*geo*-metrical) field of all other geometries, transcends this field and enters the domain of movement and metamorphosis, where rigid measure is no longer the dominating factor. The mathematicians involved in its creation since the sixteenth century have found in it a mathematical discipline of great aesthetic beauty in its clarifying effect on thought as a whole. It became clear that here was a development of geometry which would free it from the metrical concepts characteristic of Euclid, and create a unifying science encompassing the whole field.

The problem was a practical one. The great artists, such

as Leonardo da Vinci, Albrecht Dürer and many others, needed to come to terms with the laws of perspective. Together with the mathematicians they struggled in thought with problems which might be expressed in the following way. Why is it impossible to draw the forms we see, reproducing exactly their metrical qualities? How do lengths, angles and other metrical properties of a form change when one passes from the experience of *touch,* as in architecture and sculpture, to the experience of pure *seeing,* as in drawing and painting?

Projective geometry really deserves another name, for its truths are concerned as much with movement and the interplay of light and darkness as with the measurement of the earth. They are expressed in the rhythmic flow of moving, changing forms in continuity, and in the dynamic of contrast which the study of all form reveals. The aspects of form with which this geometry acquaints us are of a higher order; once known, they are to be recognized in manifold phenomena of nature, offering fresh realms of research to the scientific mind also. A discipline of thought for the pure mathematician, this "geometry" may be experienced in the soul like the sequences of colour in a Turner landscape or the drama of light and darkness in a Rembrandt. This is indeed how it must be, even in the classroom, if the tasks of the new geometry are really to be fulfilled, as were those of the old. The thought-content of the new geometry leads beyond the confines of the sense-bound experience of space and of life, which dominates science today.

The Englishman, Arthur Cayley (1821–1895) whose statement that "projective geometry is all geometry" is now classical, succeeded in showing that the Euclidean notion of distance is in reality only a particular case of a much more general definition. Perspectives and projectivities require a more subtle concept of measure, and cannot be tied to the particular Euclidean case. Moreover, they embody mobility where the Euclidean form is fixed. Furthermore, in the discovery of what is called the Principle of Duality (or Polarity) an entirely new concept has entered in, the importance of which has hardly yet made itself felt, apart from its mathematical significance. It brings the fundamental concept which underlines the statement that in projective geometry *we are*

Jean-Victor Poncelet (1788–1867)

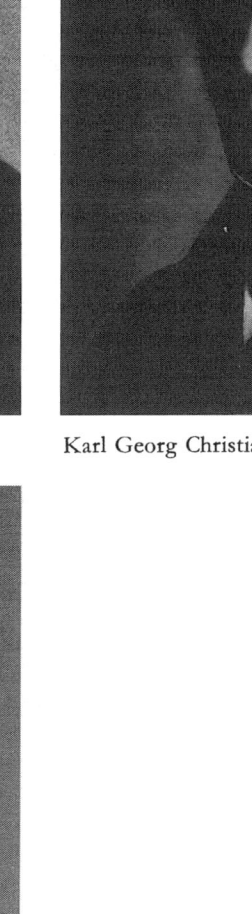
Karl Georg Christian von Staudt (1798–1867)

Arthur Cayley (1821–1895)

concerned with the whole, while the metrical geometries deal only with the *part*.

Cayley's algebraic methods were later translated by Felix Klein (1849–1925) into pure geometry, and the contributions of many other great mathematicians of the European countries in the nineteenth century, among them Poncelet, Chasles, von Staudt, Steiner, Reye, Cremona, resulted in the creation of the new geometry (12).

Rudolf Steiner repeatedly drew attention to the significance of the change from analytical to projective geometry and to the importance of mathematics as a foundation to all knowledge. In his autobiography he states clearly that mathematics formed the basis of his whole striving for knowledge. He recounts his experience when, as a child of nine years, he first discovered Euclidean geometry; how there began to dawn in him in a childlike way a conception which later grew to maturity; that the inner world exists as a sort of soul-space within, while the objects and occurrences which the senses perceive are in a space outside. Geometry seemed to him to be a knowledge which appears to be produced by man, but which, nevertheless, has a significance quite independent of him.

Then as a student, Rudolf Steiner recognised in mathematics a system of ideas entirely independent of the outer world of the senses, yet one with which it is necessary to approach that world in order to discover its laws. At this time, the concept of space caused him great difficulty. He could not come to terms with the picture of space going on and on into the void in all directions according to the prevailing theory. He describes what a momentous occasion it was for him when in a lecture on the new (synthetic or projective) geometry he learned that when a line is continued on into the infinite to the right, it returns again from the left. The infinitely distant point to the right is the same as the infinitely distant point to the left. Rudolf Steiner found this to be like a revelation in regard to the problem of space (4).

This experience of one destined to be a supreme thinker in modern time who has transcended the limits of materialism, has had, and continues to have, its echoes among lesser individuals in many a classroom. When analytical methods are kept in their place and the true spirit of projective geometry is allowed to unfold, a feeling of liberation invariably prevails and a sensing of new vistas in thought which are about to be opened up. Those who work in this field know that this is so.

The author of a classical textbook of projective geometry, J. L. S. Hatton, M.A., wrote as follows in his preface (5):

"The author trusts that this book may do something to encourage the student not to neglect the methods of pure geometry. In every other branch of mathematics analysis now reigns supreme—even in geometry it is fast gaining predominance. Twenty years' experience as a teacher of projective geometry and ten years' experience as an examiner of the University of London have led the author to regard this as a misfortune. When the great landmarks of projective geometry—the theorems of Pascal and Brianchon, of Carnot and Desargues, together with their immediate consequences—are clearly placed before the student, the author has found that even in the younger student an enthusiasm is aroused which is wanting in his study of other branches of mathematics. In the examination room it has been found that students who have mastered and absorbed the principles of pure geometry have taken a superior place to those who depend on a facility for handling analytical expressions. Such a facility with practice may undoubtedly be acquired by most pupils, but for all who are worthy to take a mathematical degree the study of pure geometry is a matter of primary importance."

Among those who took up Rudolf Steiner's indications concerning mathematics and have made significant contributions to the transformation of mathematics teaching is George Adams, M.A.(Cantab). His attention had already been drawn to this field of mathematics while at Cambridge, where he worked under the influence of such men as A. N. Whitehead, Bertrand Russell and the pure mathematician, G. H. Hardy. George Adams' researches in chemistry and thermodynamics convinced him that it was vitally necessary to overcome the prevailing tendency in science to think predominantly in terms of the *atom*—a result of the continued domination of the analytical method. It was Bertrand Russell who in this connection first directed George Adams to modern synthetic geometry.

While studying some of the earlier writings of Whitehead (12), George Adams became convinced that the profound ideas of the science of projective geometry—above all, the fundamental polar relations known as the "Principle of Duality"—were destined to have a transforming effect on the whole realm of natural science. Thus when he heard of Rudolf Steiner's indications, his own thoughts and hopes were already going in a like direction.

The tremendous impact which George Adams was then to receive in meeting Rudolf Steiner deflected him from the purely scientific career upon which he had been entering and caused him to devote the rest of his life to the manifold aspects of Rudolf Steiner's whole impulse. Nevertheless, his pioneer work in applying the thought-forms of the new geometry to problems in physics, hydrodynamics, botany and biology holds immense potentiality for science in the future. We are at the beginning of the long road towards the transmutation of the modern analytical methods which, admirably suited as they are to the study of the inorganic sciences, are insufficient where the understanding of living things is concerned.

It is a prime task in education to see to it that the analytical methods are complemented by thought-forms of quite another calibre, which will become a potent source of development and inspiration in scientific thinking.

There is a widespread urge today to explore fields of experience which are inaccessible to external sense-impressions. This is not new; it is only that in the last few hundred years it has been unscientific to regard such non-spatial supersensory realms as existent. But it is a fact of spiritual significance that we live in a scientific age, and it is no longer through mysticism, but by means of clear, activated thinking, that the break-through will come. We must call to our aid not visionary dreams but spiritual activity, a higher mode of thinking.

Rudolf Steiner pointed the way when he said (6):

"It was through synthetic geometry mainly that I brought myself to the point of consciousness concerning the process of clairvoyance. Naturally, this does not mean that someone

George Adams Kaufmann (1894–1963)

Louis Locher-Ernst (1906–1962)

who has studied projective geometry is clairvoyant, but that through it one can become clear about the process of spiritual perception. . . . He who approaches mathematics in the right spirit will find that it can be regarded as a model, a pattern, of the way in which supersensible perception may be achieved. For mathematics is simply a first stage of supersensible perception. . . ."

It is the task of the new geometry, embracing as it does the *whole* of geometry, to widen the field of man's access to knowledge. The scientific mind prides itself that it deals only with the known world, but it is in the nature of science that it explores the unknown. As the advancement of mathematics from the past to the future makes its impact felt on education, new forces will be born in the mind of man, and new ways of thinking which were not there before. This will enable the coming generations to work as creatively in a science of the living and the spiritual as has been done in the material realm. To cultivate in the classroom and beyond it vital and active qualities of thought open to receive quite new ideas is the task of modern geometry.

Cave painting, Lascaux

Experience of Form in Pre-historic Time: Movement without Perspective.

II Geometry and the Development of Thinking

From Moving Form to the Fixity of Measure

The earliest known attempts by man to express his experience of the world around him in picture form are the cave paintings of the paleolithic period. They are remarkable not only for their sensitivity of form, but for the way they express movement. In these paintings, the laws of space seem to be of little importance; form interpenetrates form, rather as in a dream-picture; it is a two-dimensional world. With all their beauty and delicacy of form, these paintings nevertheless often remind us of the early attempts of small children; there is no perspective in them and much action.

The origin of form is movement. In all organisms, the flowing processes of vital fluids are active long before the actual form appears. Even crystalline forms and the rocks of the earth are the end-result of active, turbulent forces. Man himself derives his form from the streaming embryonic movements, changing and mobile in the early stages of development. It is only gradually that the living form takes on a more permanent shape. In the last resort, fixity implies illness or death. The process of incarnation is, in fact, a gradual entry into a body which is destined to become more and more set within the three dimensions of the space of earth. Together with this process goes the development of consciousness.

It is man's destiny, in history as in each individual incarnation, to pass from a condition of unconscious creativity, through a dream-like phase, to a time when he meets with the realities of the material world, there to attain ego-consciousness. The early stages of development are sustained by exuberant forces of vitality, following which there is the true period of childhood, with its worlds of dream and fantasy; at last the powers of the intellect awaken gradually, and there is the possibility of attaining maturity. Three qualities which are basic to the being of man are here

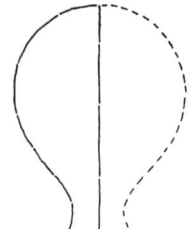

Figure 1

involved: will, feeling and thinking. The tiny child is predominantly a being of will; in the heyday of childhood, the soul lives especially in the light and darkness of many moods, in joy and sadness. With puberty there enters in the capacity of abstract thinking, the power by which man is distinguished from the rest of creation (7).

The small child's experience of form is predominantly through the will; he wants to handle the forms he sees, if possible to run around them. It is through such activity that he should first be introduced to the experience of forms. Gradually there comes then the joy of shaping forms with his hands; he models, paints and draws—freely and spontaneously, if permitted to do so—all manner of shapes in order to experience their *qualities,* straightness, roundness, the smooth and the jagged, the beautiful and the ugly.

Only later on, with the first geometry period at about the twelfth year, may an approach be made to drawings of plane figures, with recourse to instruments and actual measurements. Even then it is too early to impose upon the child abstract mathematical concepts.

Rudolf Steiner attached great importance to this gradual introduction to fixed forms in all manner of ways, for instance, in the way in which the letters of the alphabet should first be taught. Many and varied indications are to be found in his lectures to teachers (8). The practice of free form drawing previous to the actual geometry lessons and later on the simple and realistic ways of introducing projection and shadow-throwing are examples of this principle.

Free form drawing and symmetry exercises. The children are asked to draw straight lines and curves freehand. It is important that the line be really *drawn*, so that the child experiences the process in the whole arm and body—it must not be done automatically.

They are then asked to complete forms, one half of which have already been drawn on the blackboard or in their books. This is an exercise in symmetry. It is significant to note here that Rudolf Steiner also suggests the drawing of shapes from a centre outwards, giving the child the opportunity of drawing an answering shape peripherally, surrounding the first. The sketches in Figure 7 are Rudolf Steiner's own. As we shall see later on, the last two shapes are like a foreshadowing of one of the most fundamental aspects of projective geometry (see page 216).

Shadow drawing. Each child is given a candle and something with which to throw a shadow on a piece of white paper which lies on his desk. The child is encouraged to shade in the form of the shadow, and then on a second sheet, the part which is lit up, when the form of the shadow is revealed as a hollow or negative shape. Movement of the candle or of the shadow-thrower will result in changes of the forms created between the light and the darkness, and there is manifold possibility of experiencing directly how these changes come about.

Rudolf Steiner, in describing the value of such a way of approach to geometrical form, says that the activity in the symmetry exercises of completing a form already begun has a powerful moral effect on the child, while to experience and enjoy the forms, *without recourse to abstract thoughts about them,* has a healing power and brings forces of vitality to the being of the child. These forces are those which will be called upon at a much later date, when the time comes to think and learn about the forms and their laws. The memory of the forms then lies at a much deeper level, and the process of learning and memorising the mathematical facts will take place much more quickly and easily, at the time when the capacity for abstract thought really begins to dawn. But for this to happen, it is very important not to spoil the child's vital and direct experiences by giving intellectual explanations at a too early age. This applies, of course, to education as a whole and not only to the teaching of mathematics.

Experience with Measured Shapes

What follows here is by no means meant as a detailed guide to teachers, but rather as an indication of the *kind* of geometrical experience which work with certain basic constructions can give the child in the years of transition to the upper school.

When at last the first geometry period proper takes place during the twelfth year, the child is introduced to the geometrical laws which he has already experienced through the

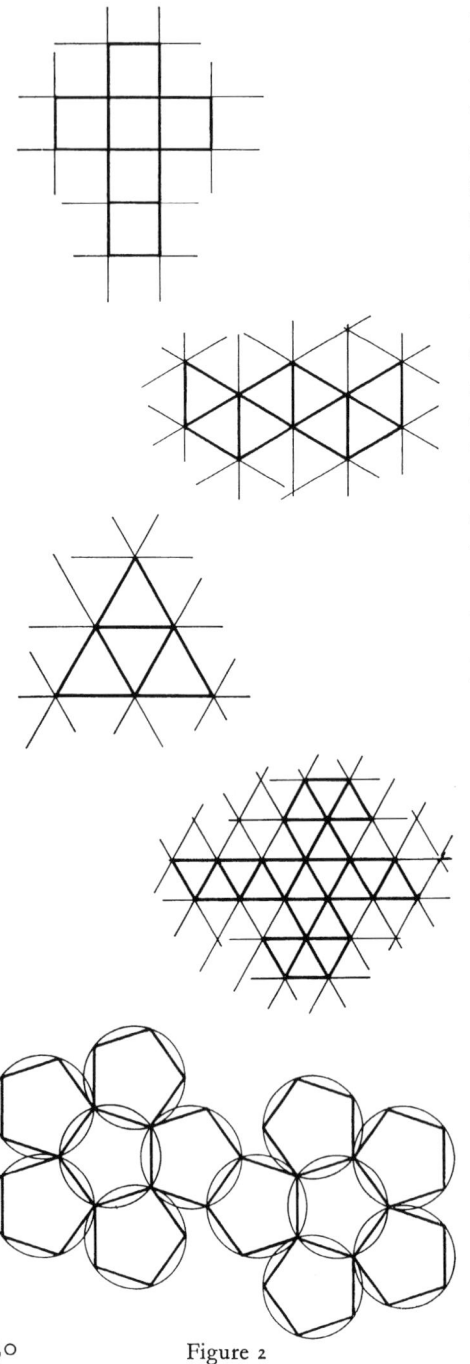

Figure 2

will both in his own body and in the world around him. He begins to learn the laws of the geometrical forms and their measurements and to make more accurate drawings. Working painstakingly, the child now begins to use the dawning light of his intelligence, but still without going into too much abstraction. In careful drawings of all kinds, with the emphasis laid on drawing, the basic geometrical concepts are approached, and it is possible to cover the first propositions and theorems of Euclidean geometry in this way, even to the Theorem of Pythagoras. (9) There is as yet no reason to attempt to prove intellectually what can be experienced intuitively first of all. When a direct experience has been achieved through the drawing and model-making, the time will come for the theoretical proof. Thus, in the gradual transition from the lower to the upper school, constructions of increasing difficulty and precision may be carried out, in the first place in two dimensions (9).

An important exercise at this time can be the making of paper models of the five so-called Platonic Forms (Figure 2). Although the first approach to these forms must be from the aspect of their finite and complete shapes, taking into account the wonderful number relationships they contain, including, in the case of Icosahedron and Pentagon Dodecahedron the Golden Number, the teacher who is conversant with projective geometry will leave doors open in thought, even unanswered questions, to which the child may return with enthusiasm at a later date (see Chapter VIII).

Quite early in the upper school the introduction of the curves which go by the name of the conic sections (circle, ellipse, parabola and hyperbola) foreshadows projective geometry. These curves are taught first—even as they were discovered—from the aspect of points and measurement. In considering them the child is presented, as humanity was before him, with the puzzle and paradox of the infinite. Here too, the teacher acquainted with projective geometry will know better when and how to deal with this question, and will also realise the importance of drawing the curves, not only from point to point, but enclosed in their envelopes (the constructions of which are made particularly easy by the use of a set-square with a right-angle constructed on the hypotenuse). The envelope or linewise curve is fundamental

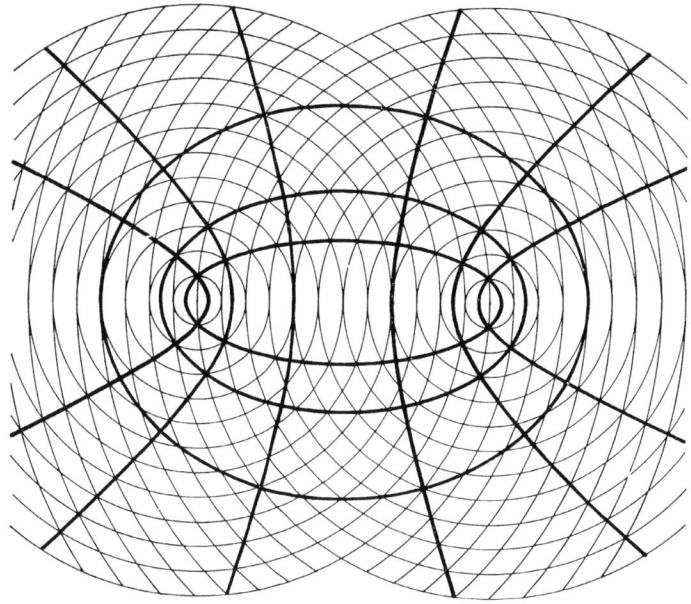

Figure 3

to projective geometry and brings an experience which complements the one-sided experience of the curves given by the pointwise construction.

The illustration included here in Figure 3 is an indication of the kind of construction which may be carried out which is easy of access to children. Numerous constructions of the kind will introduce them to the beauty and wonder of the basic geometrical shapes in a less one-sided way than is so often used, so that the experience gained from them may be drawn upon later in a systematic and theoretical approach.

A knowledge of projective geometry awakens the teacher to many opportunities to use his fantasy along lines originally indicated by Rudolf Steiner. Constructions as, for instance, the Harmonic Net (p. 55) could quite well be carried out simply as a fascinating and thought-provoking drawing, the full explanation of which will then come easily in the upper school.

Another example is the drawing of spirals. Patterns of regular concentric hexagons or squares or other shapes form a matrix upon which spirals may be drawn (Figure 4).

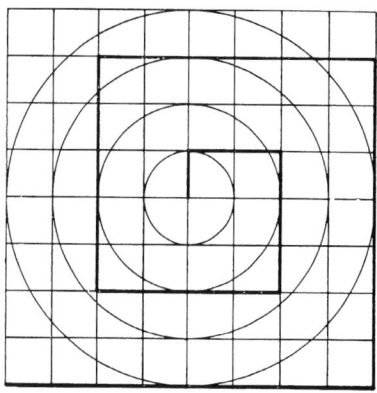

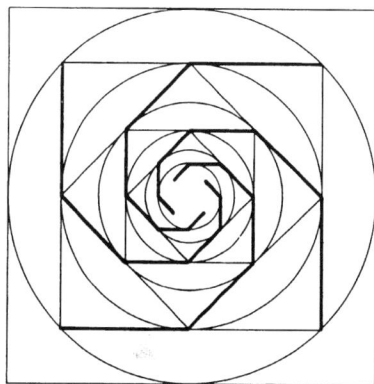

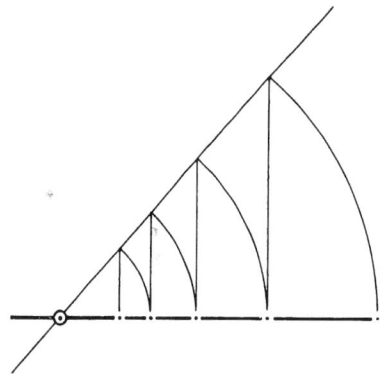

Figure 4

Figure 5

Regular concentric forms, for example, squares, hexagons, will give the kind of measure in which the equi-angular spiral may be set, both as a sequence of points, by taking the *corners* of the forms into account, and also as an envelope, the *sides* of the forms giving the direction of the spiral at each point. Having made the matrix, it is a good exercise to trace spirals on a clean sheet, seeing the difference in their shapes, according to the mutual relationship of the radial and the circular components. It will be clear that a matrix of concentric squares, triangles, hexagons, etc. will give a different progression of points and lines outward and inward, each with its particular constant proportional measure. The equi-angular or logarithmic spiral rests upon such a matrix.

Taking the outward steps between the circles in *equal* measure will result in the type of spiral which begins at the central point, the spiral of Archimedes, as it is called, the shape of which is very different from the equi-angular spiral, which never reaches the central point. It is good to experience the difference between these two types of spiral.

A measure underlying the growth progression of an equiangular spiral may also be created by drawing two lines at an angle and working alternately with arcs of circles and parallels (or only with parallels). The actual measure resulting from this construction will depend on the size of the angle in relation to the first step chosen (Figure 4).

An exercise of great value, which gives an artistic approach to something to be learned about later on, namely the degeneration of curves (30) into straight lines or other fixed Euclidean patterns, might be as follows: Starting from any given pattern of lines—say a triangle in the projective sense, or some interpenetration including circles, for example—the child learns to fill in the empty spaces with a freely drawn rhythm of curves. He learns to experience them as though they die away towards the lines like wavelets moving up the sand towards the rocks. In so doing he may have a true artistic experience of the mathematical fact that such curves keep away from the crossing points in the pattern, but tend to flatten towards the lines (Figure 5).

Such an approach to form and then to the elements of geometry is in accordance with the development of geometrical knowledge from ancient times. From the fifth millenium B.C. onwards, in the more advanced forms of society of the Ancient Orient, mathematics was an essentially practical science. Mensuration and the arithmetic of the tally stick developed into the beginning of theoretical geometry and algebra; rules of conduct were given for building and for computation, and methods were laid down for computing specific astronomical events. It is interesting to note that although mathematics began at a comparatively early time to be studied for its own sake, there was at first no attempt at logical proof. This came later, when men began to ask the question: Why?

Modern historians are puzzled to explain how the Egyptians and Babylonians knew so much mathematics and how they actually found their theorems. It is known, for instance, that they knew of the theorem called after Pythagoras, the proof of which only came later. This is not difficult to understand when one remembers that oriental mathematics was closely connected with religion. The priest-kings were the guardians and bearers of all knowledge, both religious and secular. They taught the necessary rules for practical application; the rules worked, and the ordinary man was not concerned to ask for a reason or demonstrate a proof.

In those ancient times, guided by the centres of mystery teaching, men were being directed towards the practical aspects of life, and their immediate experience of the world of the Gods was clouded more and more. The Initiate taught out of divine inspiration, of which he alone had direct experience; but he also was alone in knowing the laws of the material world in the manner in which we know them today. The ordinary man lived in a more dream-like state of consciousness, and understanding was received by him as a gift. His link with spiritual worlds was sustained for him by the mystery teachings, but through them, too, his face was set more and more towards the dark domain of earth. He was gradually to find himself alone and stark in a physical body which no longer allowed him access to the dimly remembered worlds of light. How powerfully must the weight and rigid immobility of the great Temple figures and the architecture have worked upon his soul!

This was the destiny of mankind; it is the destiny also of

Egyptian, about 2500 B.C.

Heavy Forms: Principle of the Right Angle

Egyptian, about 1050 B.C.

Parallel lines: Recollection of Infinity

every child. He must lose the wonderful dreams of childhood, turn more and more towards the tasks of earth; guided at first by parents and teachers, he must go on towards the day of his independence. The time will come when he must find his way alone—when he is able to think for himself. This should, however, be a very gradual process for every child. The divine powers guiding mankind in those olden times were wise—far wiser than we men are today. The wrong which is done to a child by exposing him at a too early age to a merely intellectual education is however becoming abundantly evident today; there is a growing awareness of the fact that the premature introduction of too much abstraction only stunts the child's natural and vital interest in learning.

With the decline of Egypt and Babylon and the rise of Greek culture a great change took place for mankind. The spirit of enquiry grew in men's minds, and the desire to understand the place of man in the universe according to rational principles. Early Greek mathematics, from the time of Thales, who is said to have travelled, like Pythagoras, to the eastern centres of learning, initiated the foundations of modern science upon mathematical reasoning. What had hitherto been accepted as a practical mathematical fact was now approached in the spirit of understanding rather than of utility, and mathematics became the foundation of the logical and exact thinking of Plato and Aristotle. The Greeks, with their fine feeling of form, cultivated mathematics more in the spirit of geometry than algebra, and their work culminated in the setting down of all that was then known of geometrical facts. Euclid wove elementary plane and solid geometry into the close deductive system that was to become the standard material in education and the basis for scientific thought for over 2,000 years. His "Elements" have had an unimaginable influence upon the thinking of mankind, reaching even right into the sphere of social life.

Euclid's theorems have led the thought of man into the material realm and have taught him to look at the world from his own point of view. This type of geometrical thought is clearly reflected in the forms assumed in social relationships. The "Euclidean" circle, formed from its centre, the point of a compass, is like a picture; each individual man is alone in his own centre and the rest of the world is around him, outside.

Greek, end of the sixth century B.C.

Step Measure

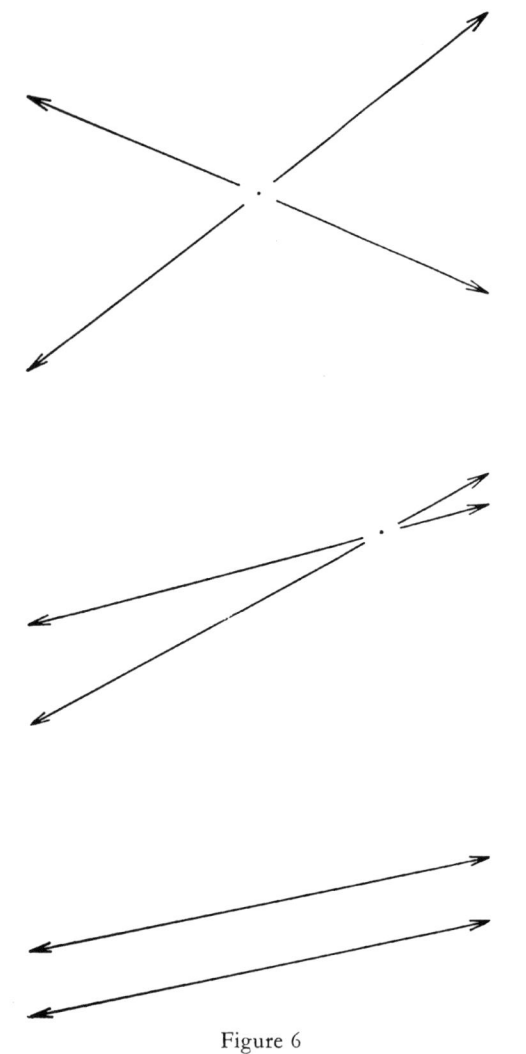

A king on his throne, his subjects at his feet, a centralised government or a dictatorship are social forms of the past. Today men seek for associative relationships in which all may play their part, mutually sustaining one another and interwoven, like the tangents of another kind of circle (see page 188).

The Question of the Infinite

Euclidean geometry deals only with forms in finite, i.e. measurable space. Euclid rounded off his system so effectively with his postulate concerning parallel lines that it has stood, like a great bulwark against any thought of infinitely distant elements as they occur in modern geometry. Well may it be that men's minds have thus been channelled overwhelmingly towards the facts of the earth and the tasks arising from it. Although the puzzle of the infinite in mathematics occupied the mathematicians and philosophers through the centuries, the kind of thoughts about it which stem from ancient Greece could give no answer to the problem.

The Greeks took cognisance of the problem of the infinite in a variety of ways, but they did not solve it. Stimulated by their awareness of the theory of proportions and of irrational numbers, and by what were called the paradoxes of Zeno, which were preserved by Aristotle, they were led towards the idea of infinite divisibility, but they abhorred the idea of an infinite without (10). Their conception of measure led them to define a line as the shortest distance between two points, a finite length; area as a part of a plane bounded by lines, and volume as a limited space bounded by planes, like a box or a room. Their thought led them to the notion of the "geometrical atom" (the point of no dimensions) and they tried to conceive of a line, area or volume as being built up of a large but finite number of these indivisible "atoms".

Thus space for the Greeks was finite, like a vast container, capable of holding a large but nevertheless limited number of physical atoms. They imagined the earth to be a finite plane;

Figure 6

Aristotle denied the existence of the actually infinite. "Infinitum actu non datur."

The discovery of the five regular solids and of the conic sections took place in Greek times. We have given an indication as to the way in which the Platonic forms may be introduced to the child at first. The following indications concerning the conic sections may serve as a guide to the way these curves may be taught at this time; they are approached purely from the metrical point of view, as distinct from the way we shall experience them in projective geometry. The Greeks knew the curves as loci of a point. The problem of the tangent, at which they worked—and with it the question of the envelope—was only really understood centuries later. In the conic sections as well as in the Platonic solids, the paradoxes involving the question of the infinite are unavoidable. Here a fundamental question arises to which a satisfactory answer can only be given later on.

Rudolf Steiner introduced the study of the conic sections from their metrical aspect around the fourteenth or fifteenth year. The curves arise when a double cone is cut by a plane. The cone opens out in two directions from its apex and the sections made by a plane which cuts it at various angles will give a curve. This curve will be a circle, ellipse, parabola or hyperbola, according to the orientation of the plane with respect to the cone (Figure 7). In this way, different variations of the curves will arise, the metrical aspects of which may then be dealt with in detail, by plotting the points of the curve in a plane.

Figure 7

Previously, in Figure 3, the ellipse and the hyperbola are shown in relation to two fixed points; in the case of the ellipse the *sum of the distances to either point of any point on the curve* is a constant number, while in the case of the hyperbola it is the *difference* of these two distances which remains constant. The use of concentric circles with a repeated measure of growth which is the same in both families gives a simple way of illustrating these facts. In Figure 8 we have taken a family of circles related to one another in a proportional growth measure, together with a set of parallel lines in the same measure. From the illustration it is easy to see that any point on a curve is a certain distance from the centre of the circles and a certain distance (measured perpendicu-

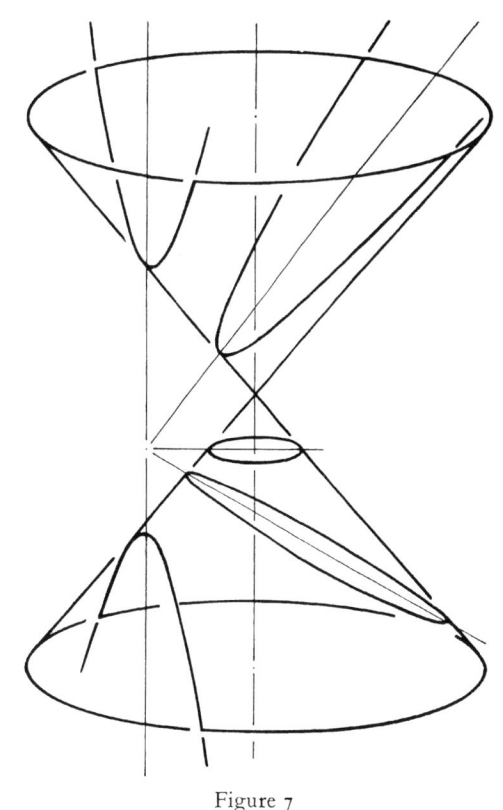

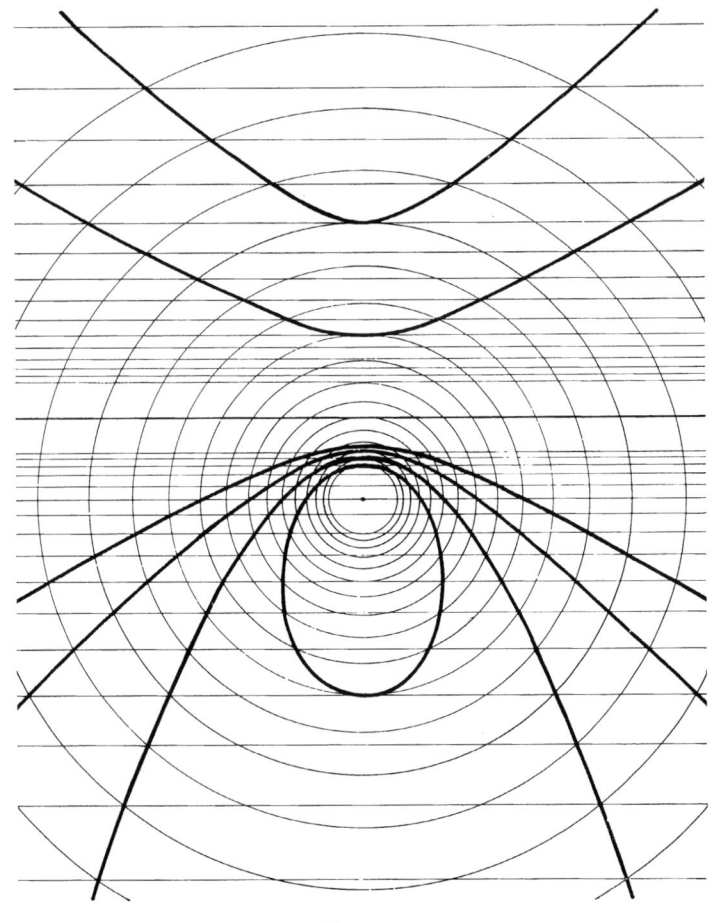

Figure 8

larly) from the "central" line of the parallel lines. These two distances related *proportionally* to one another will determine whether a point belongs to ellipse, parabola or hyperbola. If the proportion expressed by these two distances is a number which is less than one (that is to say, if the point is nearer to the centre of the circles than to the line) the curve will be an ellipse; if greater than one a hyperbola, if equal to one a parabola. The constant line is called the directrix and the constant point the focus. Inherent in all this are the questions involving distance (finite or infinite?) and also the way the curves change into one another.

Greek mathematics, and with it Greek architecture and sculpture reflects the past, while at the same time containing all the seeds of the future. The beautiful forms are the expression of a wonderful balance; well set and fashioned according to the laws of earth measurement, they seem, however, to be poised between earth and heaven.

So, too, the healthy child between the years of nine and twelve. He is well poised in his body and rhythmical in his movements. Behind him is the time when he has been led down with kindness and understanding to his present situation on the earth; before him the world begins to open its doors more and more, beckoning him on towards independent investigation.

Gradually he begins to question authority; his young mind seeks to answer his own questions independently, he begins to need the certitude of objective proof, and on entry into the upper school the child stands at the threshold of life. By this time he should have had a kind of introduction to almost all the fields which will now be his to enter into with his thinking. Now is the time to draw upon the rich wealth of experience which should have been acquired and absorbed in all the earlier years of artistic approach to knowledge, and to deal with it with clear, abstract thinking. By now the young individual's own sword of thinking should have been well forged in metal which, first molten, has become keen and strong. The tool must now be his, upon which he may rely, so that he may acquire clear understanding and therewith real love for knowledge. If this is so, he will enter with zest into the requirements of the upper classes, for example, into the geometrical theorems and their proofs, from which he will derive a wonderful feeling of assurance.

To continue the historical comparison: the decline of the School of Alexandria and the antique culture, and the sudden growth of Islam, caused the development of mathematics to cease practically for some centuries. It was preserved and cultivated in India and Mesopotamia without anything of importance being added creatively. When it returned to Europe in the Middle Ages, it had taken on a strongly Arabic form, a fact which has been of deep

Spanish, thirteenth century

The Cross: Infinity has come to rest in the Centre

Hans von Tübingen (about 1400–1462)

Three-dimensional Space; above it hovers the Cosmic Plane

significance for the development of scientific thinking ever since. To the physical quality of Greek thought was added the abstraction of the Arabic; mathematics was cultivated especially in algebraic form and was destined to become more and more abstract.

Already in the Middle Ages, the artists concerned themselves with the problem of space and perspective drawing. Is it possible to draw just exactly what the eye sees in space? Will the shepherd who climbs to the top of the mountain reach the sky, the end of the universe? What "space" will he reach if he pushes his head through?

The scholastic philosophers in Europe were those who continued the study of mathematics and the question of the continuum and of infinity, and it was especially St. Thomas Aquinas who accepted Aristotle's "infinitum actu non datur", but who considered the continuum as potentially divisible ad infinitum. "Ex indivisilibus non potest compari aliquod continuum" (A continuum cannot consist of indivisibles). This manner of thinking, writes Struik in his History of Mathematics, was not without influence on the inventors of the infinitesimal calculus in the seventeenth century. When, around the turn of the fifteenth century, with the dawning of the Renaissance, the West awakened to its cultural tasks and the development of mathematics was resumed in earnest, the work of the University of Bologna saw the beginning of modern mathematics.

Rudolf Steiner refers to the fifteenth century as the time when man was most deeply of all immersed in the physical world, when his mental conceptions were drawn entirely from this world. At that time, man's consciousness, far from being dreamlike, had grown clear in respect to his grasp of material entities and processes. The more active his intellectual forces became, the more the world of spiritual Imaginations became unreal, a world of fiction. It is the time of the great explorations, of discoveries and inventions in the physical world; the spiritual world has become remote. Religion is for the individual a matter for dispute, religious conviction resting on faith. Man experiences himself alone, and only if he himself wills it so, is he a seeker after truth.

The young individual, during and after puberty, is preoccupied on the one hand with the outer aspects of life and his growing desire for complete independence; on the other, he is idealistic, although he may try to hide this and, especially today, in his search for inner spiritual values, he may experience deep uncertainty. It is now that he will begin to use and be thankful for an active power of thinking which is not merely superficially intellectual, but which allows him to begin to approach life realistically yet with a deep moral force. He must find his own way, but he must have been enabled to acquire the forces with which to do so.

The Birth of Modern Geometries; Two Pathways

Whereas in ancient times the wisdom of mathematics was given to man as a teaching from the mystery centres of learning, now, at the dawn of modern time, it is characteristic that men were activated in their search for knowledge by an individual fire of personal inquiry. One only need read the lives of such men as Leonardo da Vinci, Kepler, Copernicus and many others, to see with what tremendous intensity they exerted their own individual forces of thought towards discovery and perfection in their various fields of activity.

Resulting from the work in mathematics and all that developed from it, two pathways were opened up by man in his search for knowledge. One pathway has been well trodden, with tremendous practical effect for science; the other has been very well laid down but far less often trodden. It is this second pathway which we shall be taking in this book. Even today the realisation has not generally dawned that the two paths are really one. What appears to be the first path, and to some people the only one, is actually incomplete without the other. This is a fact of fundamental importance both scientific and philosophic; it lies at the root of all Rudolf Steiner's work and is of great significance for education.

At the entrance to these two pathways in mathematics may be written *analysis* on the one hand, *synthesis* on the other. With this the situation is, in fact, neatly described, for it would be a truism now to say that the synthesis, which is the whole, contains the part, which is what is reached by analysis. Synthesis is a true description of the part played by projective geometry among all the new geometries. For, as already quoted, "Projective Geometry is all Geometry".

It was the work of two friends, both Frenchmen, which gave the leading impulse—one in one direction and one in the other—for the further development of mathematics: René

Descartes (1596–1650) and Girard Desargues (1593–1662). Descartes was a philosopher, living in Paris, Desargues an architect of Lyons. It is interesting that the great Descartes thought so highly of his practical friend's opinion that, in regard to his own work, he looked first to him for criticism. Concerning his "*Méditations Métaphysiques*", Descartes said of Desargues: "Je me fie plus en lui qu'en trois théologiens"—a saying which reflects the situation of science at the time.

Of the work of these two friends, Descartes' was destined to be taken up immediately. Analytical geometry, with the calculations based on the cartesian system proved of great practical value in the application of mathematics to physics. The works of Desargues seemed to be so much less important, and in fact they were set aside and lost for two hundred years. Even Descartes himself did not perceive the significance of Desargues' work, so powerful was the force which was leading men's minds towards technical achievements and atomic conceptions.

The steps in geometry which were taken at this time went beyond Euclid, opening the way to the so-called non-Euclidean geometries and then to projective geometry. At the same time, mathematics took on a more and more abstract form, for it was Descartes' monumental contribution to mathematics, that he showed the way to the competent union of algebra and geometry. It rested with the individual mathematicians and the qualities and propensities of thought characteristic of each one, whether mathematics was pursued in the guise of algebra or in real geometrical construction or pictorial imagination. Projective geometry itself is often cultivated analytically; analysis as we have seen carried the day. It is usually easier to prove a theorem by analytical methods and the algebraic proof apparently inspires the mathematician with greater confidence than a visual proof derived from diagrams.

It is symptomatic that the actual break with the Euclidean tradition was made more or less simultaneously by three famous mathematicians of different European countries. The new geometry arose in the minds of men all over Europe, and quite often without actual collaboration. The three mathematicians who first developed a geometry in which Euclid's parallel axiom was denied were Gauss

(1777–1855) of Germany, the Hungarian Bolyai (1802–1860) and the Russian Lobaschevski (1793–1856). All three worked independently of one another. It was a first breakthrough, and it inspired the development of what is called hyperbolic non-Euclidean geometry. Following this, Bernhard Riemann (1826–1866) and still later Felix Klein (1849–1925), both of Germany, developed aspects of elliptic non-Euclidean geometry (11).

It would exceed the task of this book to name the many mathematicians and philosophers concerned with these problems or to touch on their work, from the time of Gauss until the appearance in 1910 of the *Principia Mathematica* by Bertrand Russell and A. N. Whitehead (12). Suffice it to say that despite Euclid's great genius and the part his work has played, and still does play, his geometry does not satisfy the present-day requirements for logical rigour. His system became too narrow, above all due to the axiom of parallelism. To quote the English historian Heath: "When we consider the countless successive attempts made through more than twenty centuries to prove the Postulate, many of them by geometers of ability, we cannot but admire the genius of the man who concluded that such a hypothesis, which he found necessary to the validity of his whole system of geometry, was really indemonstrable." (13)

As a result of the long and intense mathematical research for a set of axioms which are independent, complete and consistent, it was discovered that non-Euclidean geometries may equally well be created. Euclid's geometry is one among a number of geometries, each of which has its own set of axioms.

It was as though, in Euclid's closed system, which was designed to bring systematic thought to bear upon finite forms—one might say, upon the physical forms of the earth—the one possible opening towards the infinite, or the cosmic, was destined to remain closed. The question might well be asked; was not this necessarily barricaded by the setting up of the postulate, in order that the consciousness of humanity might for a time be directed mainly towards the tasks concerning earthly life?

At the beginning of the modern era, however, the mathematicians at last brought the powerful light of a newly acquired clarity of thought to bear upon this momentous question, and consciously attained to the thought of the infinitudes of cosmic space. Since the seventeenth century, the mathematicians began to speak of the "infinitely distant" elements of Euclidean space.

"Projective Geometry is All Geometry"—Cayley

Thus the Euclidean notion of distance was widened. It was Kepler (1571–1630), in his work on the conic sections, who first felt his way towards the idea of an infinitely distant point. The first to bring the infinitely distant elements into geometry in a conscious and explicit way was Desargues, followed by his famous pupil, Blaise Pascal (1623–1662). Yet it was not until in the late nineteenth century that Arthur Cayley (1821–1895) was able to show the much more general definition of distance of which Euclid's is a particular case, and to prove that the last stage in the further development of the Euclidean and the non-Euclidean geometries—namely projective geometry—plays a paramount role in all geometry.

Euclid's geometry still holds a central place for all that concerns measurement on the earth. It is however recognised today that it is only a first approximation to the geometry of the universe as a whole. Euclid is concerned almost exclusively with metric concepts, such as lengths, angles, areas and volumes. Non-metric projective geometry is free of all these. The inclusion of the concept of the infinitely distant entities—point, line and plane—opened up the way to the creation of a geometry based solely on the *inter-relationships* of the three basic geometrical entities, without recourse to measure at all. *Projective geometry is not merely a geometry of created forms, but the geometry of the relationships between form-creating entities.*

Thus, the culmination of the development of the non-Euclidean geometries was the creation of projective geometry, fathered, one might say, by the great Frenchman Poncelet (1788–1867), tended by men of almost every European nationality, and brought to its maturity by such men as Christian von Staudt (1798–1867), Cayley and also Felix

Klein, who translated Cayley's algebraic methods into the language of pure geometry.

With this culmination in projective geometry, Euclid is, as it were, gathered up into the fold and given a right place in the synthesis, the whole. This is a very important fact, to be remembered today, when piecemeal attempts are being made to reform the mathematics curriculum. For this, above all, we need the understanding of the whole.

One reason, perhaps, why this understanding has not yet dawned is because in the hands (or heads) of the mathematicians, from the time of Desargues and Descartes—the more so since the mathematicians joined forces with the logicians—mathematics has become so complicated and is often conducted in such a rarified atmosphere of symbolism, that the normally intelligent but mathematically uninitiated individual, even if he is not averse to being lifted off the earth in a cloud of formulae or of refined logic, finds it hard to follow.

The cry which is being heard all over the educational world today is the cry of the modern soul of man seeking synthesis amid all the diversity, demanding an education in which to be made whole again.

It is for this reason that our entry into the beautiful realm of pure geometry, which is in reality near and accessible to every human soul, will be as simple as can be. We shall address ourselves wherever possible to the whole human being, to the will and the feeling, as well as to the thinking, for projective geometry is indeed the geometry of the whole human being.

A word about drawing: bearing in mind that it is the task of the geometrical concepts we are considering to lead towards an understanding of the non-spatial, mobile and ethereal processes which take place in and around the material world of the senses, we try to reveal this finer quality in our drawings.

To bring a finished construction into movement in thought is a most valuable educational exercise. Instead of simply looking at a ready-made configuration, one brings to bear upon it one's own creative faculty of imagination and will, and brings movement into the picture, thus experiencing a considerable element of freedom.

A drawing, however well executed, which gives an impression of too much solidarity and fixity, or reveals only the finite aspect of a form, will not contribute to the experience which projective geometry wants to give; on the contrary, just that quality which is unique in projective geometry will be lost.

For this reason we draw fine lines, wherever possible avoiding leaving broken ends on the page, and we commend to the reader the delicate use of colour. A good way of indicating movement among the elements of a drawing is to use the sequence of the colour circle. To show polarity, we can use either complementary colours—particularly "peach-blossom" and green—or else the contrast between blue and red. Contrasting techniques in the use of the colours—for instance, concentration of the colour in the centre and an arrow-like radiation outward from the one pole, and shading becoming more intensive as it moves inward, to give a peripheral impression in the other—will do much to awaken a real feeling for the qualitative nature of polarity. (See the coloured illustrations in "Plant, Sun and Earth"[1].)

In this way, an abstract thought is approached through artistic feeling and we are led beyond the bounds of the materialistic and spatial towards a truer understanding of the thought.

III First Steps in Projective Geometry; Movement

The first steps towards the new geometries were made possible through the postulation by Kepler of the idea of infinitely distant entities in geometry—the infinitely distant point, line and plane were gradually assumed. Two theorems basic to projective geometry could then be enunciated: Desargues' two triangle theorem and Pascal's theorem concerning the hexagon and the conic (pages 78 to 87). With this step, the rigidity of form imposed by fixed measure, where the right angle plays an essential part, was overcome. A tremendous freedom of movement was opened up.

Euclidean concepts, as we have seen, concern the laws of finished forms; laws which it is necessary to know in order to build the form. Thought is thus united with an experience of the will, in the sphere of touch. The new concepts are necessary in order to understand exactly the changes which forms undergo through perspective. Here thought is brought to bear concerning the activity of seeing. We never *see* a cube, for instance, according to the Euclidean laws inherent in it, which in thought we know to be true. In fact we see many different pictures of it, but never this one. When movement arises between the viewer and the object, there ensues a manifold possibility of alteration in the form seen. How then can the Euclidean experience through the will and the perspective experience through seeing be brought together? Only when this is done can it be truly said that we apprehend the whole.

First let us use the following constructions and considerations as an approach to the understanding of the part played by the so-called point, line and plane at infinity.

Miniature from "De Sphaera", Moderna, Biblioteca Estense

Saturn Sphere and Earth Perspective

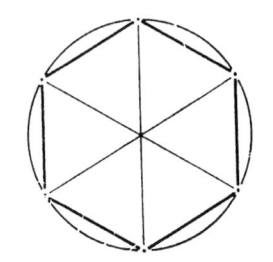

Figure 1

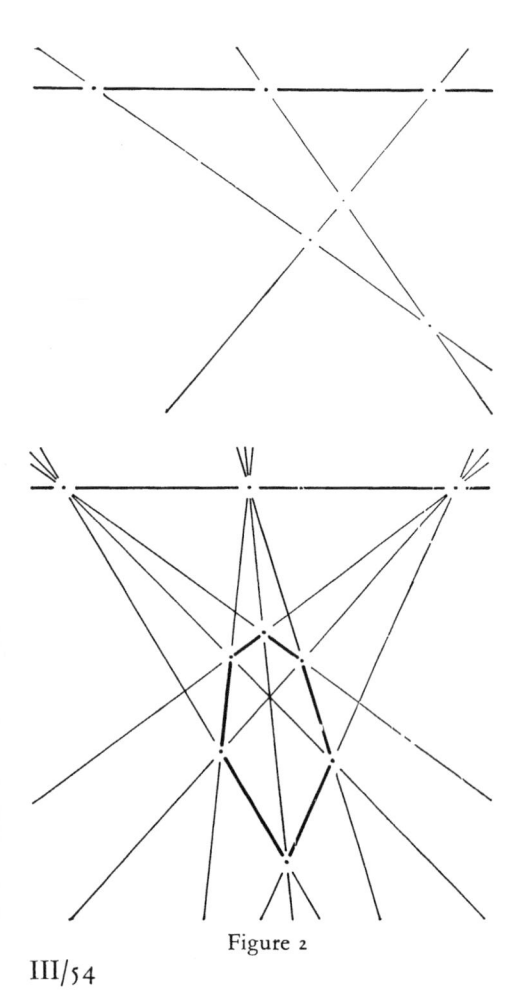

Figure 2

The Line-woven Net in Step-measure

According to Euclid, a hexagon may be drawn with its points on a circle, by making the length of the radius six times round the circumference. This procedure relates *points* of the hexagon to a *central point,* according to a predetermined measure (Figure 1).

In projective geometry, where measure is not primary, *any* six points or *any* six lines in a plane create a hexagon. We may call a hexagon simply a "six-point" (or a "six-line") and we always think of the lines as being of infinite extent. (When we use the word "line" we will always mean a straight line).

Let us now proceed to draw a hexagon without regard to measure or symmetry, commencing, not with a central point, but with a line which we will think of as picturing a horizon, like a "vanishing line" in perspective drawing (Figure 2).

Choosing freely any three points on this horizon-line, we will draw through each of them a line (and we will allow these three lines to make a triangle). All the further lines we shall now draw from the three points on the horizon-line are determined and can no longer be freely chosen, for we can only draw lines through points which are already there. The three new points which result from the first three lines are joined each to the one of the three horizon-points to which it is not as yet joined. Continuing thus, with six further lines the hexagon is completed. It is surprising to see how the last line is predetermined by two points which fall in line with one of the horizon-points, thus giving rise to the last side of the hexagon and completing the form. We only have to draw carefully, using fine lines and choosing to position them so that all the points of the hexagon can be found within the limits of the paper. Any one of the three points forming the first triangle may become the point of intersection of the three diagonals of the hexagon.

The miraculous way in which this form arises merely in the interweaving of lines will be experienced still further if the process be continued so that a whole field of hexagons arises in the plane (Figure 3).

A hexagon network will thus be woven throughout the entire plane, all the lines proceeding from the three points on

Figure 3 ▶

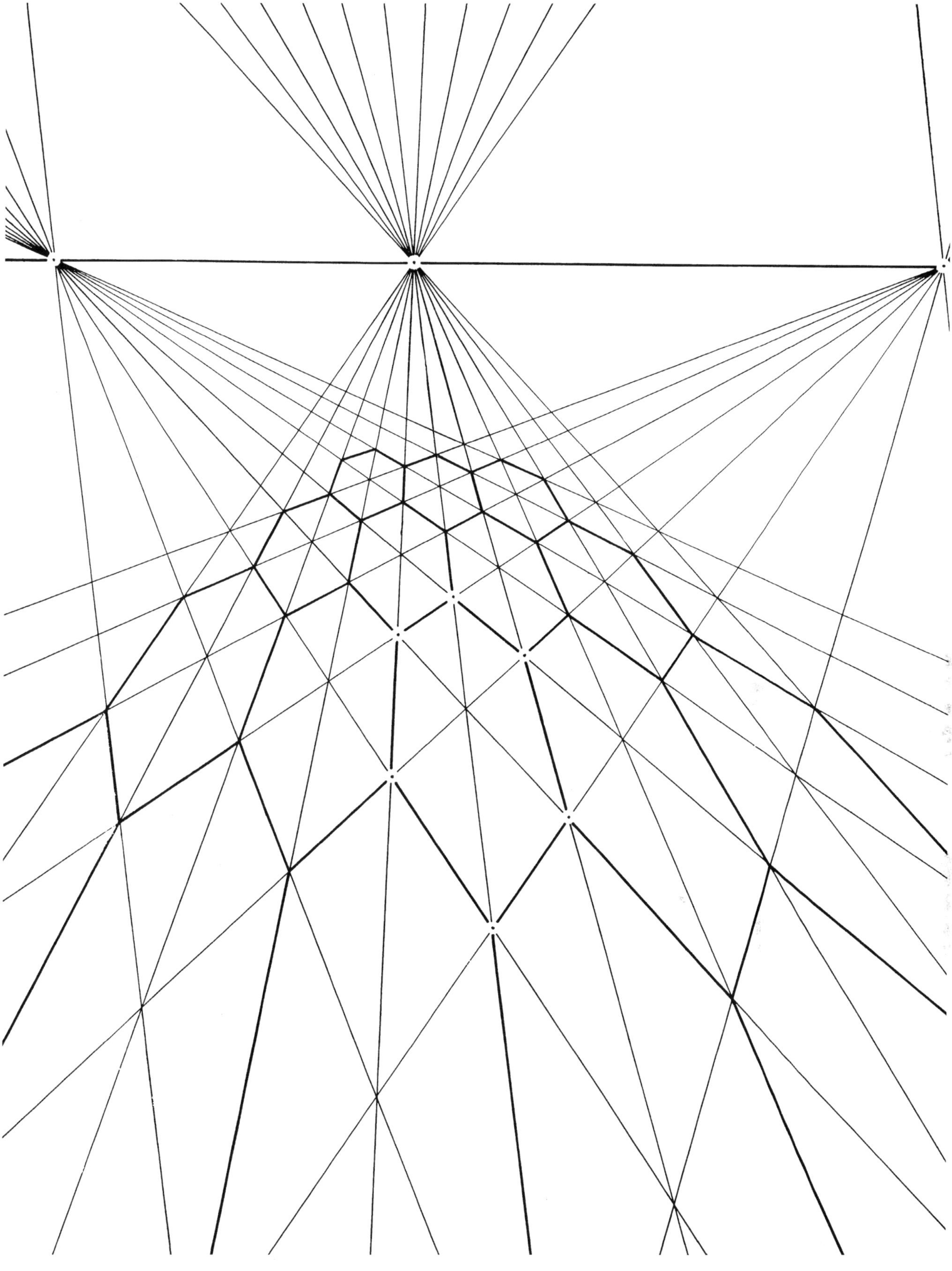

the horizon-line. No one hexagon is similar to another, and they lie side by side in never-ending sequence, becoming smaller and smaller as they recede towards the horizon-line, which they will never reach. Here, the law of symmetry and of angular and linear measure, given originally by the circle, has disappeared; yet order is maintained, as though, it seems, by some hidden law.

Move any or all of the three original points into any of the infinite number of positions on the horizon-line, and the network will always arise, each time with a different form and measure. The practical experience while drawing, when points seem to fall in line like magic, gives remarkable insight into the way the lines interweave to create ordered form without recourse to any preliminary measure.

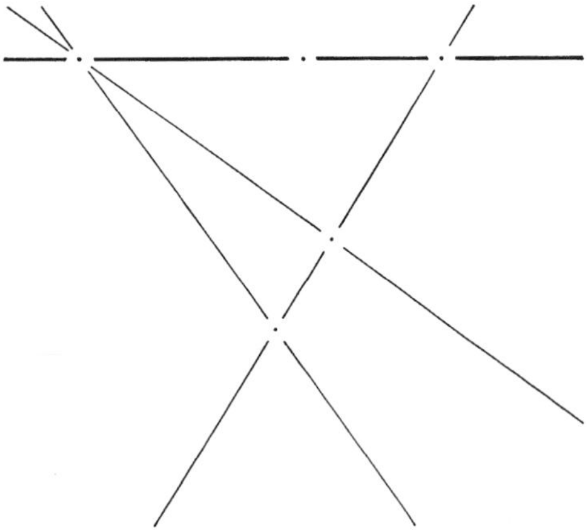

The network is like a matrix in which other interpenetrating nets are to be seen, among them triangular forms and quadrangles. The triangular net is basic to all the others. The regular triangle and the quadrangle (square), as well as the hexagon, have a simple relationship to the circle. In rather the same way as the circle gives a basis for all these regular forms, so the net which arises from the points on the horizon-line becomes the common ground for the projective triangle, quadrangle and hexagon.

In order to draw a quadrangle and then use it as a beginning for a whole net of quadrangles instead of hexagons, we begin as before with three freely chosen points. Now, however, the lines from the three points will play different parts in the figure. In the hexagon net, each line served alternately as side and then as diagonal. In the quadrangle net, two points must be different from the start. If one point sends forth two lines giving sides of the quadrangle, another will give a diagonal. The two points of intersection of the first three lines, drawn freely from the freely chosen points on the "horizon-line", will form two "corners" of the quadrangle, which is therefore already fixed by these two points, for the third point on the horizon-line will pair with the point giving sides and will supply the other two sides. There remains a sixth line to be drawn (the second diagonal), which, when added, completes the figure and determines a fourth point on the horizon-line. This fourth point then forms a pair with the point from which the other diagonal has already been

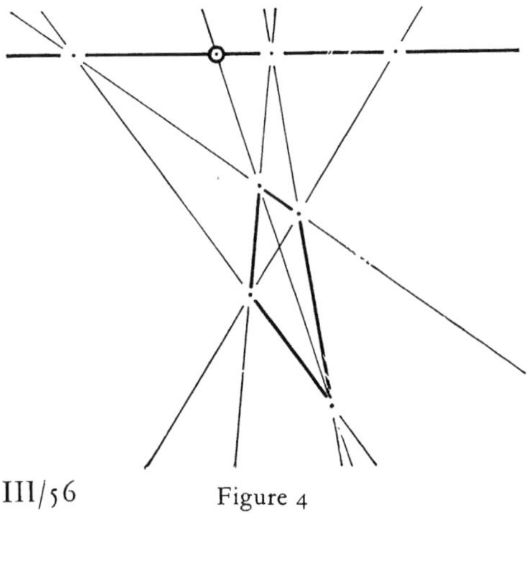

Figure 4

drawn. The point-pairs on the horizon-line will mutually separate one another (Figure 4).

This figure, called the complete quadrangle, will play a prominent part in our subsequent considerations. For the moment, we simply note the remarkable fact shown by the construction, that *any* other quadrangle drawn from the three given points will *always* have its sixth line (or second diagonal) passing through the *same* fourth point. Not only is this true of all quadrangles in the plane of the paper in which we are working, but also, as we shall see, of any quadrangle in

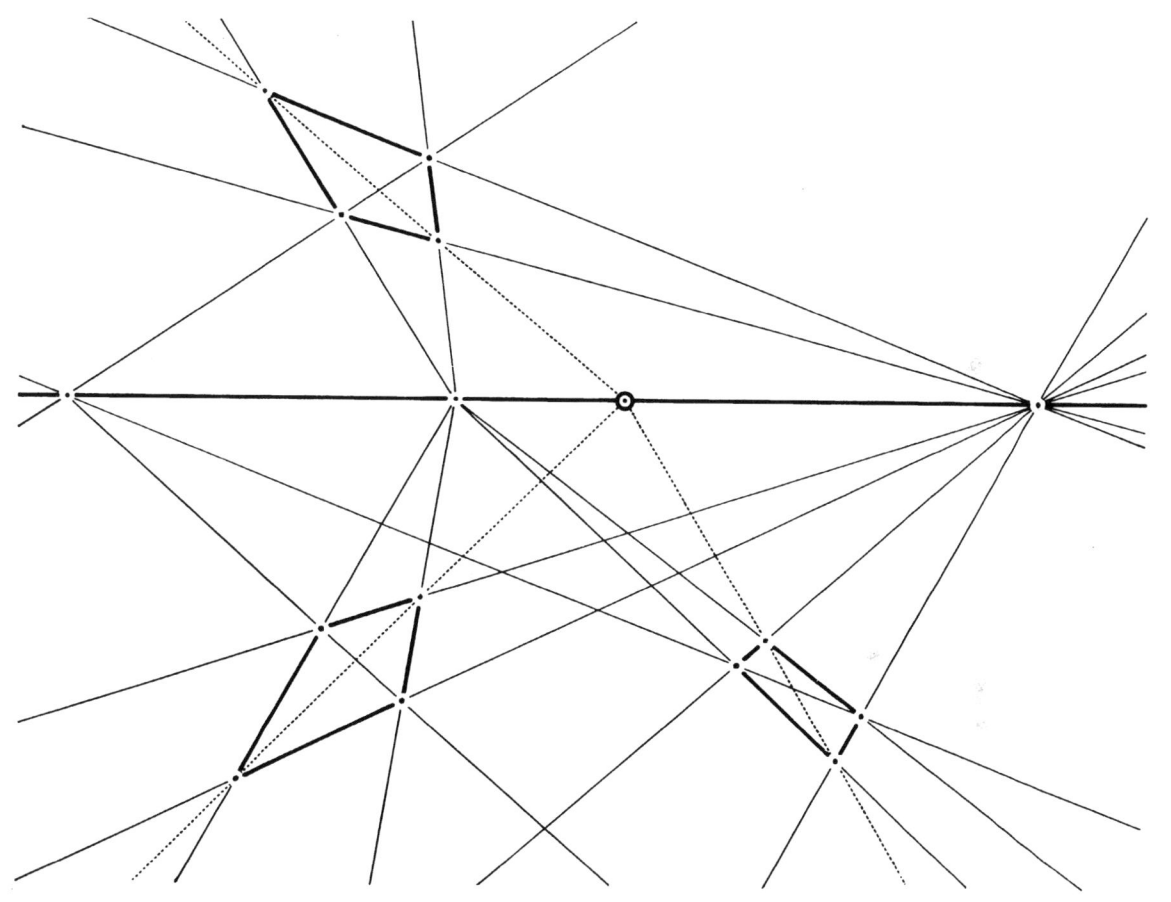

Figure 5

any other plane of the line as well! (Figure 5, see also Figure 14.)

As with the hexagon, so with the first quadrangle, the beginning has been made for a whole field of quadrangles to be drawn side by side. Such a field is called a Harmonic Net or Moebius Net (Figure 6).

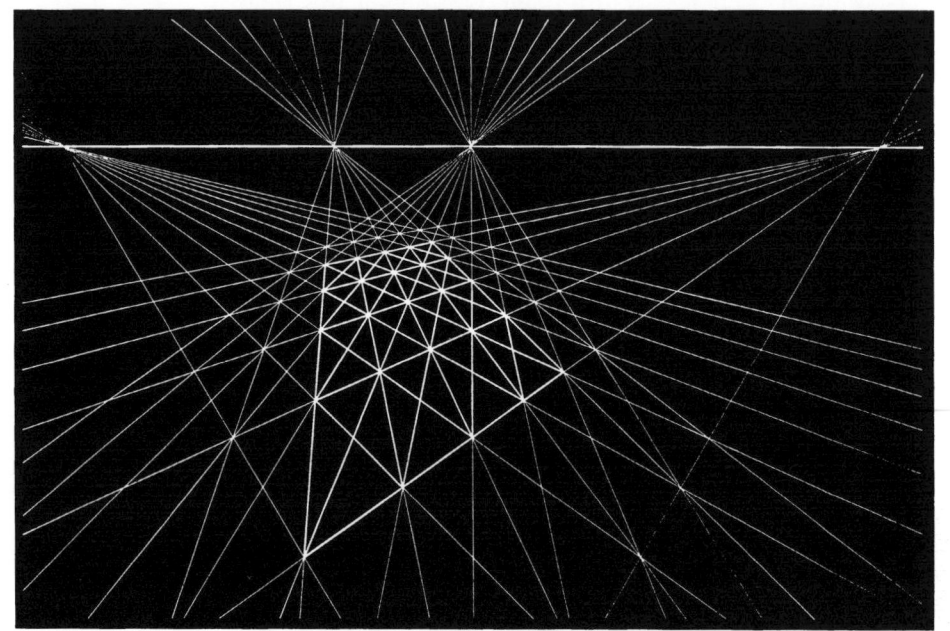

Figure 6

The Infinitely Distant Point of a Line

It should be clear from the foregoing (and further practice will show) that if one of the points moves out along the horizon-line, its pair will change position in relation to the other two. We must practise seeing the construction in *movement,* realising that the situation we happen to have chosen is only one among infinitely many. Let us choose the moment when one of the points on the horizon-line has moved infinitely far away. We can *use* this infinitely distant point of the line as one of our chosen three. What does this mean?

We accept the hypothesis that parallel lines, like any other two lines in a plane, have one and only one point in common, namely, an infinitely distant point. Figure 7 may help to illustrate this fact. Consider a line turning in a point and another line not in that point. The common point of these two lines moves out to one side or to the other, according to the direction in which the moving line is turning. At the moment when the turning line is parallel with the other, their common point is considered to be infinitely distant. As the turning line moves round, the common point of the two lines immediately returns from the opposite direction. *Thus, to draw a line from the infinitely distant point of any line to a point not on that line, means to draw a parallel to the line.*

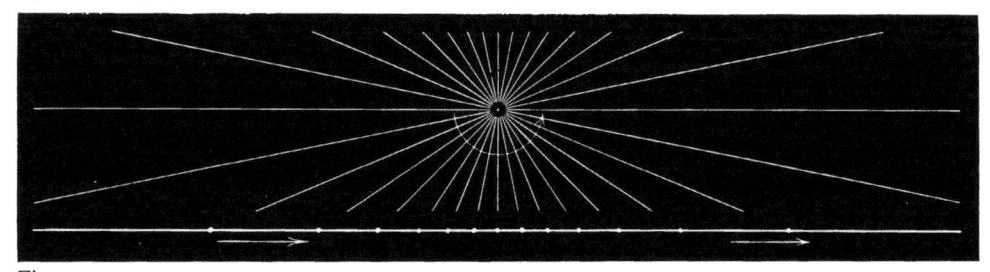

Figure 7

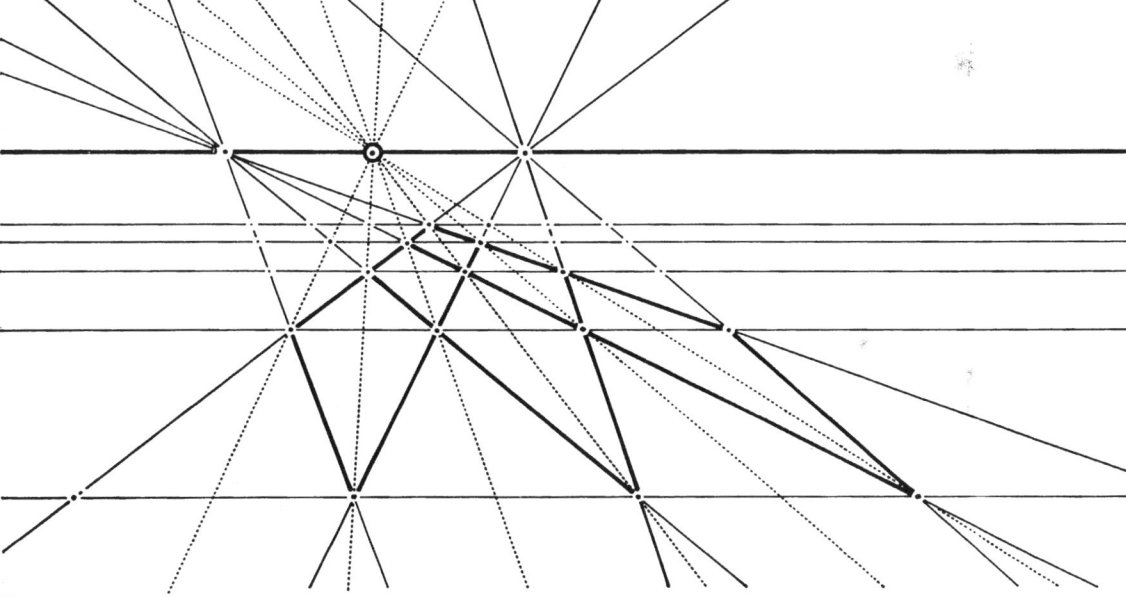

Figure 8

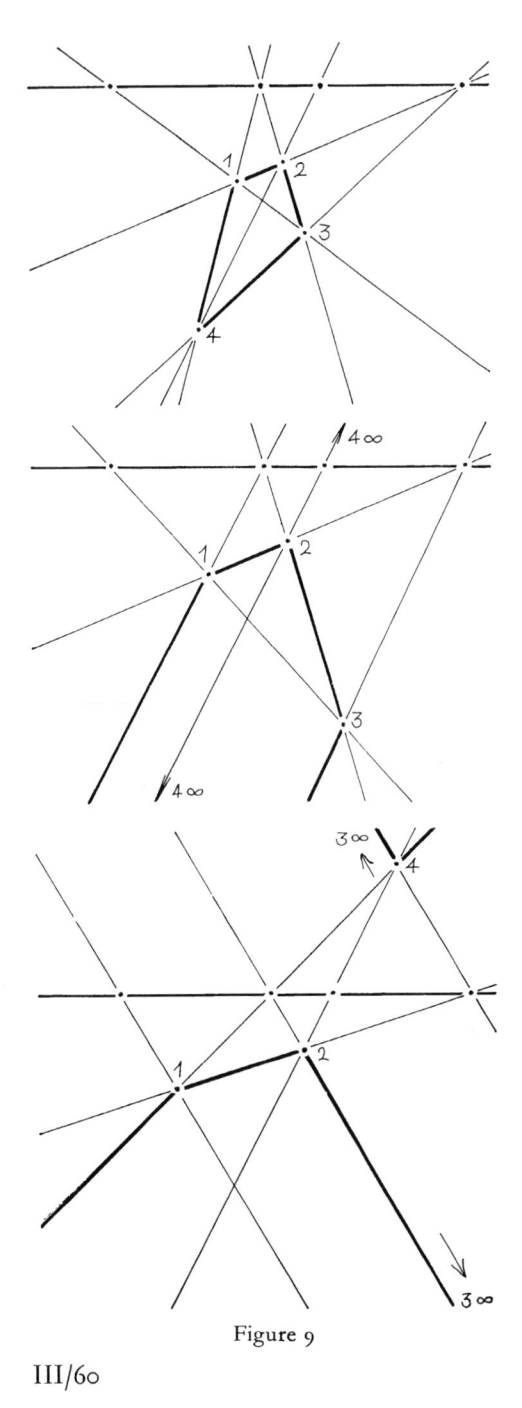

With this in mind, we now proceed to draw the Quadrangle Net as before, but this time using the infinitely distant point of the horizon-line as one of the freely chosen points. It is immaterial whether sides or diagonals be drawn to this point (Figure 8).

The result is remarkable! Take a ruler and measure distances between points. It will be seen that the point on the horizon-line which is paired with the infinitely distant one has placed itself exactly midway between the other two! With all the freedom given in choosing the points, this will always happen. Moreover, a glance at the construction shows that along all the lines parallel to the horizon-line the points are set at equal distances! Steps of equal measure have appeared along all the lines coming from the infinitely distant point of the horizon-line. Evidently, the setting of a point at infinity on the horizon-line brings an element of symmetry into the picture, and a certain fixed measure. The points are in the so-called arithmetical progression, the measure of equal steps. We will call it "Step-measure" (26).

In this drawing we have used the infinitely distant point of the horizon-line; but what of the infinitely distant points of the many other lines we have drawn? We must realise that every line has its infinitely distant point, remembering that we took our start from the fact that two lines in the plane always have a common point and two points a common line. (It follows from this that the community of all the infinitely distant points of a plane have the characteristics of a line and may rightly be called the infinitely distant line of the plane.)

Now look at one of the quadrangles and note that its "bottom" point is formed by two lines coming from two of the four points on the horizon line. Bring movement into the picture (Figure 9). Take hold of this bottom corner and pull! Move it out, for instance, along the diagonal, and watch how the two sides which have this point in common swing open, until the moment comes when they are parallel. At this moment the corner of the quadrangle has disappeared into the infinite.*

If the two sides of the quadrangle continue to turn, behold, this corner will appear again from above! For projective geometry the quadrangle is still there; it has

Figure 9

* The infinite is denoted by the symbol ∞.

merely changed its shape as the result of a continuous movement of one of its points. To recognise it, we must become free of the old, habitual point of view, that a quadrangle must be some sort of fenced in area, with definite measurements. It is simply four points and the six lines common to them.

What now is inside and what outside? The question becomes less important. It is, however, certainly helpful at this stage to use the "crutch" offered by Euclidean geometry and to shade in the quadrangles according to familiar patterns based on Euclidean concepts.

Practice in drawing the nets and allowing them to extend out through the infinite below and return again from above the horizon-line will be a good though fairly exacting exercise, requiring the use of a large sheet of paper, in order to include distant points (Figure 10).

In making these nets the reader might do well to start with a symmetrical situation, with one of the four harmonic points in the infinite, when it is easier to follow each line out through the infinite to make sure that the forms are correctly linked as they open out below and return from above. The possibilities of movement in these constructions brings an element of time into the experience of the forms.

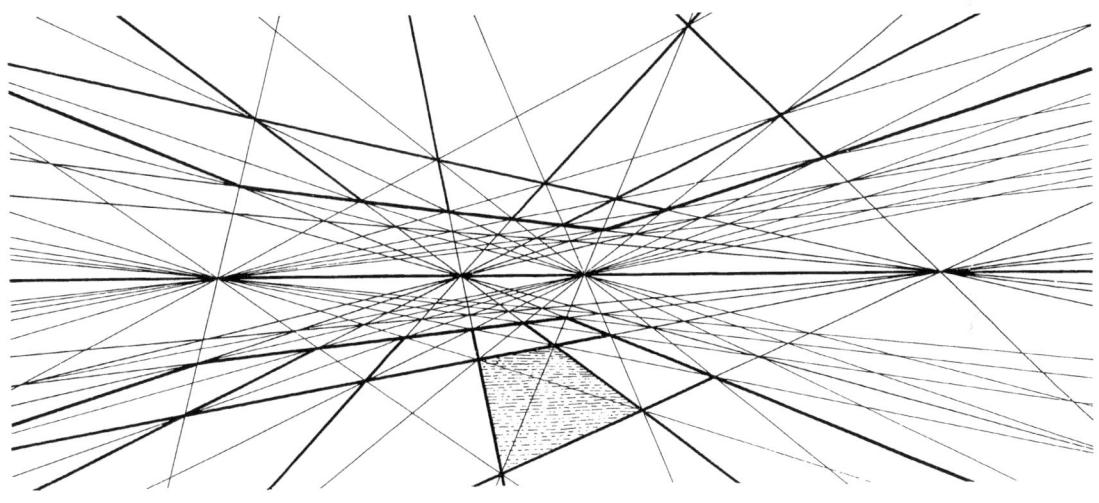

Figure 10

The Infinitely Distant Line of a Plane

A rather more difficult conception is the *one* infinitely distant line of the immeasurable plane. The totality of all the infinitely distant points of a plane has the property of a line. Consider the quadrangle net and let the horizon-line move further and further out. There must come a moment when not only one sheaf of lines from a horizon-point becomes a parallel sheaf, but the same happens to all four. This is an idea which we may find difficult to reconcile with our normal consciousness of space, but which is, nevertheless, a perfectly clear and exact thought. It is a thought with which we must work. All lines of a plane have their infinitely distant point on a line which we call the infinitely distant line of the plane.

Like all other lines of a plane, the infinitely distant line is straight. Yet it does seem to have some sort of circling quality about it. The following example will help to explain this, albeit in a rather simplified way (14).

We feel the need to compare this all-encompassing, infinitely distant line to a very large circle. We should, however, have to think of an immeasurably large one, which is, in fact, a straight line. Think of the curvature of a circle. A very small circle has a very strong curvature; if the circle becomes still smaller, there will be a yet stronger curvature until the whole form contracts into a point. If the circle grows larger it becomes less curved until it loses its curvature altogether and becomes a straight line (Figure 11). In like manner, we can conceive of a circle which has grown until it has turned into the infinitely distant line of its plane.

We will now (Figure 12) construct a quadrangle net starting from three points of the infinitely distant horizon-line—the line at infinity of the plane in which we are working. We choose freely any three infinitely distant points, determining them by the situations in which we set the first three lines. (Each line gives the direction in which one of the infinitely distant points would be.) These first three lines will determine three of the points of the first quadrangle and two of its sides. Two more lines drawn parallel, one each to two of the three, will complete the quadrangle, except for the second diagonal, which, when drawn in, leads to the fourth point

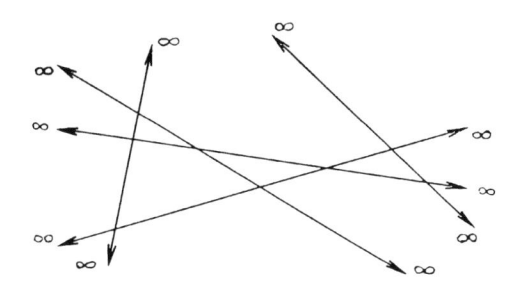

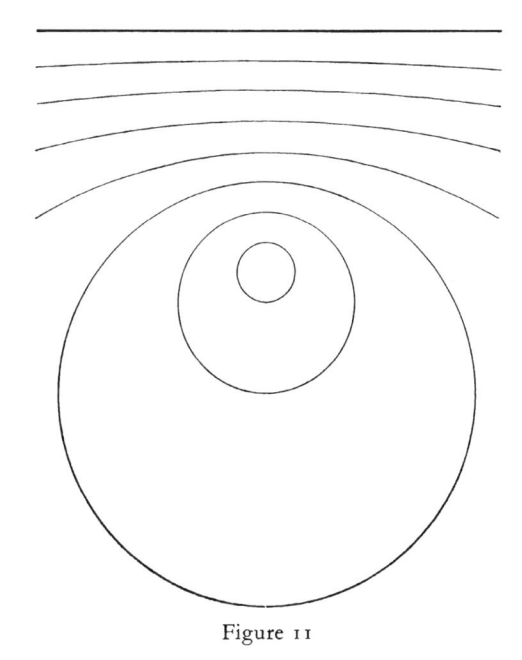

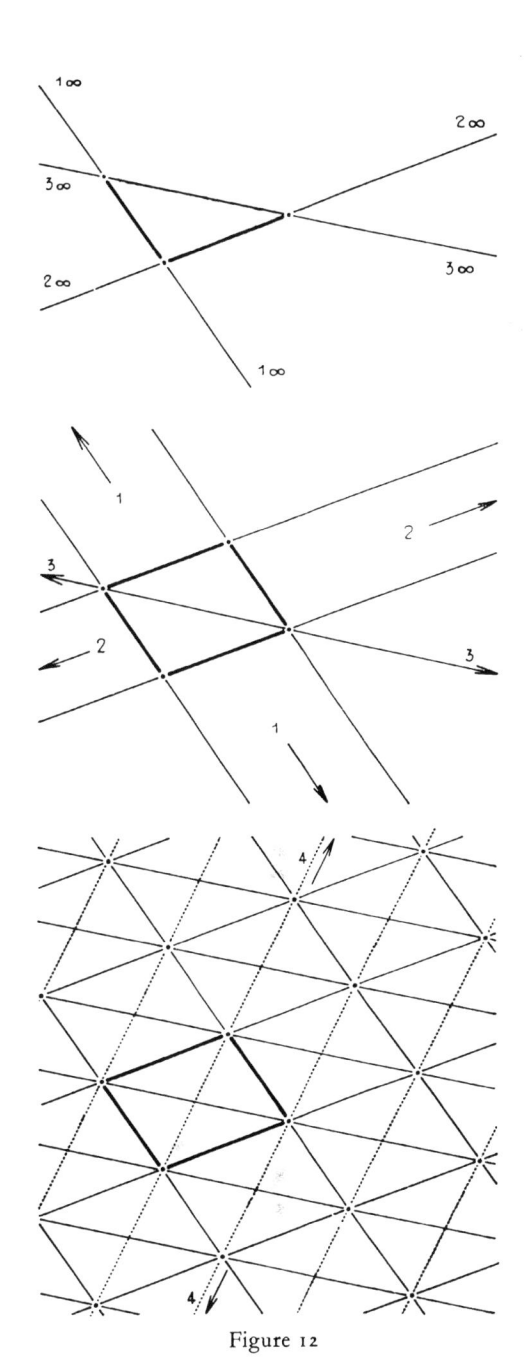

Figure 11

Figure 12

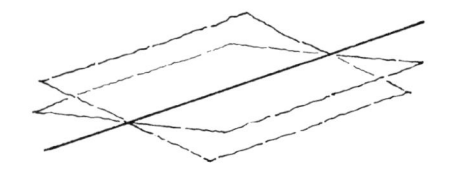

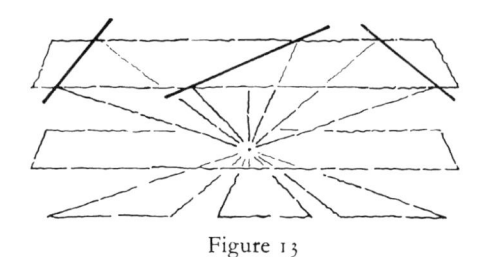

on the infinitely distant line, the pair to one of the original three. Proceeding consistently with the quadrangle construction as before, we find that the sheaves of parallel lines raying from the points on the infinitely distant line give rise to a net of quadrangles (Figure 12).

More regularity appears in this net; we note that Stepmeasure has arisen along all the sets of parallel lines. It is clear that had we happened to choose the first three infinitely distant points at the appropriate angles, the net might have been composed of squares. This would be a special case. Thus, to work from the line at infinity of the plane appears to take away a certain freedom and results in a more set and regular pattern.

Another way of coming to terms with the conception of an infinitely distant line—and one which will help us in our next step—is to consider the following.

Picture two planes and their common line. Any two planes have a line in common, even two parallel planes.

Think of any two planes in space; let one of them rest immobile, while the other has the freedom to pivot on a fixed point. At every moment the two planes have a line in common, even if it is very far away. When the planes become parallel, their common line slips out to the infinite in one particular direction. A line may move out to the infinite in *any* direction and become the infinitely distant line of a plane (Figure 13).

Figure 13

The Infinitely Distant Plane of Space

There are many ways by which to help one's imagination to compass the mathematician's concept of the plane at infinity—the infinitely distant plane, as we would perhaps do better to call it. This plane, like any other plane, is flat! It is as flat as one must think an infinitely large sphere to be; one which has grown so vast that it has no more curvature left and is therefore a plane. Bound as one's imagination is to the spatial, it boggles at this. Here is a threshold where *thought* can enter, but not an imagination used only to physical things (15).

Figure 14, 15 ▶

III/65

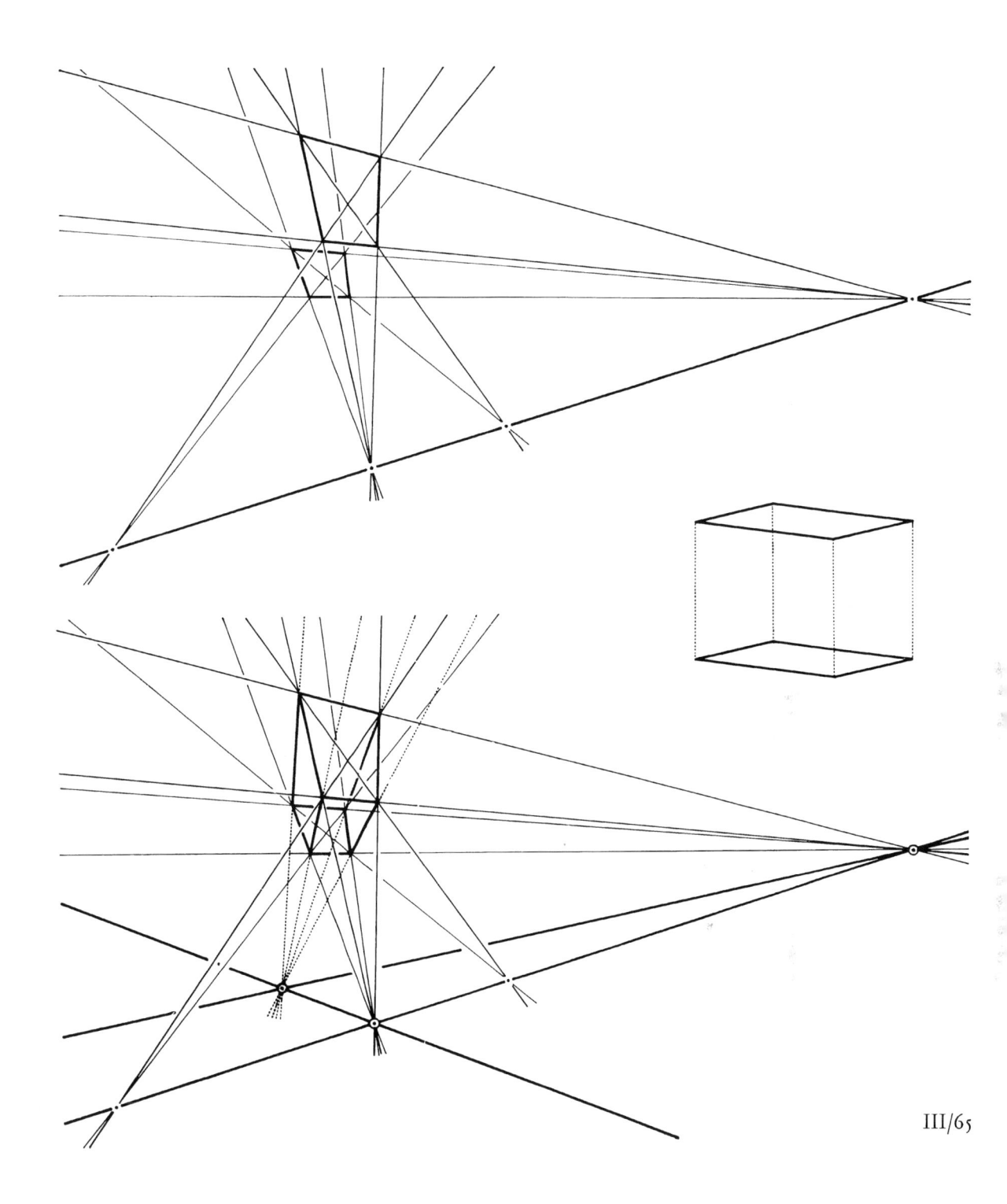

Let us return to the quadrangle and allow it to reveal yet another of its surprising qualities. This time we will draw two quadrangles, but we will think of them as being in two *different* planes of the same line. Having chosen the first three points freely on the "horizon-line", as we know, the same fourth point will be determined by both quadrangles. A little manipulation of the lines which we imagine to pass behind the others will make quite a realistic picture. The fact that we can choose to draw the horizon-line in any situation we like, helps us to free ourselves from the domination of rigid symmetry and measure (Figure 14).

Now, thinking of the two quadrangles as though in different planes, let us ask: What will happen if we regard them as the "top" and the "bottom" of a three-dimensional form, such as the cube, and join corresponding corners, as though adding the third dimension?

With all the freedom of construction in the quadrangles, it is amazing to find that the four lines joining corresponding corners all *meet in a point*! This new point may be anywhere—near or far, above or below. It may even be infinitely distant, should the four lines chance to be parallel, which is perfectly possible. The new point plays a part similar to two of the original points on the horizon-line. From *three* points the three times four lines spring, which, weaving together create a three-dimensional form. This form, apart from the special right-angled measure, has all the characteristics of a cube. It has twelve edges, eight corners and six faces.

The *points* of the triangle ray forth lines which, four by four, create the twelve sides of the form, while the *lines* of the triangle send forth, two by two, its six planes. The form's eight points arise in the interweaving of the twelve lines and the six planes. Even the diagonal lines of the four new faces of the form will be found to meet in pairs on respective sides of the triangle, if our drawing is sufficiently exact! (Figure 15).

All is harmoniously woven together, each single part contributing towards the creation of the whole form. The lines proceeding from the points of the triangle and the planes proceeding from its lines all co-operate to create an ordered whole. Once the form is finished, we see that the triangle is its source and origin. The triangle in the plane is

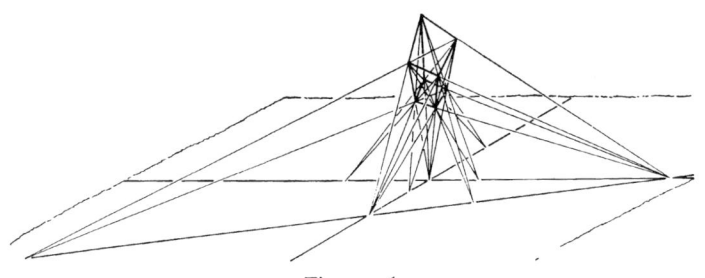

Figure 16

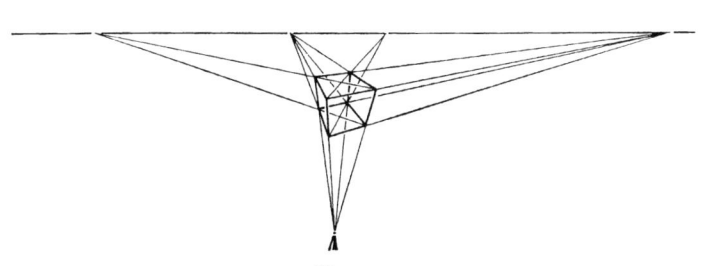

Figure 17

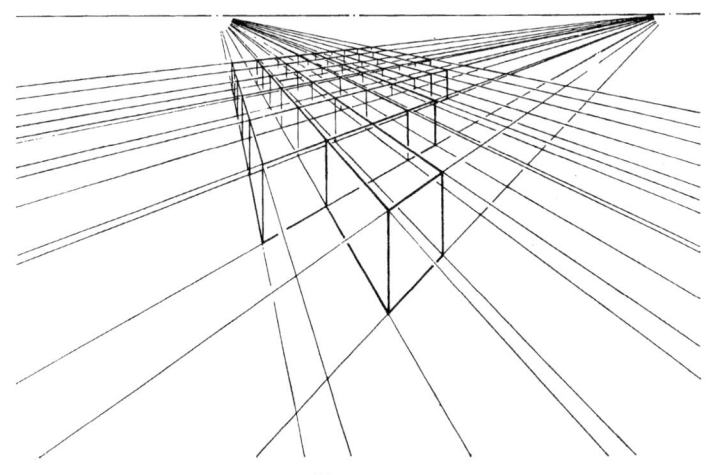

Figure 18

III/67

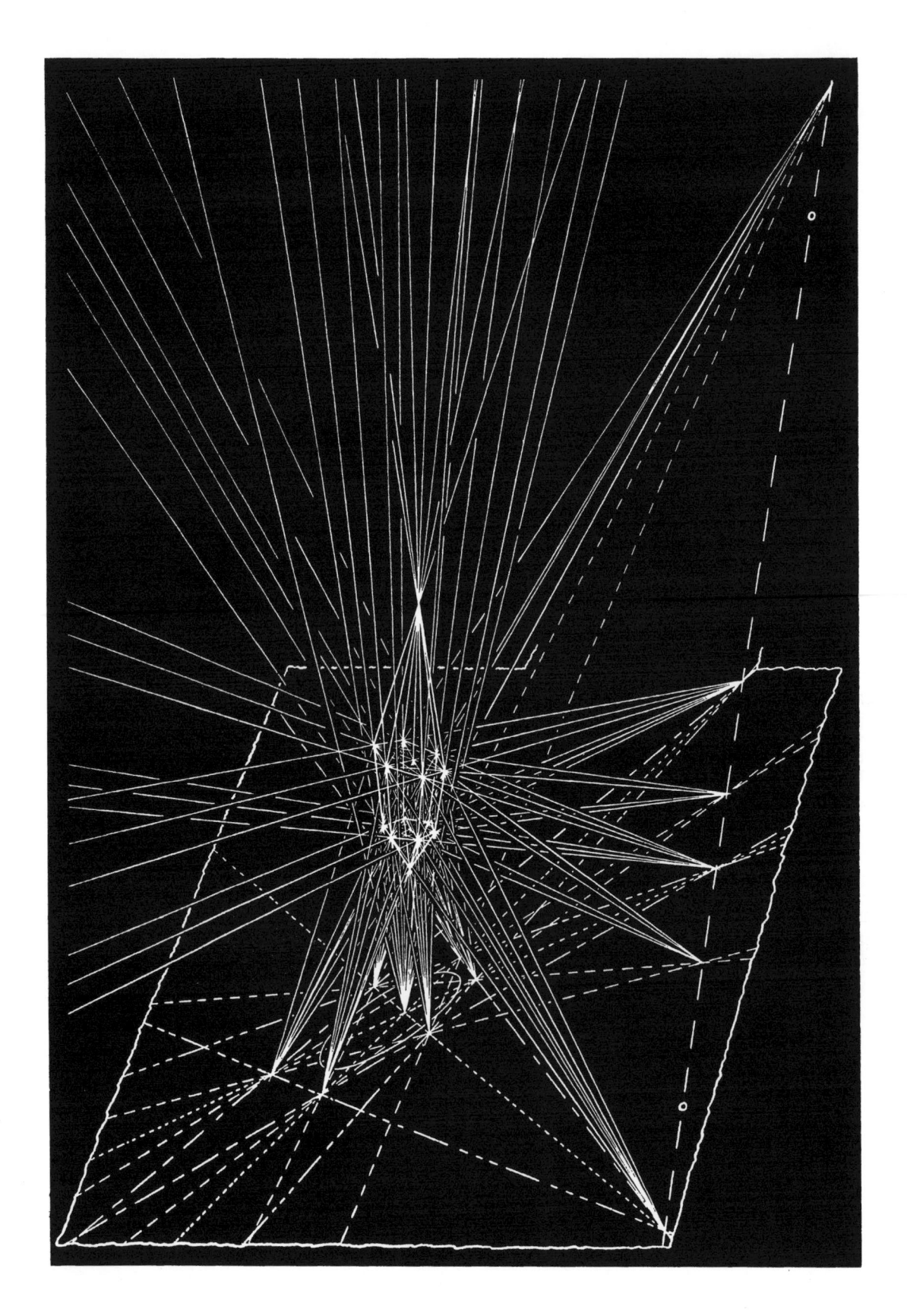

III/68

like an archetypal pattern of origin for the three-dimensional form, which arises simply through the interplay of the six planes, which spring in pairs from the three lines of this archetypal triangle.

We have to think actively, bringing will into our thought and imagination in regard to this configuration. We shall then perceive the manifold possibilities of *movement* inherent in it. To move only one part will change the picture. The thought underlying the construction—the basic concept—compasses a whole gamut of possibilities. The underlying thought may find its external manifestation in a multitude of forms, all of which are variations or metamorphoses of the original one (Figures 16, 17, 18 and 19). Figure 18 is like a crystal lattice; one point of the archetypal triangle is infinitely distant. In like manner other crystal forms may arise from other archetypal forms in a plane. Figure 19 shows the quartz crystal construction.

And now it is necessary to take hold in thought even of the archetypal plane itself and move it about, trying to picture the changing form as we do so, until at last we send this "creative" plane to infinity. The thought is perfectly clear, though the picture is difficult to imagine. Just as in a plane we saw a line slip out to infinity and become the infinitely distant line of that plane, so now we see the archetypal plane recede into the distance in one or the other direction until at last it becomes the *one* infinitely distant plane of all space—the mathematician's "plane at infinity". Now it is on all sides at once, above and below, to the left and to the right, and yet it is still a single plane! We reach with our thought beyond the confines of measure.

At this moment the archetypal plane, which has given birth to the crystal form, becomes that plane, which for Euclid was non-existent, the all-encompassing, infinitely distant plane of space. For projective geometry this is an exact thought. Cayley called this plane the "Universal Absolute".

It is precisely at this moment that our form assumes the parallelism and the metrical characteristics of the measurable world. It appears like some crystal we may hold in our hand, such as the parallelepipedon form of the Iceland Spar crystal, for example. As a special case it might appear as an actual cube (Figure 20).

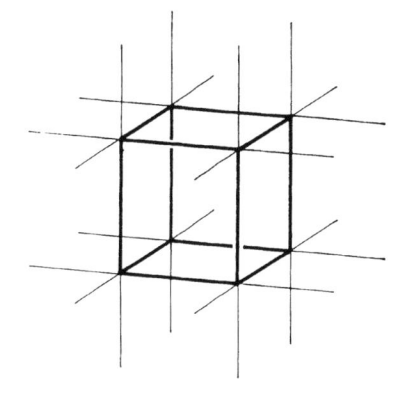

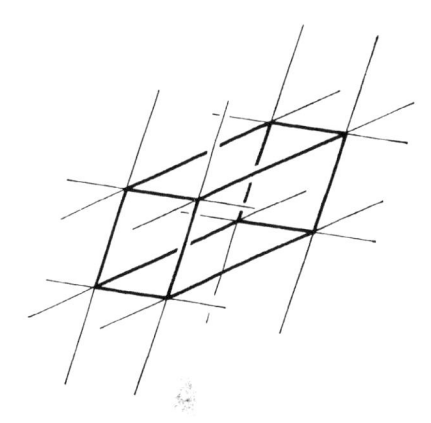

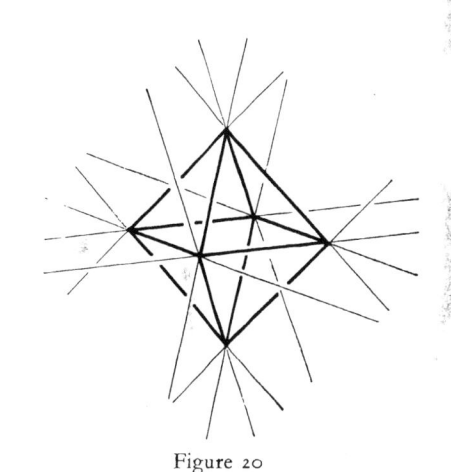

Figure 20

We see the form arising through the inward raying and interweaving of planes and lines from the infinite periphery of space. Like raying light, the lines and planes create a form. Once it is there we may check its measurements; but this comes as a last step. In the process of construction we have learned more about the form than is given merely by its measurements.

Real or Unreal?

Of the physicist Max Planck it is quoted: "That which can be measured is real." We, with our construction and all that it entails, have evidently by means of our thinking pierced through to a non-spatial (perhaps one might say super-spatial) concept, which, however, has a special bearing on the metrical aspect of the form. Maybe there is more to a form than its metrical aspect? At least we may go as far as to say that in addition to the Euclidean idea of a cube, which sees it created about a centre in three-dimensional space, in accordance with the law of the right angle and with a fixed measure, we have glimpsed the possibility of another creative idea underlying it, which complements the first.

The crystal is real and permanent here on earth. Science investigates the laws of the crystal lattice by the analytical and atomistic method. Without setting aside the true spirit of modern scientific enquiry, based on clear mathematical thought, we have reached a less one-sided, more complete understanding of the crystal form.

In England in 1924, Rudolf Steiner described the crystal in a way which a science restricted by atomistic forms of thought, finds it impossible to understand. He said: "We must look right through the earth, as though she were not there; and because she is not there we can look further. In connection with the minerals we must see what is above us and what is all around us in the surrounding cosmic periphery. The earth must be as though dissolved away; we must see the same below as above, in the west as in the east. . . . If you find a quartz crystal in the mountains, it is usually attached to the rock below it; that, however, is only because it is disturbed by the earthly situation. . . . In reality it is formed by the spiritual element which streams in from all sides and is mirrored within itself, so that you see the quartz crystal floating freely in universal space. Each single crystal, which forms itself perfectly on all sides, appears like a small world of its own. . . . There exists not only one world, but as many spatial worlds as there are crystals in the earth. We look into worlds immeasurable. Looking at a crystal of salt we may say: Out there in the universe is Being; the salt crystal is for us the manifestation of something which permeates the whole universe in the form of Being; so too the pyrites crystal, be it cube or dodecahedron. . . . In the manifold crystal forms is revealed to us a great world of Beings, which live and find expression in mathematical-spatial form. Looking at the crystal we see the Gods" (16).

Through the geometry of the ancients we can experience the fact that we belong to the realm of the Spirits of Form, while modern geometry is inspired by the Spirits of Movement. In modern geometry we are enabled to take a first step beyond the boundary of the sense-perceptible and to reach out towards the creative Thoughts of divine worlds. The infinitely distant plane of modern geometry is not perceptible through the senses; *but it is a clear and exact thought*. As such, it comes from the side of natural scientific and mathemetical research to meet spiritual scientific research. The time is at hand when through active scientific thought mankind must find the way in clear consciousness to the worlds which lie behind the sense world, from out of which the sense world is born. The spirit of man has forged the way to the world of the atom; so, too, will the spirit of man discover the worlds which abound with the forces of creation, the raying formative processes which give birth to worlds (17).

Fra Bartholommeo (1472–1517)

Point and the Upward-Bearing Plane

Point, Line and Plane

Our considerations so far have led us to see how forms can arise in the interplay of the three basic geometrical entities, point, line and plane—formless and ultimate—considered in their entirety and without regard to measure. To begin with, we have met only with straight-edged and flat-surfaced forms, but later on we shall see that point, line and plane, uncurved in themselves, are the progenitors also of all plastic forms.

We will now enter in thought, but also with imagination, into the different properties and qualities inherent in these three, and then consider systematically their relationship with one another in space.

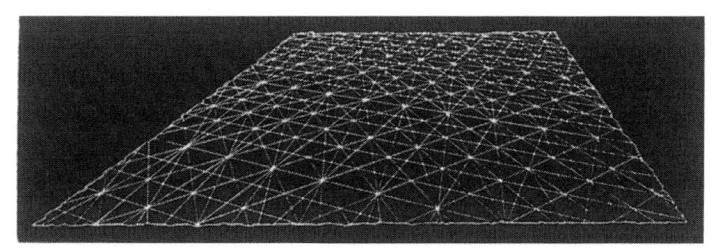

Figure 21

The ideal Plane: it has no thickness, is always perfectly flat and extends into the vast distances of space. Imagine, from the mast of a ship, the vast expanse of sea, spreading away into the blue to meet the sky in the distant horizon—the horizon which is, however, infinitely distant on all sides. It is a picture which approaches in mood the quality of the ideal plane. In drawing the plane, we must give it an edge, draw less of it than is really there, simply in order to make it visible; but this is arbitrary, for in reality it has no shape nor size. It is an expression of infinite expansion; its organs are lines and points (Figure 21).

The ideal Point: has also neither shape nor size; it is sharp and contracted, enclosed in itself. Perhaps it is like the sharp red point of pain. In order to make it visible in a drawing, we must, of necessity, draw too much of it. The ideal point is an expression of infinite contraction; its organs are planes and lines (Figure 22).

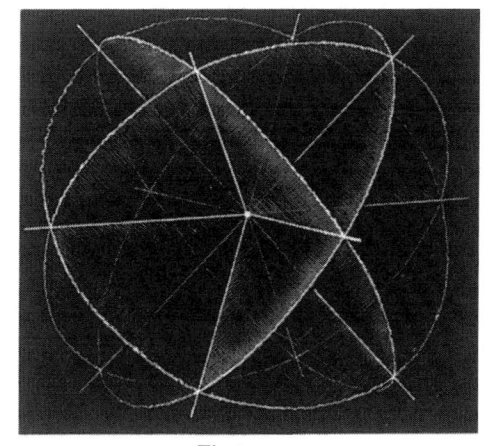

Figure 22

The ideal Line: has no thickness, but it is of infinite length and perfectly straight.* Thus it has an aspect of expansion and also of contraction. It mediates between plane and point (Figure 23).

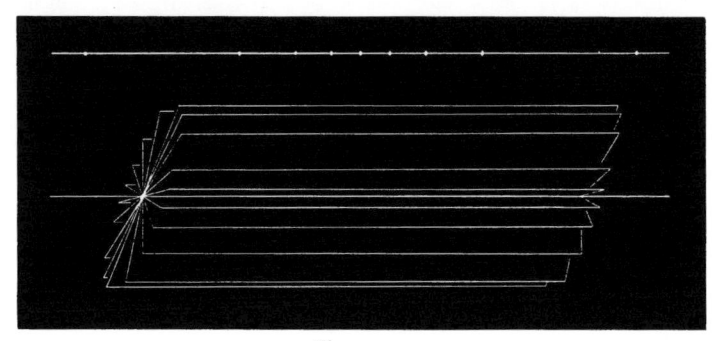

Figure 23

Thus points, lines and planes are thought of as entities in themselves. In their mutual interplay, however, they create one another, and together they are the formative organs of space.

As, however, the very task of modern projective geometry is to enable us to transcend the limits of Euclidean forms of thought and to spiritualise our conceptions of space, we must learn to free ourselves from the old spatial habits of thought. Thus it is necessary to speak exactly about the relationships between point, line and plane, using words which have of necessity a spatial connotation, without restricting their meaning to the old spatial thought-forms. Modern geometry contains, for example, the thought: *a point is in a plane*; but equally, the thought: *a plane is in a point*. The mathematician has technical terms which we shall, however, leave aside, for they hinder rather than help the process of becoming less tied to physical concepts in space. We shall therefore speak of *lines in a point* or of *planes in a point* without being any the less exact than the trained mathematician, who speaks of "bundles" and "sheaves" (18. See also page 218.)

* We use the word "line" always to mean *straight* line, as distinct from "curve" (curved line).

Point, Line and Plane in Mutual Interplay

A plane may be regarded as a manifold or community of lines and points. It may be determined by any three points not in line, by a point and a line not in the point, or by two lines if they have a common point (Figure 24).

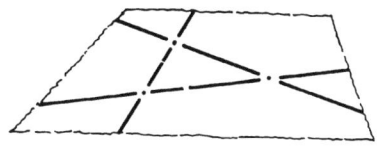

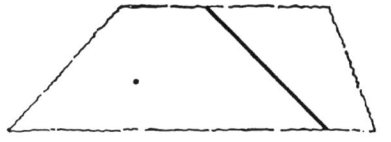

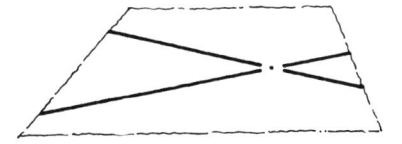

Figure 24

A Point may be a manifold of lines and planes. It may be determined by any three planes if they are not in a line, by a plane and a line not in the plane, or by two lines if they have a common plane (Figure 25).

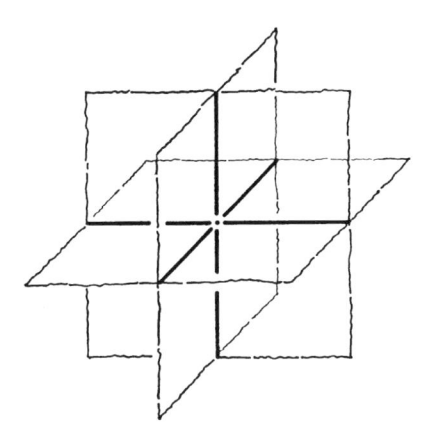

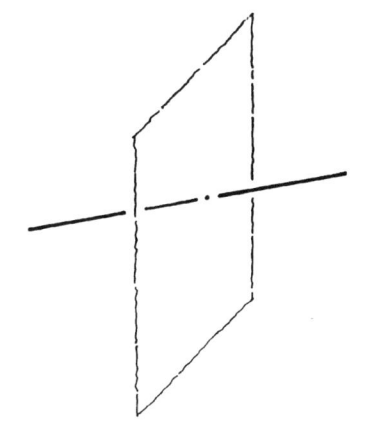

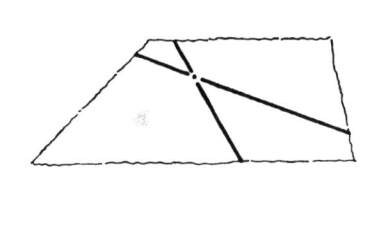

Figure 25

A Line may be a manifold of points or of planes. It may be determined by two points or by two planes.

Add a *third point* not in line with the two which determine the line and you will determine a *plane*. Add a *third plane* not in line with the two which determine the line and you will determine a *point*.

For the modern geometrician the mutual symmetry of these statements is complete through the inclusion of the concept of infinity. A plane, for example, is determined by three points, whether they are in the finite or whether one, two, or even three are infinitely distant. *Any* two lines lie in a plane, *if* they have a common point, and *any* two lines have a common point, *if* they lie in a plane. The statement is true without exception only if parallel lines are included. Any two planes have a line in common; two (or more) parallel planes have an *infinitely distant line* in common (Figure 26).

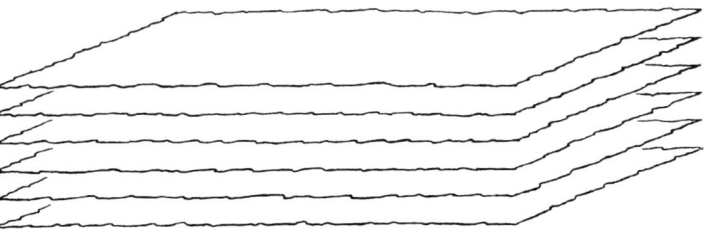

Figure 26

The (Axioms of) Community of Point, Line and Plane

Any two planes have one and only one common line. This line contains all the points which the two planes have in common.

A line and a plane always have a common point. If they have more than one, then the whole line with all its points lies in the plane.

Any three planes have a common point. If they have more than one point in common then all three lie in a line.

Any two points have one and only one common line. This line contains all the planes which the two points have in common.

A line and a point always have a common plane. If they have more than one, then the whole line with all its planes lies in the point.

Any three points have a common plane. If they have more than one plane in common, then all three lie in a line.

Two lines either have *both* a point *and* a plane in common, or they have *neither* a point *nor* a plane in common. In the latter case we call them "skew" (19).

This statement is so formulated as to make the distinction between "a line as such" and "all the points of a line". In fact we distinguish between *three* aspects of the line: the line as an undivided whole; the line as a manifold of points; the line as a manifold of planes.

The archetypal interdependence in the relationships between the points and the planes of space comes to expression in all spatial forms. If through any set of laws a configuration of planes is created, it will always have a sister-form in which the roles of the points and the planes have changed places. Only when a form contains points and planes which are equal in number and function will the form correspond to itself in the sense of this mutual relationship between point and plane. It is then its own sister-form. With the word "form" we do not merely mean a fixed and rigid form, but also a transmutable one, in other words a "type". In Chapter VIII we shall consider in detail this important aspect of the world of forms.

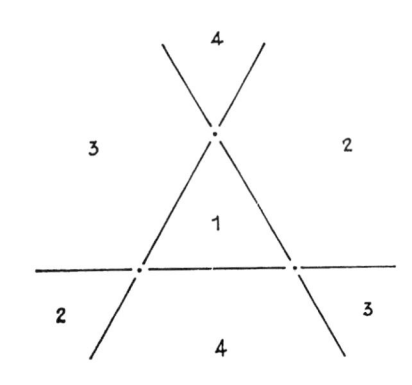

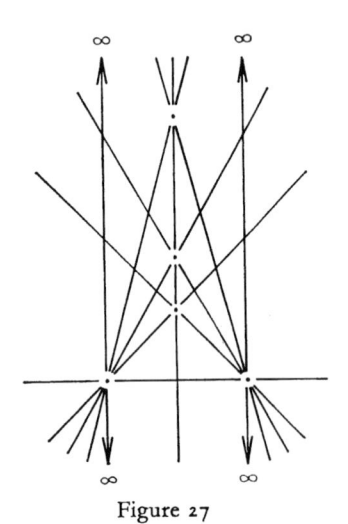

Two Basic Theorems: Desargues and Pascal

We will bring to a conclusion our considerations concerning the first steps taken at the dawn of modern time in the overcoming of the old, fixed and limited geometrical conceptions by describing two important theorems upon which these first steps rest: Desargues' two-triangle theorem and Pascal's theorem concerning the hexagon.

Girard Desargues (1593–1662), architect of Lyons and friend of René Descartes, was the first to use Kepler's idea of elements at an infinite distance. In considering the line as a whole, including its infinitely distant point, he widens the concept of a triangle. Behind all Euclid's definitions of a triangle lies the projective one which includes them all.

In projective geometry a triangle is defined as follows:
Any three lines in a plane, but not all in a point; or:
Any three points in a plane, but not all in a line.

Here all the points and lines of the plane play a part, including the infinitely distant ones. The thought is exact, and its external manifestations are unlimited (Figure 27).

We have to overcome our usual spatial conception of a triangle, as we did for a quadrangle. The three lines which form the sides, when considered in their entire extent, enclose not one but four areas of the plane, *all of them triangular*. Besides the area enclosed by the Euclidean triangle, we recognise three more, each of which opens out and stretches away through the infinite, returning on the opposite side. Follow two of the sides outwards, through their infinitely distant points and back again until the third side is reached from the opposite direction. For projective geometry this area equally represents a triangle. Moreover, a triangle still retains its identity, if one or two of its points or one of its lines are infinitely distant.

Figure 27

Desargues' Two-triangle Theorem

Draw any two triangles and relate them by pairing their points and their sides (the points AA′, BB′, CC′; the sides aa′, bb′, cc′). Then look for the three points in which the pairs of sides meet and the three lines common to point-pairs. It will usually happen that the lines joining AA′, BB′, and CC′ will somewhere form another triangle of lines (remembering that we take the whole extent of the lines into account) while the three points determined by the paired lines (AB, A′B′; AC, A′C′; BC, B′C′) will also form a triangle (of points) (Figure 28).

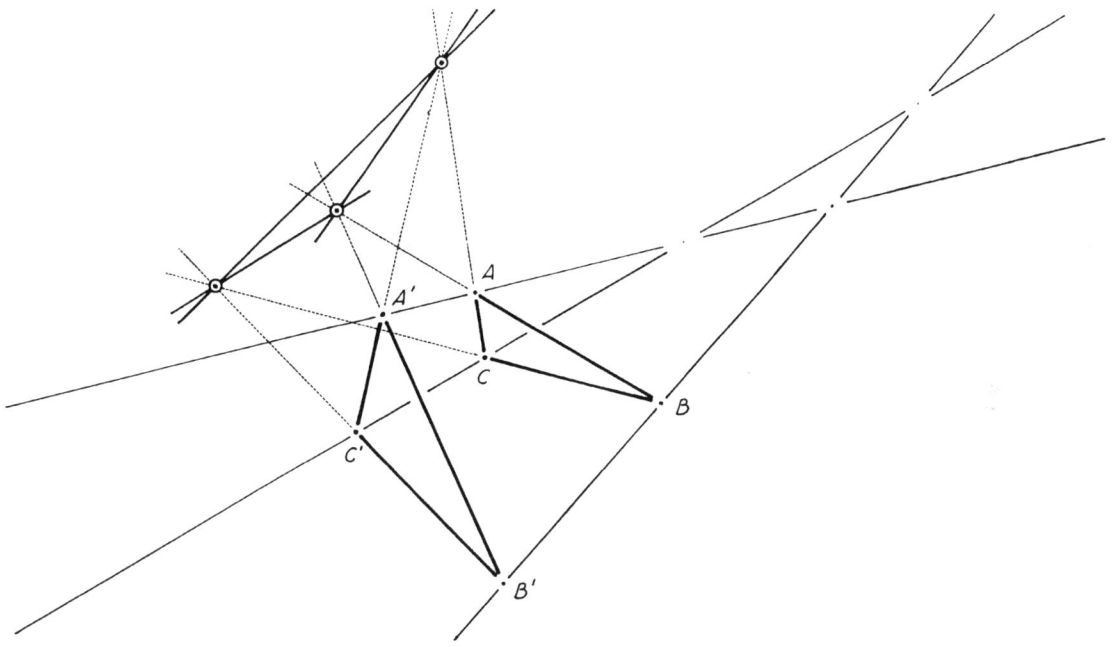

Figure 28

Desargues' Theorem shows the special case of this general situation—the simple yet wonderful fact that *if* the triangles are so arranged that the three lines given by the points of the two triangles *lie in a point,* then the three points given by the pairs of lines of the two triangles will *lie in a line,* and vice versa. Moreover, this holds true for *any* two triangles, whether they lie at random in a plane or anywhere in the whole of space!

Figure 29

Figure 30

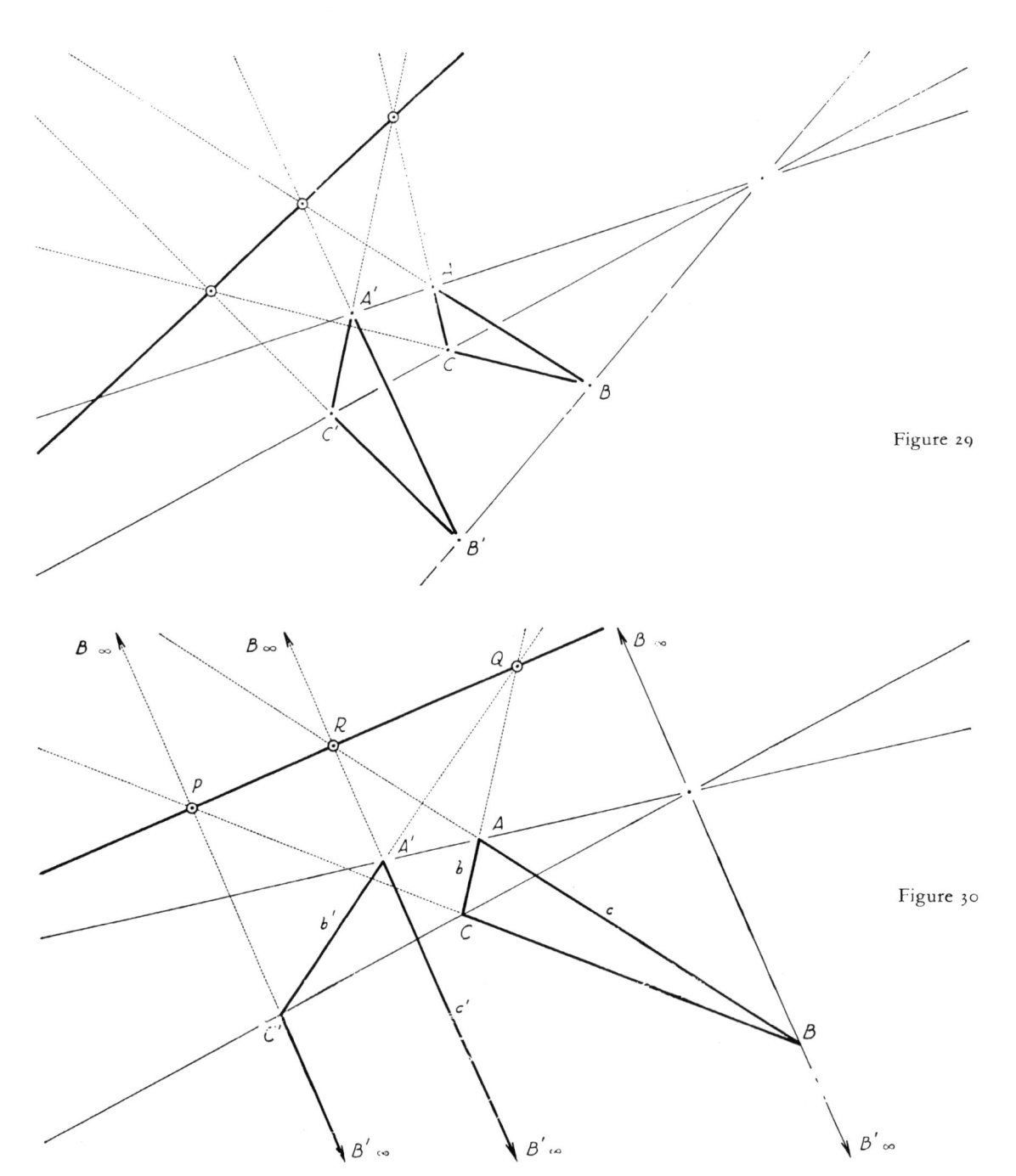

III/80

Statement of Desargues' Two-triangle Theorem: If two triangles are such that the lines common to corresponding points meet in a point, then the points common to corresponding lines lie in a line, and vice versa (Figure 29).

In drawing the triangles, the only rule to be observed is accuracy; apart from this there is complete freedom. The drawing will, of course, serve its purpose best if it is so arranged that where necessary all points lie on the paper. It belongs, however, to this exercise, that we learn to use infinitely distant elements and come to see that they play exactly the same part as those that are in the finite. In Figure 30, for example, the point of one of the triangles (B') is infinitely distant.

We practise the construction in various forms, taste the freedom of expression which the statement allows and see how the infinitely distant elements not only play their part with all the others, but are indispensable, if the statement is to be complete.

In the example, Figure 31, A'B'C' is a triangle with the

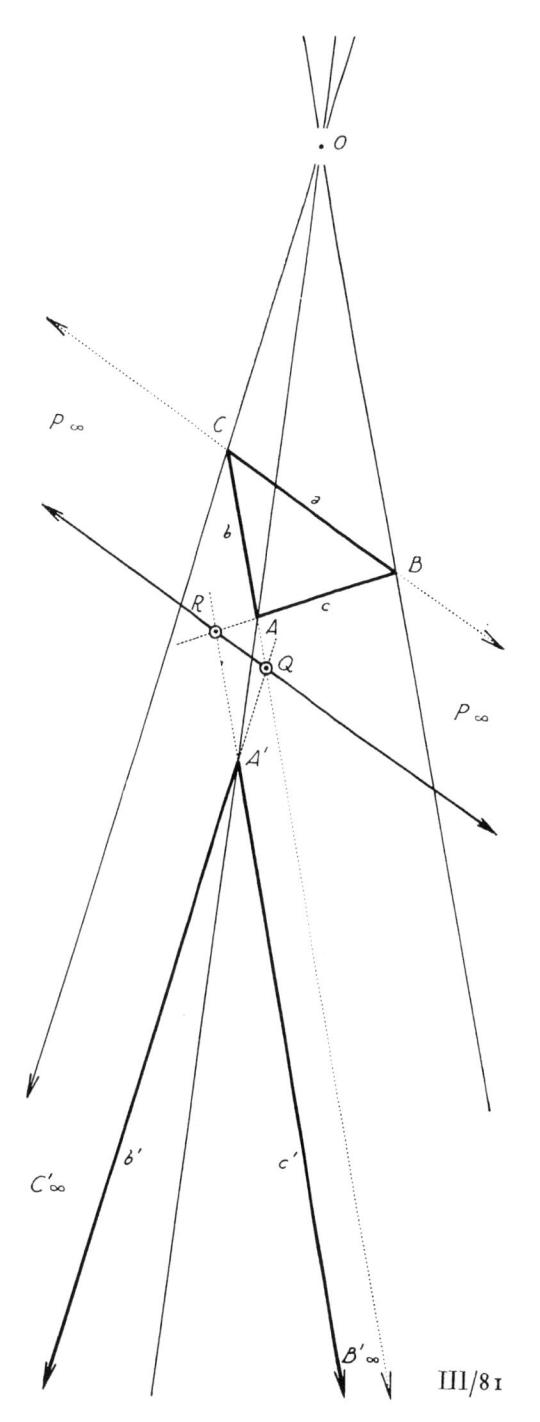

Figure 31

points B' and C' (and therefore the side a') infinitely distant. Choose a point O and draw in it the three lines to the points A', B', and C'. (As we have seen, the lines joining O to the infinitely distant points B' and C' on the line a' of the triangle will have to be drawn parallel to the sides c' and b' respectively.)

A second triangle ABC may then be constructed in *any* situation, provided that its points lie one on each of the three lines of O (that is, A on OA', B on OB', C on OC'). It will be seen that the points P, Q and R in which the paired lines meet respectively will *always* lie in a line.

In our case, the lines c and c' meet in R; b and b' in Q. We then discover that the line a lies exactly parallel to the line determined by R and Q! These two lines therefore meet in an infinitely distant point, namely in P. P is therefore on the infinitely distant line of the plane in which the two triangles lie, *which is also the side a' of the triangle A'B'C'*. The corresponding sides aa' of the two triangles meet therefore in P, which is in line with Q and R, in the line O. (In such manner further examples may be understood and carried out.)

As long as the two triangles lie in a single plane, it is not possible to prove the theorem with projective methods alone; when, however, they lie in two different planes, i.e. freely in space, the statement becomes almost self-evident.

Picture two triangles lying in two different planes, such that their corresponding sides meet in pairs in three points of the line o common to the two planes. It will follow that each corresponding pair of triangle sides will also determine a plane and the three planes thus determined will form a trihedron or three-sided pyramid, whose sides converge in the point or apex O (Figure 32).

Had we started with the opposite aspect and considered the triangles to be plane sections of a trihedron with apex at O, it would have followed that corresponding pairs of triangle sides each determine one of the trihedron planes; each pair of sides must therefore have a common point which can only lie in the line o common to the planes of the two triangles.

On the basis of this situation the proof, resting as it does on the Axioms of Community of point, line and plane, is

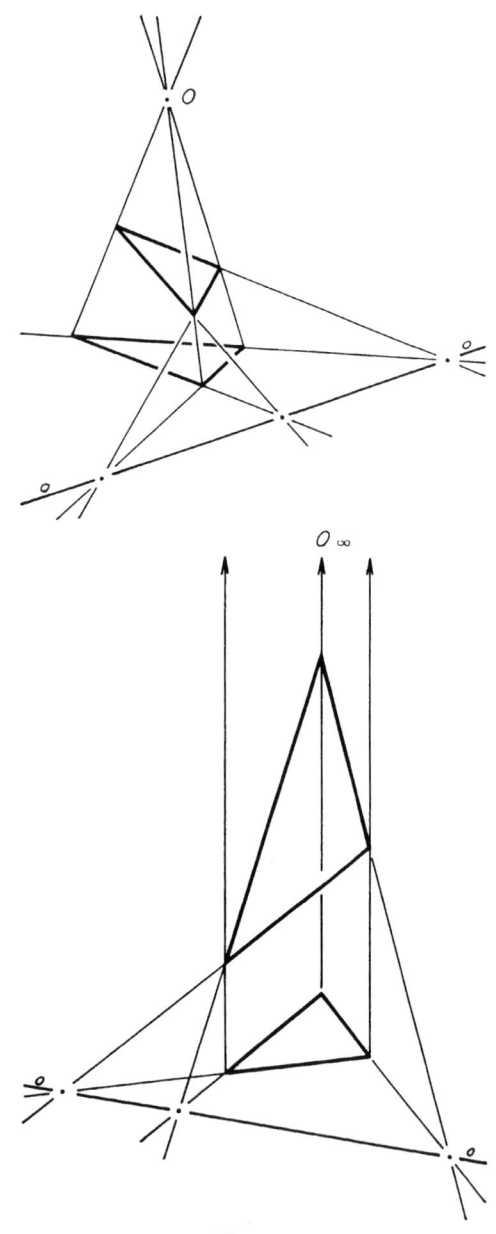

Figure 32

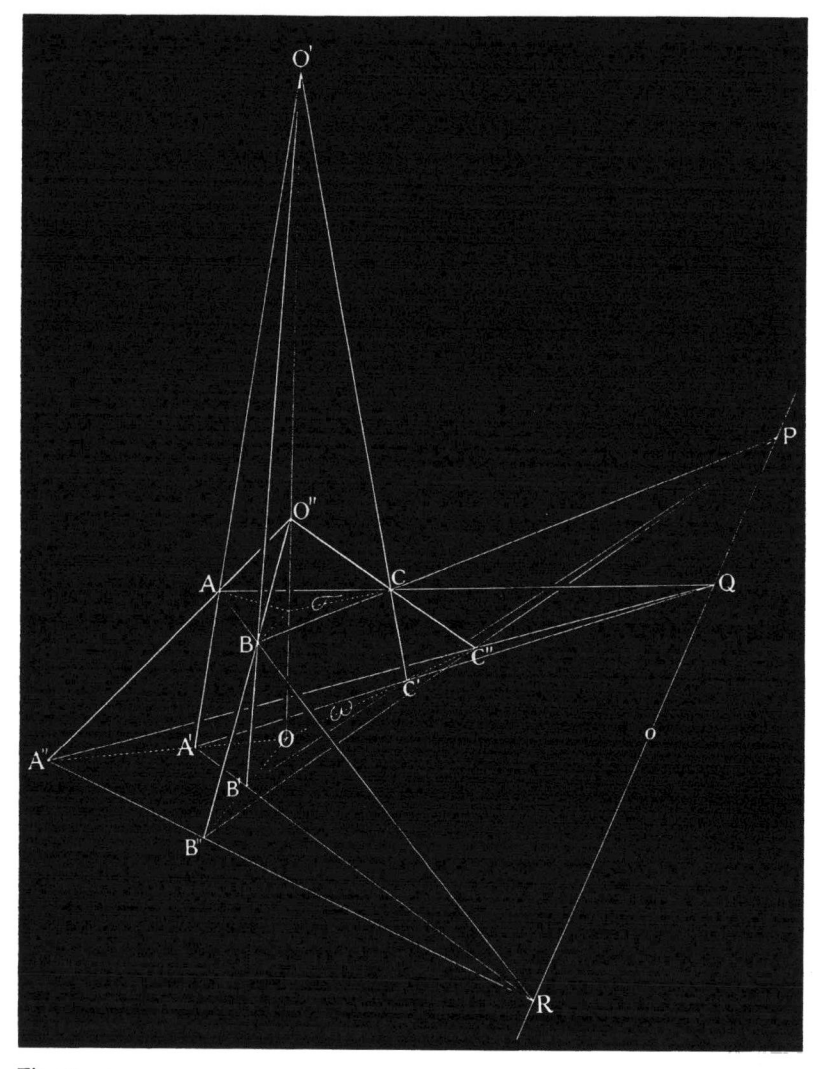

Figure 33

straightforward (20). Varying the construction, we experience its manifold possibilities, including the cases when infinitely distant elements function just as well as those in the finite.

The theorem shows that the quality of incidence is far more fundamental than any question of size and measurement. The all-important factor is *relationship*—the related situations of the various members. The movement of any one or even of all the members is possible without in any way impairing the underlying harmony of the whole.

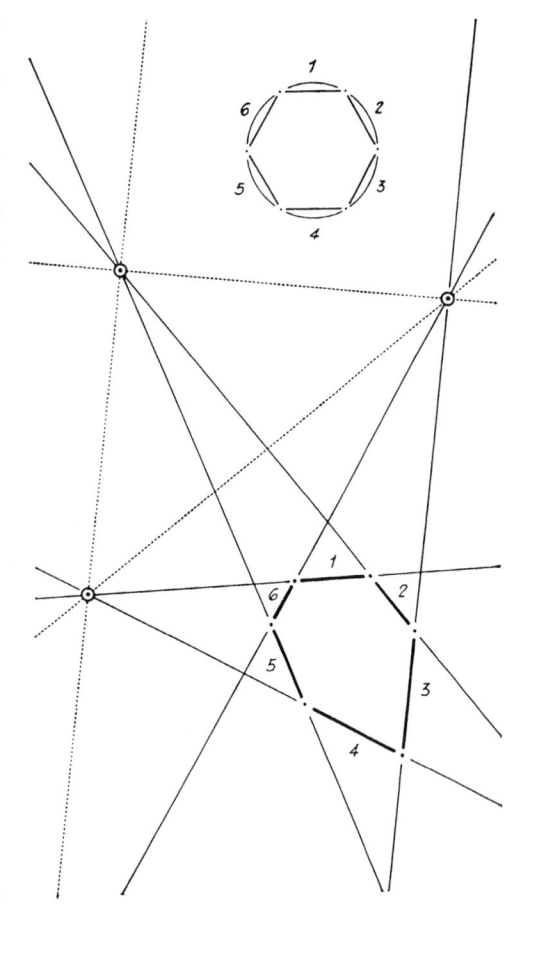

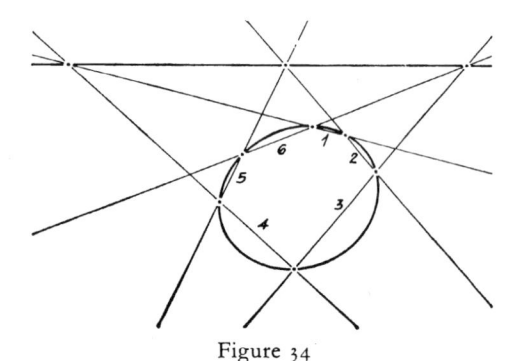

There are altogether ten points and ten lines and they are so arranged that in every line there are three points and in every point three lines. Thus all lines and all points are of equal value. Every line can play the part of o and every point the part of O. The choice of O, however, determines o and also the two triangles. Figure 33 shows the complete configuration for Desargues' theorem.

Pascal's Theorem

Blaise Pascal (1623–1662), Desargues' celebrated pupil, contributed substantially to the theory of the conic sections. He discovered his famous theorem in 1640, at the age of sixteen. It deals with six points of a plane; these form a hexagon, but our conception of a hexagon must be wider than the fixed form of regular measure. Given any six points, we can draw the six lines that join them in successive order, returning to the point with which we started. The lines follow one another from one to six, and we can call "opposite" pairs (from the Euclidean point of view) the first and fourth, second and fifth, the third and sixth respectively.

Opposite pairs meet in three points, and these as a rule will form a triangle. Pascal tells us however that if these three points do not form a triangle, but *are in line,* all six hexagon points will lie on a conic. Six arbitrary points will not generally lie on a single conic; they will only do so if Pascal's condition is fulfilled (Figure 34).

Statement of Pascal's Theorem: If a hexagon be drawn through any six points of a conic, the three meeting-points of "opposite" pairs of lines will be in line. Or more precisely: *The points of intersection of the opposite sides of a hexagon inscribed in a conic are collinear.*

The remarkable thing is that given *any* conic (circle, ellipse, parabola or hyperbola), we can choose the six points freely, *anywhere* on the curve, and can join them in *any* cyclic order. Thus from the ordinary point of view our hexagon will look quite "jazz", but in the light of projective

Figure 34

geometry will still be a hexagon. The points in which "opposite" pairs of lines meet will always be in line! This line is called the Pascal Line of the hexagon. We know from Euclid how to draw a conic section. Take a ready-made curve and prick it through to a clean sheet; then choose on it quite freely any six points and find the Pascal Line. What mysterious law lies behind this freedom? No wonder Pascal called this the Hexagrammum Mysticum! (Figure 35).

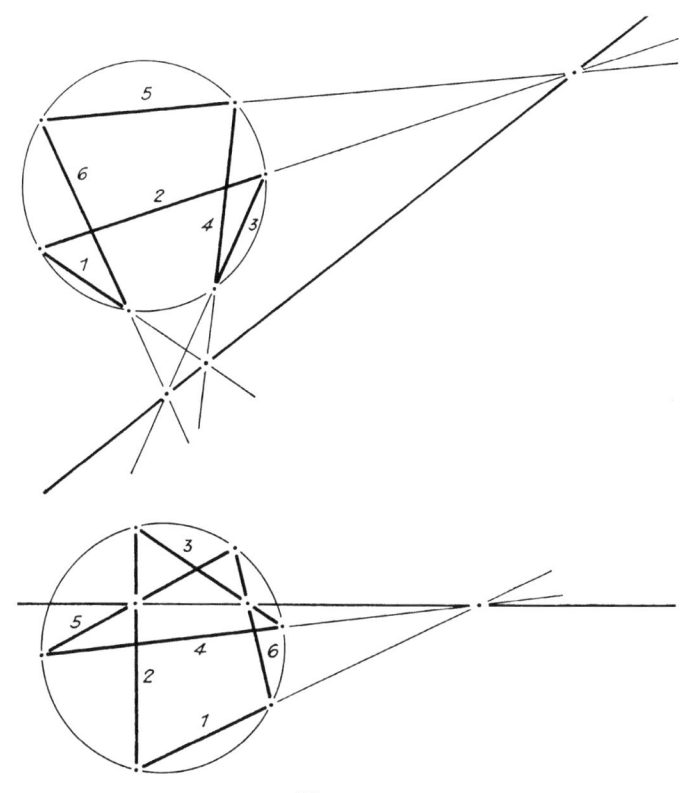

Figure 35

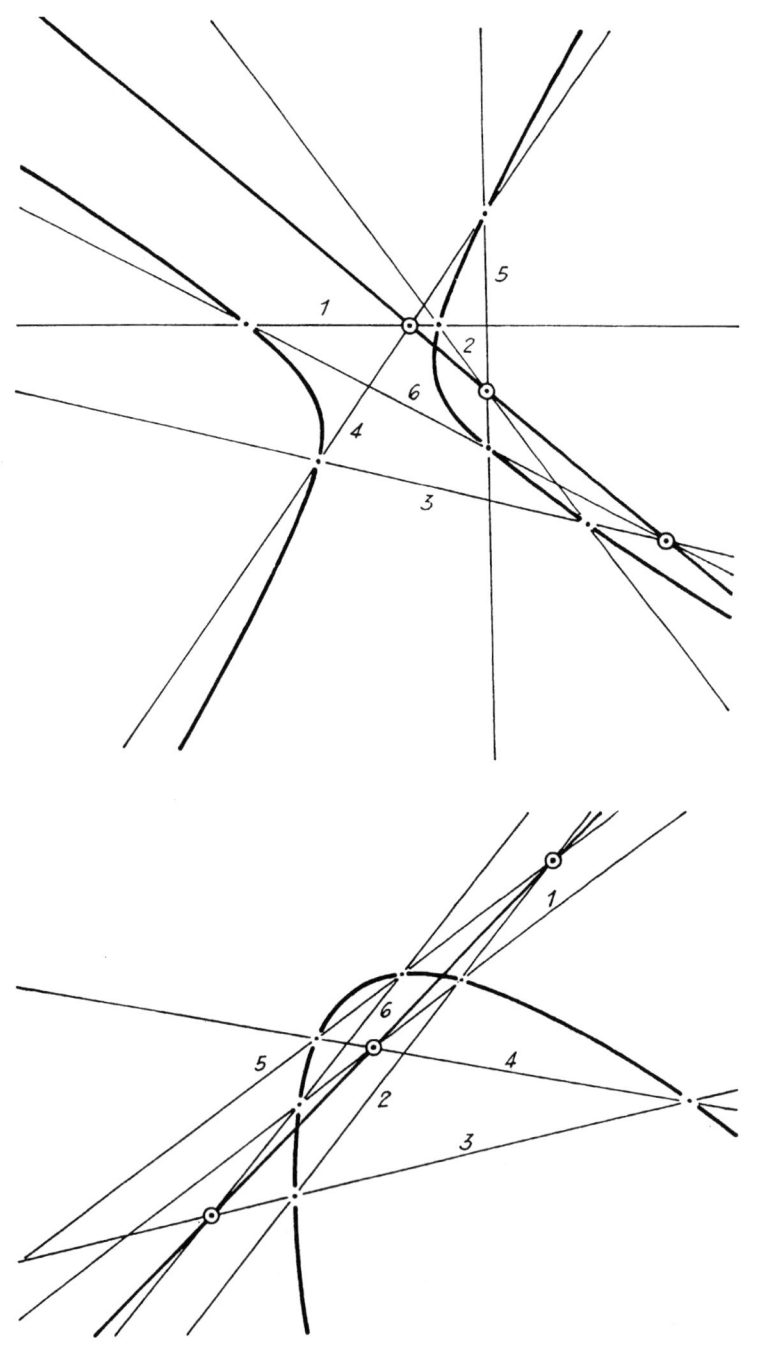

III/86

Figure 36

This Pascal construction holds good for all conics, i.e ellipse, parabola, hyperbola as well as circle, and in all cases the cyclic order of the six points can be chosen at random (Figure 36).

It is interesting to realise that for any six given points there will be sixty Pascal Lines, according to the way we join them, though if the points are very regularly situated, some of these lines will prove identical with one another.

There are in fact twelve types of hexagon, determined by six regularly spaced points of a circle. Considering the number of distinct figures given by rotation, the number of hexagons for each type is given here, the overall total being sixty. It is a useful exercise to find the Pascal Line for each of the twelve types, some of which will have one of the three points in the infinite, and some all three. Thus, for the regular hexagon inscribed in a circle, the Pascal Line is the infinitely distant line of the plane in which the hexagon lies (Figure 37).

Pascal found and proved this theorem by Euclidean means, using proportions, but in fact measurement plays no part in it at all. The constructions rest entirely on the relationships between points and lines, that is, on the quality of incidence. We shall see later on how the essential quality of these two theorems is truly projective; that is to say, *position* is the factor upon which they entirely depend and not measurement.

The beginning of the new way of thinking had been laid down, but it remained overshadowed for centuries by the tremendous impetus given to natural science by men like Galileo and Newton, who knew well how to use the tools provided for them by Descartes.

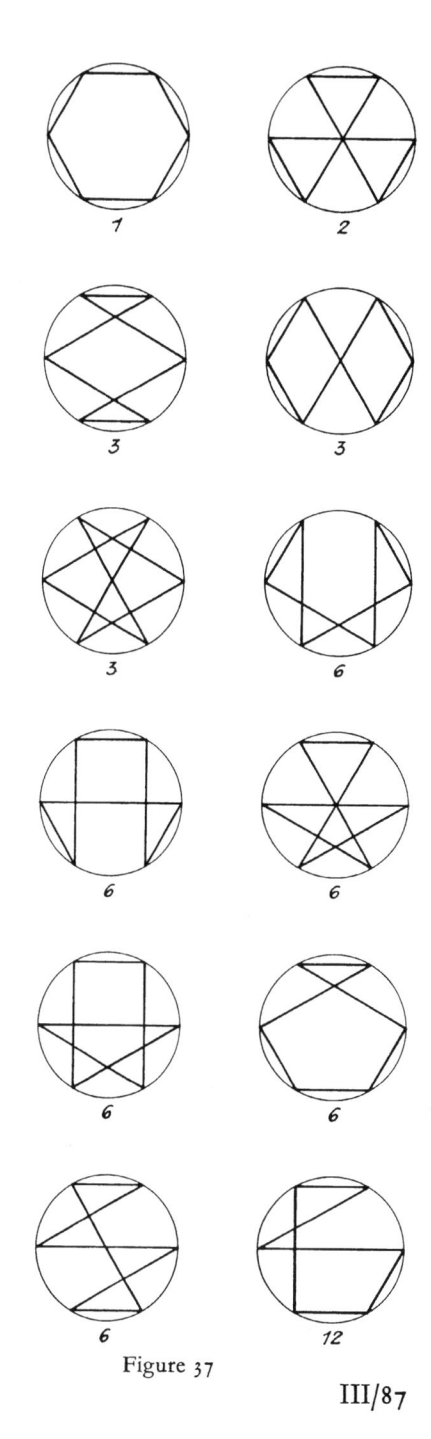

Figure 37

The world of ideas on the one hand and of sense-perception on the other had become the field of action for the individual. It had become possible for men to think their way into the laws of natural phenomena as never before; to contemplate, for example, the phenomenon of the falling stone in a way which left the soul free, untouched by the actual elemental forces involved. Galileo's laws of falling bodies, Newton's theory of gravitation, these were the outcome of the intelligent use of the abstract mathematical formulae, which could be applied to whatsoever phenomenon presented itself. Of the two friends, Descartes and Desargues, who worked together at the dawn of a new age in the evolution of mankind, it is Descartes' name which has become famous, while Desargues has remained unnoticed.

IV Further Discoveries; Duality and Projection

It must not be thought that all the mathematical problems at which men were working were new: many may be traced far back into antiquity. But, matured in the minds of great thinkers, the problems became resolved and reduced to that form of simplicity which is the mark of a true scientific theory.

In the decades following the death of Desargues and Pascal in 1662, while men continued to develop the vast field of mathematics in particular towards utilitarian ends, many a spirit, moved by the sheer love of pure mathematical thought and discipline, turned also to the seemingly less useful but often far more beautiful pathways of mathematical thought. The Serbo-Croatian, R. J. Boskovic (1711–1787) published in 1757 his work on the theory of conics, which deals brilliantly with problems concerning transformations through the infinite (21). Then, towards the end of the eighteenth century many men were born in the different countries of Europe whose work, published mostly in the first half of the nineteenth century, led to the unfolding and blossoming of pure projective geometry. Among them was Jean Victor Poncelet (1788–1867).

Poncelet, a Frenchman, contemporary of Lambert, Monge, Brianchon, Legendre, Carnot, the Swiss mathematician Jacob Steiner and the German, Gauss, was a lieutenant of the French Army. Twenty-four years old, he was left lying among the dead on the battlefield of Krasnoi, during the ghastly retreat from Moscow in 1812. A prisoner and—as he describes it himself—after terrible ordeals and sickness, recuperating in the gentle sunshine of the Russian springtime, he felt the strong urge towards some spiritual activity. He undertook, without books or belongings, to rethink and rediscover through his own initiative what he had known of mathematics. During the following two years he created, as though out of nothing, what in fact became the firm foundations of the new geometry, in which the ideas of metamorphosis, of the infinite widths of space and of the line- and plane-woven creation of forms are the leading thoughts.

His work, published in 1822 in Metz, some years after his return to France, *Traité des propriétés projectives des Figures,* states clearly that not only has every line its infinitely distant point, but the infinite distances of every plane must be thought of as a straight line and the infinite distances of space as a single plane (22).

The "Principe de Continuité", which underlies all Poncelet's considerations, opens the way to the clear understanding of the part played by the infinitely distant elements in the transformations of the conic sections one into the other.

We will attempt, step by step, to approach these truths of projective geometry, leaving aside (for the sake of simplicity) much which the trained mathematician might well deem it necessary to include and towards which he will find some guidance in the *Notes and Bibliography*.

The reader may find our next steps a little arduous, but his perseverance will be well rewarded later on. He will appreciate at a deeper level the reasons for all the freedom of construction already experienced in the Line-woven Nets and the theorems of Desargues and Pascal.

Michelangelo (1475–1564)

Man between Weight and Lightness

Projective geometry gives a twofold approach to the creation of forms, and this gives access to a truer and more complete understanding of all form. We learn to see that all form is in fact the outcome of the interplay and union of opposites. In any single instance it may be the one or the other quality which is more in evidence, giving the form its individual characteristics.

In a truly balanced form the two extremes are in wonderful equilibrium as in the perfect human body, poised as it is between the dark, unconscious realm of will, in the gravity of earth, and the powers of the light of consciousness, whereby man stands erect.

In our geometrical considerations the two extremes—they are polarities—are represented by the point and the plane (in the geometry of the plane it is the point and the line). The point with its centric quality and its characteristic way of moving, and the plane or the line, which move with a sweeping, moulding gesture—these are the progenitors of all forms. We attain to a deeper appreciation of form if we add a further conception to that way of thinking which sees a form as an assemblage of points or atoms. We learn to experience the modelling, plastic activity which is characteristic of planes and lines as they contribute to the creation of a form from outside.

Together with the great artists and mathematicians at the time of dawning awakening at the beginning of modern time, we seek the artistic experience and the clear understanding of the upward-bearing, planar forces of the living world (see Chapter IX). The step now taken in geometrical thinking allows of the conception, not only of forms but also of forces in polar contrast to one another. The time will come when in the light of man's understanding it will dawn upon him that the very forces with which he stands upright and walks the Earth are the forces of the Resurrection.

Resurrection of Christ. Tombstone about 1550. Laufen, S. Germany

Victory over the Three Dimensions of Earth

The Principle of Duality

This distinctive and vitally important characteristic of projective geometry gives it its quality of *wholeness*, not shared by the Euclidean system nor by any of the metrical geometries. Any statement concerning incidence of points and lines in the plane gives rise to a second statement obtained from the first by interchanging the words "point" and "line". Each statement has its dual.

We have seen that a line can be thought of as a single and undivided whole, and likewise a point. But we know also that a line may be an organism of points, a line-of-points, and that a point may be an organism of lines, a point-of-lines. Further, we are conversant with the concepts: plane-of-points and point-of-planes.

The fundamental relations of space may be expressed by saying: "As plane is to point, so is point to plane." This is generally referred to as the *Principle of Duality*. (The word "*Polarity*" is however a more apt description). Within the two dimensions of the plane, the fundamental relations can be expressed by saying: "As lines are to points, so are points to lines." This, too, is called the *Principle of Duality*.

In this book we shall make a clear distinction, for we believe it to be valuable to avoid a confusion of terms in connection with this very important theme. We shall call the relations between planes and points in space the *Principle of Polarity,* and shall reserve the term *Principle of Duality* for the dual relations between the points and lines of the plane, and also (as will be seen later on) between the lines and planes of the point. In the polar-reciprocal relations of plane and point in space, the line—the third element—plays an intermediary role. It would, in fact, be more truly applicable to use the word "Trinity" in this connection (23). In the following chapters we shall be repeatedly practising the Principle of Duality *in the plane*, as it plays its part between point and line.

Perspective and Projective Relationships

We shall first consider the elementary facts concerning perspective and projection in the plane—that is, between point and line—which will lead us to the projective construction of plane curves.

People often experience difficulty at first in thinking out the dual constructions. We are so unused to this peculiar way of thinking geometrically. The Principle of Duality is, however, the very quality which gives to modern geometry its task for the future. An inner initiative in thinking is demanded, which cannot rely on the support given by the pictures of the sense world. The abstract conceptions of the old geometry rest, for example, on such pictures as a square field or a cubic box or building. In modern geometry we practise a mode of thinking for which no such sense-perceptible pictures or habits of thought exist. The Principle of Duality turns the geometrical picture into its opposite in a very subtle way; it is a turning inside-out which has no ordinary spatial counterpart. The activity of thinking thus involved is of great value in education. Furthermore, there are many good reasons why the thought processes of this geometry should in the future permeate and educate man's spiritual activity of thinking.

It will be a help to realise that in seeking the dual of the point common to two lines (namely, the line common to two points) we have to *draw in the line* joining the two points. It is characteristic that points remain separate until joined by a line, while the common point of two lines is always there; we only have to find it.

The fundamental process is the establishment of one-to-one correspondences in the plane between points and lines, as follows (Figure 1):

Figure 1

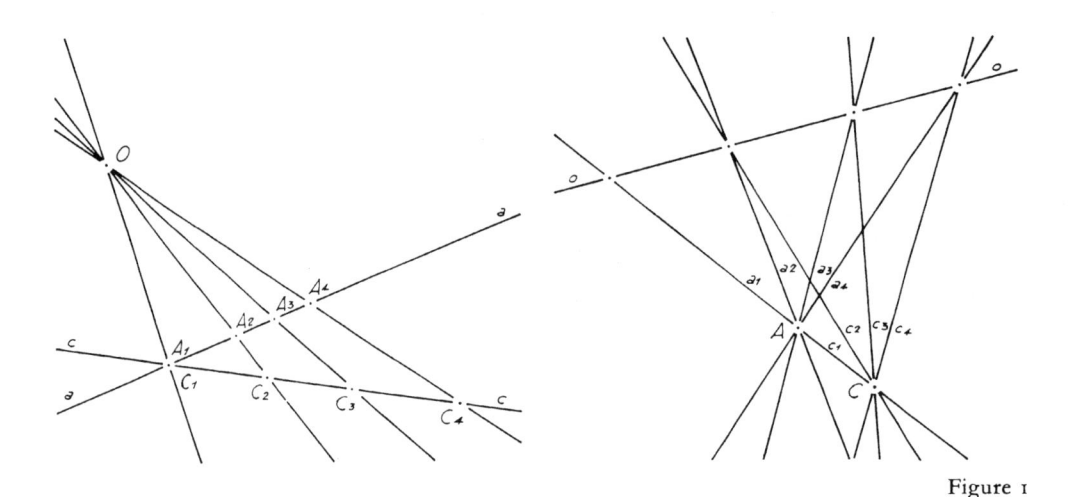

Perspectivities

Two lines are put in direct perspective with one another, point by point, by an intermediate point—as it were an eye. Every point of the one line thus corresponds with a point of the other, inasmuch as they lie along the self-same ray of the eye (or centre of perspective) which overlooks them both. The meeting point of the two lines, considered as a point of either line, will correspond to itself.

We write: $A_1A_2A_3A_4 \overline{\wedge} C_1C_2C_3C_4$, which means that the points $A_1A_2A_3A_4$ are in direct perspective with the points $C_1C_2C_3C_4$. Or, when the centre of perspective is O:

$$A_1A_2A_3A_4 \underset{\wedge}{\overset{O}{=}} C_1C_2C_3C_4.$$

Two points are put in direct perspective with one another, line by line, by an intermediate line—as it were a common horizon. Every line of the one point thus corresponds with a line of the other, inasmuch as they meet in the self-same point of the horizon (or horizon of perspective) which relates them both. The common line of the two points, considered as a line of either point, will correspond to itself.

We write: $a_1a_2a_3a_4 \overline{\wedge} c_1c_2c_3c_4$, which means that the lines $a_1a_2a_3a_4$ are in direct perspective with the lines $c_1c_2c_3c_4$. Or, when the horizon of perspective is o:

$$a_1a_2a_3a_4 \underset{\wedge}{\overset{o}{=}} c_1c_2c_3c_4.$$

Here and in all that follows, the reader should practise putting the one or the other point or line in the infinite, realising that this will make no fundamental difference to the construction. Thus, for example, the centre of perspective O may be infinitely distant, or the horizon of perspective o may be the infinitely distant line of points in the plane of the construction.

A direct perspective (also called a central perspective) is wholly determined when *two* corresponding pairs are given. It is important to note that in a perspective there will always be a self-corresponding element, namely, either the meeting-point of the two lines or the common line of the two points which are being related one to the other. This is in fact the distinguishing factor between a direct perspective and a projection.

Projectivities

A correspondence between elements may also be engendered indirectly, by two perspectives or by a sequence of perspectives. There may be any number of intermediate steps, but the one-to-one correspondence is maintained as between points-of-lines or lines-of-points. Such a relation is projective; we call a sequence of perspectives a projectivity or projective transformation.

The projective idea is more general than that of perspective; a perspectivity is a special case of a projectivity. In normal speech, we tend to use the word "projection" for what is really a perspective relationship. We "project" a picture on a screen, whereas it is really "perspectived" there, though this would be an awkward term to use. It will, however, be helpful to keep this fact in mind.

A projectivity can be expressed as the product of a sequence of two or more direct perspectives. Thus, for example (Figure 2):

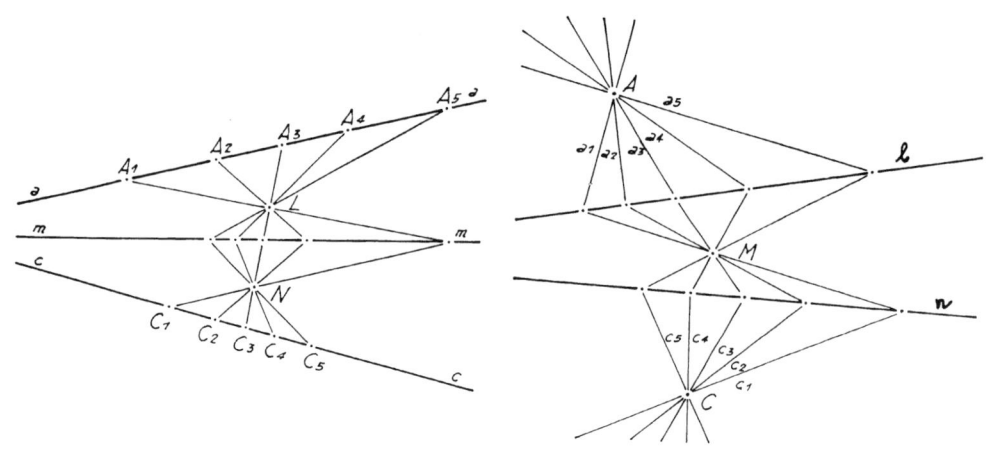

Figure 2

Two lines a and c are related in perspective with one and the same intermediate line m. From the point L, a is seen in perspective, point by point, with m; from the point N, m is seem in perspective with c. By means of two points and one intermediate line, a is thus related point by point with c.

The points A and C are related in perspective with one and the same intermediate point M. By means of the line l, A is in perspective, line by line, with M; by means of the line n, M is in perspective with C. By means of two lines and one intermediate point, A is thus related, line by line, with C.

We write:
$A_1 A_2 A_3 A_4 A_5 \,\bar{\wedge}\, C_1 C_2 C_3 C_4 C_5$

We write:
$a_1 a_2 a_3 a_4 a_5 \,\bar{\wedge}\, c_1 c_2 c_3 c_4 c_5$

This means that the points of the line a are in projection with the points of the line c, and that the lines of the point A are in projection with the lines of the point C. Thus a correspondence between elements is a projectivity, if such a correspondence is made by a sequence of perspectives.

Note that in using the sign $\bar{\wedge}$ the projection is stated without inserting the intervening steps (perspectives). In working out a projection, it is necessary to follow and sometimes to state the steps by which it has been brought about. We would then, in this case, write as follows:

$A_1 A_2 A_3 A_4 A_5 \stackrel{L}{\wedge} M_1 M_2 M_3 M_4 M_5 \stackrel{N}{\wedge} C_1 C_2 C_3 C_4 C_5$ or

$a_1 a_2 a_3 a_4 a_5 \stackrel{l}{\wedge} m_1 m_2 m_3 m_4 m_5 \stackrel{n}{\wedge} c_1 c_2 c_3 c_4 c_5$

Opposite kinds of element may also be related in perspective and projection; for example, the points of a line may be in projection with the lines of a point, and vice versa.

The Projection of Three into Three

Time and again the number Three is decisive in projective geometry. A projectivity is governed by this number and the Fundamental Theorem rests upon this fact (p. 102). If any three pairs of elements (points of two lines or lines of two points) are chosen arbitrarily, they will always project into one another. We will demonstrate this in drawings.

Three pairs of corresponding elements are given (Figure 3). To continue the projectivity, i.e. to find further corresponding pairs of elements, it is necessary to carry out a sequence of perspectives which brings about the given projectivity. There will be very many ways of doing this; we will undertake two of them—simple ones. We choose two active elements and an intermediary one. That is to say:

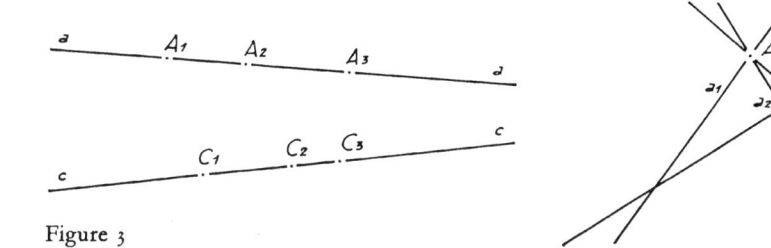

Figure 3

$A_1 A_2 A_3$ may be projected into $C_1 C_2 C_3$ by means of two projecting points L and N and one intermediate line m.

$a_1 a_2 a_3$ may be projected into $c_1 c_2 c_3$ by means of two "horizon" lines l and n and one intermediate point M.

Having chosen any two of these three elements (L, N, m, or l, n, M) the third will be determined according to the situation of the other two. We note that there will be a self-determining point (or line) in each of the two perspectives which bring about the projection. In our first example we choose m and L (M and l) first; N (n) will arise from this, and we see that the common points of m with a and c (the common lines of M with A and C) will be the self-corresponding ones (Figure 4).

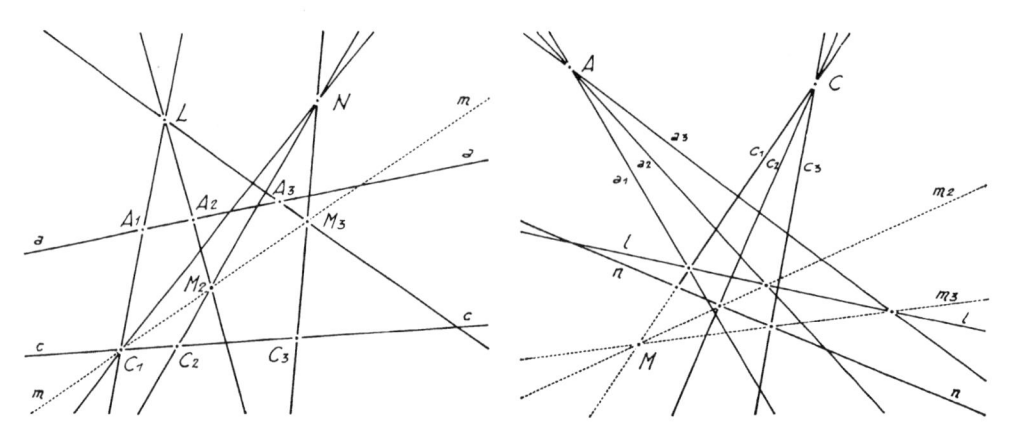

Figure 4

The points $A_1A_2A_3$ and $C_1C_2C_3$ lie on a and c respectively.

Through C_1, for example, draw any intermediate line m and take any centre of perspective L which lies on the line joining A_1 and C_1. From L the points $A_1A_2A_3$ are in perspective with $C_1M_2M_3$ on m. The common point of the lines joining M_2C_2 and M_3C_3 respectively will be N. From N the points $C_1M_2M_3$ are in perspective with $C_1C_2C_3$.

By this sequence of the two perspectives, first from L and then from N, via the line m, the projection of the points $A_1A_2A_3$ into $C_1C_2C_3$ has been effected.

The lines $a_1a_2a_3$ and $c_1c_2c_3$ lie in A and C respectively.

In c_1, for example, choose any intermediate point M and draw any horizon of perspective l which passes through the meeting-point of a_1 and c_1. By means of l the lines $a_1a_2a_3$ are in perspective with $c_1m_2m_3$ in M. The common line of the meeting-points of m_2 and c_2 and of m_3 and c_3 respectively will be n. By means of n the lines $c_1m_2m_3$ are in perspective with $c_1c_2c_3$.

By this sequence of the two perspectives, first by means of l and then by means of n, via the point M, the projection of the lines $a_1a_2a_3$ into $c_1c_2c_3$ has been effected.

Now we will make the construction to bring about the projection in the other way (Figure 5).

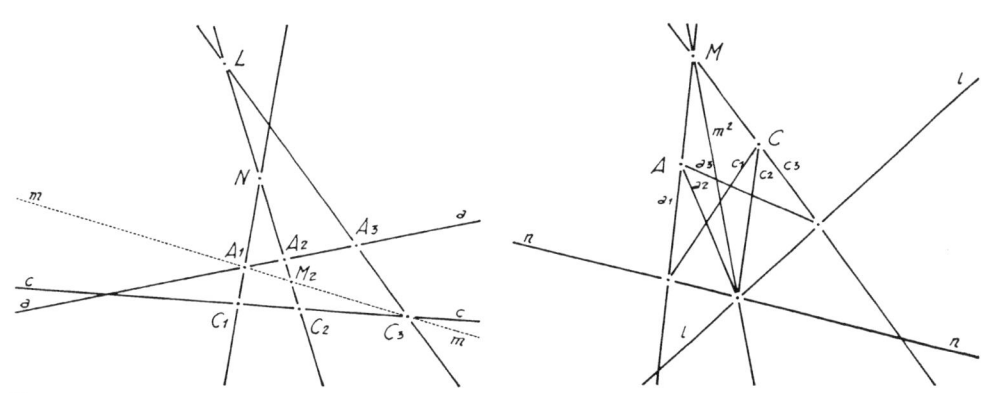

Figure 5

Draw the three lines joining the point-pairs. Let L and N be any two of the three points determined by them; from this will arise the situation of the intermediate line m. (Here it is the line joining A_1 and C_3).	Find the three points in which the line-pairs meet. Let l and n be any two of the three lines determined by them; from this will arise the situation of the intermediate point M. (Here it is the meeting-point of a_1 and c_3).
L being the centre of perspective for a and m, then $A_1A_2A_3 \stackrel{L}{\barwedge} A_1M_2C_3$. The centre of perspective for m and c, being N, then $A_1M_2C_3 \stackrel{N}{\barwedge} C_1C_2C_3$.	l being the horizon of perspective for A and M, then $a_1a_2a_3 \stackrel{l}{\barwedge} a_1m_2c_3$. The horizon of perspective for M and C being n, then $a_1m_2c_3 \stackrel{n}{\barwedge} c_1c_2c_3$.
Thus, by the sequence of the two perspectives: $A_1A_2A_3 \barwedge C_1C_2C_3$.	Thus, by the sequence of the two perspectives $a_1a_2a_3 \barwedge c_1c_2c_3$.

We said before that a perspectivity is a special case of a projectivity, thus, in the aforegoing illustrations:

If m were to pass through the common point of a and c, the self-corresponding points of the two perspectivities would coincide and the result would be a perspectivity between $A_1A_2A_3$ and $C_1C_2C_3$.	If M were to lie in the common line of A and C, the self-corresponding lines of the two perspectivities would coincide and the result would be a perspectivity between $a_1a_2a_3$ and $c_1c_2c_3$.
It would also be a perspectivity if the three lines joining corresponding points A_1C_1, A_2C_2 and A_3C_3, instead of making a triangle, were to meet in a point. (Compare the construction on p. 104).	It would also be a perspectivity if the three points in which the pairs of corresponding lines a_1c_1, a_2c_2 and a_3c_3, instead of making a triangle, were to lie in a line. (Compare the construction on p. 104.)

The Fundamental Theorem

Thus we can make a construction whereby for any freely chosen fourth point, its pair can be found (Figure 6).

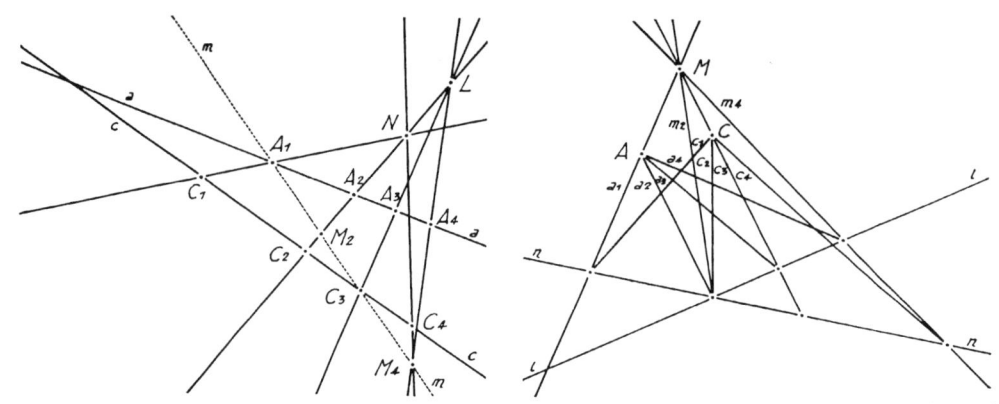

Figure 6

Let L and N be the two centres of perspective establishing the projection $A_1A_2A_3 \,\overline{\wedge}\, C_1C_2C_3$ between the points $A_1A_2A_3$ of line a and the points $C_1C_2C_3$ of line c.

The auxiliary line m used in the process is drawn through a pair of non-corresponding points, say A_1 and C_3 of lines a and c, and M_2 is the projection of point A_2 into a point of m.

Take any fourth point A_4 of the line a and project it from L into M_4 on the line m; then project this point M_4 in turn from N into C_4 on the line c. It is found that A_4 and C_4 correspond to one another in the sequence already established by the first three pairs of points. $A_1A_2A_3A_4 \,\overline{\wedge}\, C_1C_2C_3C_4$.

Let l and n be the two horizons of perspective establishing the projection $a_1a_2a_3 \,\overline{\wedge}\, c_1c_2c_3$ between the lines $a_1a_2a_3$ of point A and the lines $c_1c_2c_3$ of point C.

The auxiliary point M used in the process is the meeting-point of a pair of non-corresponding lines, say a_1 and c_3 of points A and C, and m_2 is the projection of line a_2 into a line of M.

Take any fourth line a_4 of the point A and project it by means of l into m_4 of the point M; then project this line m_4 in turn by means of n into c_4 of the point C. It is found that a_4 and c_4 correspond to one another in the sequence already established by the first three pairs of lines. $a_1a_2a_3a_4 \,\overline{\wedge}\, c_1c_2c_3c_4$.

Put in a quite general way, this theorem says: *A projectivity is fully determined by three corresponding pairs of elements. Through this correspondence the partner of any freely chosen fourth element is predetermined.*

The Theorem of Pappos

It is characteristic of projective geometry that we could have approached it from many sides, and now, having begun in the way we have, there are many ways of proceeding, any of which would lead in the end to the same well-knit whole.

We will now choose to consider another theorem in relation to what we have just seen—the Theorem of Pappos. It is a direct result of the Fundamental Theorem.

This theorem goes back to antiquity. Pappos of Alexandria discovered and proved the following: choose at random any three points on each of any two lines; pair them, and then join the points which are *not* paired. It will be found that in whatever way the construction is carried out, the points in which corresponding line-pairs meet will always fall in line! (Figure 7).

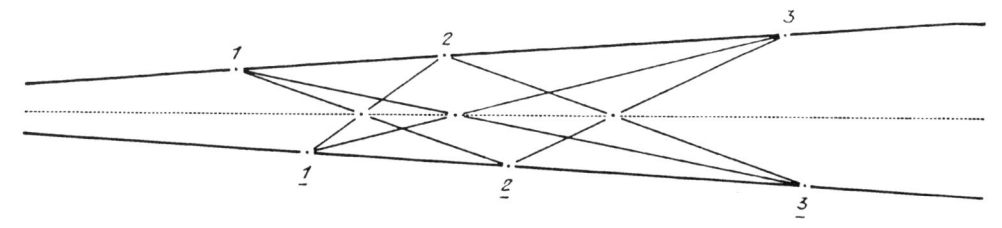

Figure 7

This theorem, too, is fundamental to projective geometry; it contains absolutely no concept of measure. It is significant that centuries had to pass by before the time came for the dual of Pappos's theorem to be discovered (Figure 8).

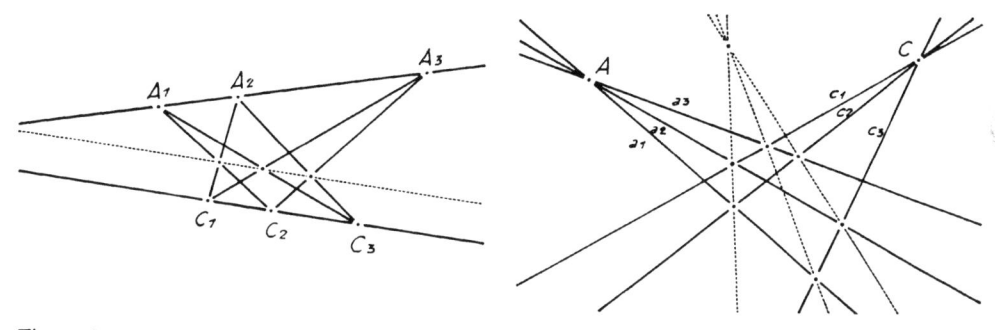

Figure 8

If $A_1 A_2 A_3$ are any three points of a line, and $C_1 C_2 C_3$ any three points of another line, then the three points in which the line-pairs $A_1 C_2$ and $C_1 A_2$, $A_1 C_3$ and $C_1 A_3$, $A_2 C_3$ and $C_2 A_3$ meet respectively are collinear.	If $a_1 a_2 a_3$ are any three lines of a point, and $c_1 c_2 c_3$ any three lines of another point, then the three lines determined by the point-pairs $a_1 c_2$ and $c_1 a_2$, $a_1 c_3$ and $c_1 a_3$, $a_2 c_3$ and $c_2 a_3$ meet respectively are coincident.
The line in which the three new points lie may be called the Pappos Line.	The point in which the three new lines meet may be called the Pappos Point.

The Pappos construction is clearly understandable in the light of the Fundamental Theorem. Any three pairs are bound to be in projection; in the case of the point-pairs, the Pappos line is obviously the intermediary line, while in that of the line-pairs it is the Pappos point which plays the part of the intermediary point.

Let us look for another construction which will bring about the projection as given in the Pappos construction. We shall need in the one instance two projecting points in addition to the intermediary line, which is already there, and in the other instance we must choose two horizons of perspective in addition to the intermediary point, which is already there. Then we can go ahead.

First, however, looking at the Pappos construction, we notice that it might involve a special case: in the pointwise construction it might chance to occur that the three lines containing the points are coincident, while in the linewise

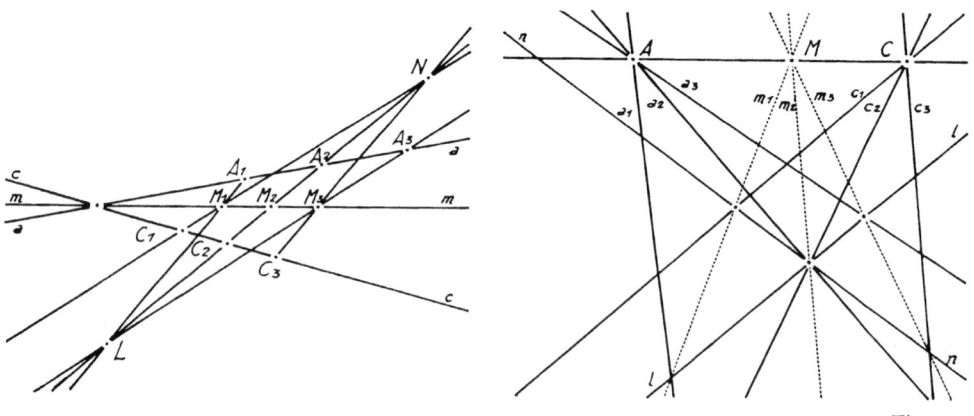

Figure 9

construction the three points containing the lines might be collinear (Figure 9). In this case the points of a and c (or the lines of A and C) would be in direct perspective.

When, however, the three lines a, c and m (the three points A, C and M) form a triangle and we are concerned with the more general case of a projection and not at the same time with a direct perspective, then: $A_1A_2A_3 \barwedge C_1C_2C_3$ or $a_1a_2a_3 \barwedge c_1c_2c_3$. The sequence of perspectives may be brought about as in Figure 5, but choosing L and N rather differently this time (Figure 10).

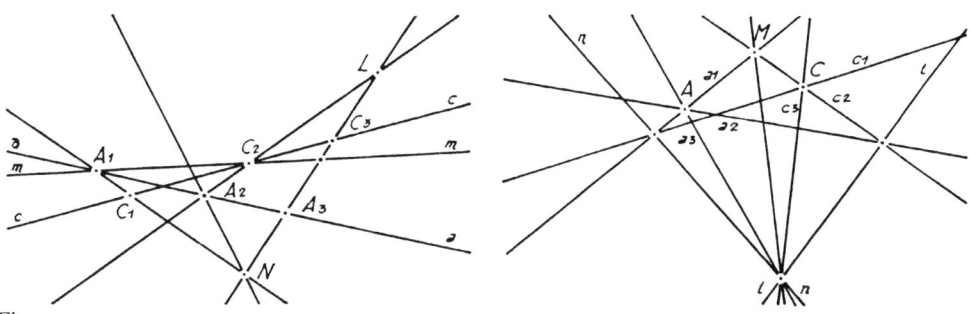

Figure 10

If now we let L move to C_3 and N to A_3, we get the familiar Pappos figure, which can also be regarded as a sequence of perspectives from a to c, in which, however, each point pair can take on the function of L and N (Figure 11). The dual case would be carried out correspondingly.

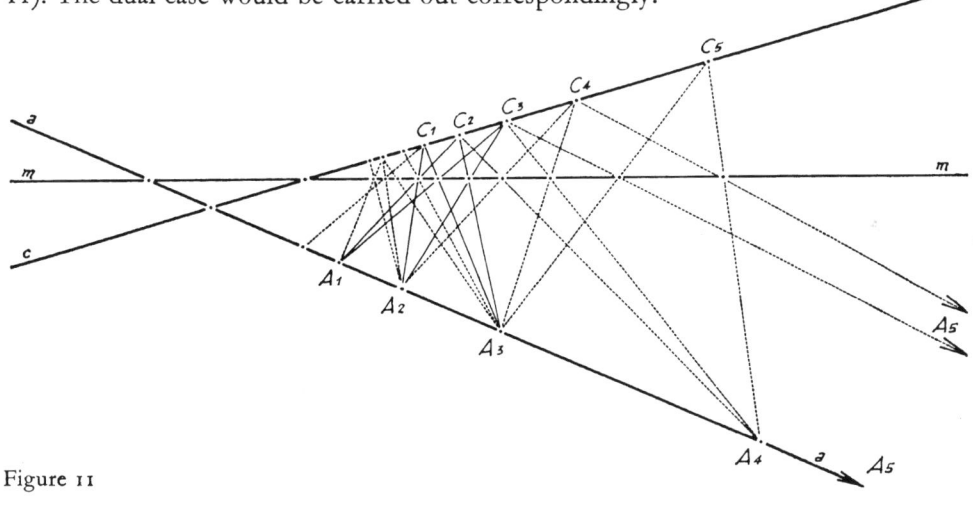

Figure 11

The Pappos construction pointwise. The lines a and c are chosen arbitrarily. The three arbitrarily chosen points $A_1 A_2 A_3$ are then related by the Pappos construction to the three arbitrarily chosen points $C_1 C_2 C_3$. By this construction the line m is determined.

Now choose a point at random on either of the two lines a or c, say A_4 on a, and look for its partner with the help of the Pappos construction. Taking any one of the point-pairs on a and c, say A_2 and C_2, as the projecting points, make the projection of this new point A_4 of a into a point of c, using the intermediary line m.

It will be seen, for example, that a line of C_2 will take the freely chosen point A_4 into a point of m; this point of m will then determine the ray from A_2, which in turn determines C_4. Thus we can say: $A_1 A_2 A_3 A_4 \barwedge C_1 C_2 C_3 C_4$ (Figure 11).

The process can be continued at will, and the strange thing is that for any newly chosen point A_5 the same point C_5 would have been determined had the projection been carried out by any other of the three original pairs of points.

In other words, the line m, which arose like magic from the Pappos construction on the original point-pairs, will continue to act in the same way for any other sets of points on the two lines a and c. This line m is in fact the *axis of projection* for the projectivity between the line a and the line c.

Given the number Three, an elemental, archetypal process is set going which results in an ordered and self-sustaining whole—a projection of all the points of one line into all the points of another line.

The Pappos construction linewise. The points A and C are chosen arbitrarily. The three arbitrarily chosen lines $a_1 a_2 a_3$ are then related by the Pappos construction to the three arbitrarily chosen lines $c_1 c_2 c_3$. By this construction the point M is determined.

Now choose a line at random in either of the two points A or C, say a_4 in A. Taking any one of the line pairs in A and C, say a_2 and c_2, as the projecting lines, make the projection of this new line a_4 of A into a line of C, using the intermediary point M.

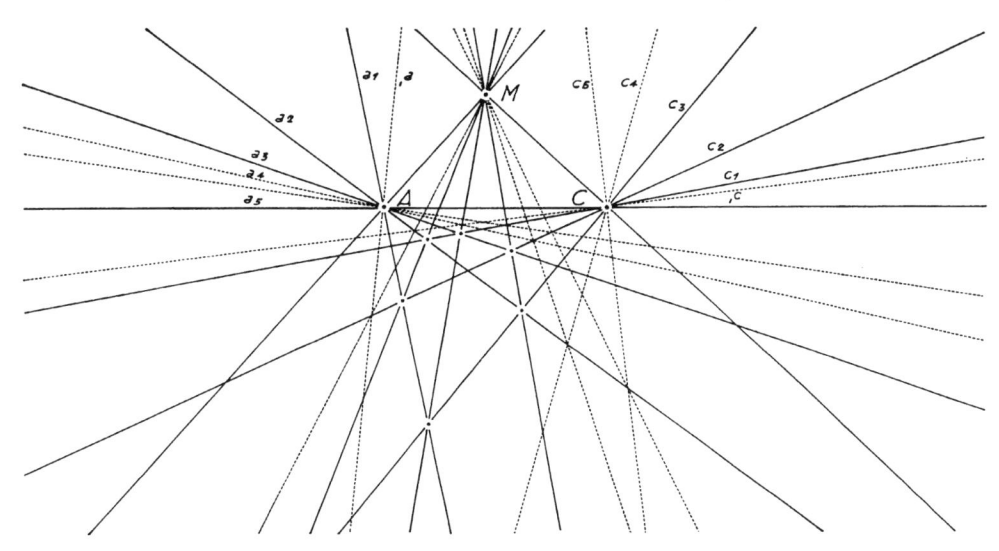

Figure 12

It will be seen, for example, that a point of c_2 will relate the freely chosen line a_4 with a line of M; this line of M will then determine the point of a_2, which in turn determines c_4. Thus we can say: $a_1 a_2 a_3 a_4 \barwedge c_1 c_2 c_3 c_4$ (Figure 12).

Once more, the process can be continued at will, and the important law comes into force: If we look for the partner in the projection of any particular line, we come to the same result, irrespective of which of the already existent line-pairs we use in the construction.

In other words, the point M which arose like magic from the original line-pairs, will continue to act in the same way for any other sets of lines in the two points A and C. This point M is in fact the *centre of projection* for all the lines of the point A into all the lines of the point C.

The reader will find this construction far more difficult to think through than the pointwise one. It is all the more valuable to make the effort. The difficulty arises through the fact that the linewise construction has no familiar spatial comparison with sense-perceptible events, such as the moving of a point-like object from one place to another. In the linewise case it is a matter of moving the infinitely long line from one position into another; it is a rotation, an angular movement of a line which is held in a point.

Projective Creation of Curves—The Rainbow

If now in the point-wise Pappos construction we seek for the *lines common to projectively related point-pairs* and in the linewise construction we seek the *points common to projectively related line-pairs,* a wonderful surprise will be in store for us; unexpectedly, like a rainbow, a beautiful curve will arise.

Indeed, the likeness is not far removed. Amid the weaving of the rays, in seeming chaos and yet in secret order, the curve arises. It is the outcome of the rhythmic interplay of line and point—symbols of light and darkness.

The drawing will best be carried out in the mood of rhythm. We keep going to and fro between line and point—line-of-points, point-of-lines, line-of-points, point-of-lines—in rhythmic alternations. It matters not where we begin, nor how we continue; the curve will always be there in some shape or form, be it ellipse, parabola, hyperbola, or even the circle itself. It will gradually be formed as we go on drawing

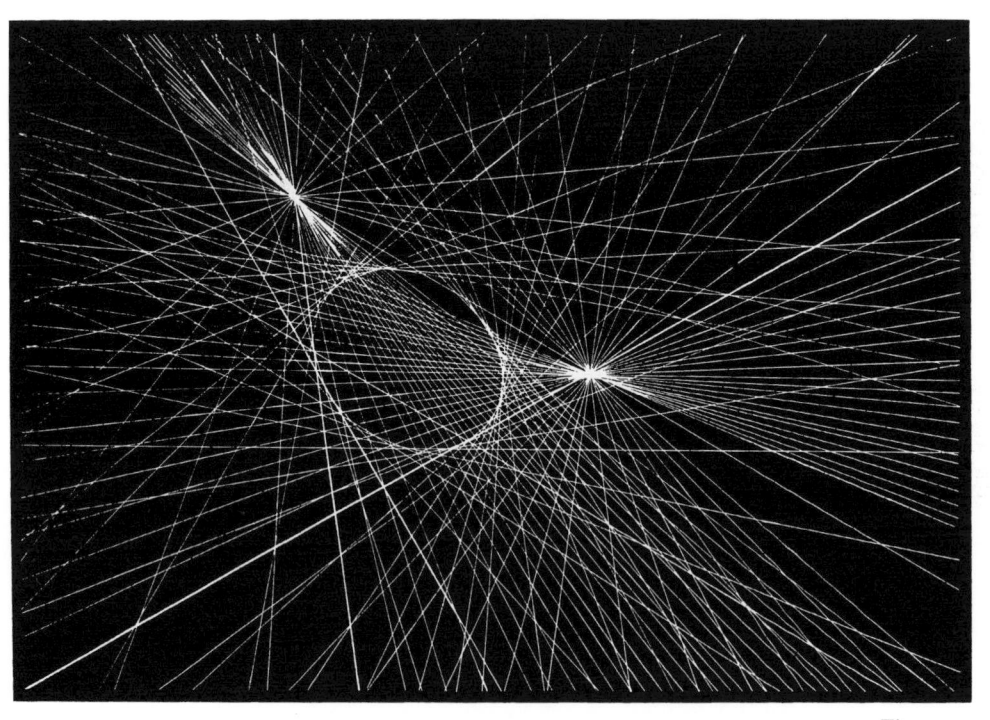

Figure 13

lines through points, or finding the meeting-points of lines. We call it the *"circle-curve"*—a more beautiful and, for projective geometry, far truer name than *conic* or *conic-section*. According to its setting, the curve will be some particular variation of the circle (18, 24).

As with the rainbow, however, we shall only be able to *see* the curve if the right conditions prevail. The lines and points must interweave in such a way that its form, or at least part of it, is on the page. But we may be sure that whenever a complete Pappos construction is faithfully carried out, with all the lines considered in their infinite totality and not merely as segments, the curve will arise somewhere, whether close by, on the page, or away out beyond its boundaries in the far distances of the plane.

It will be as follows: When the projective construction is linewise, that is to say, when *any three lines* of a point are in projection with *any three lines* of another point, the curve will arise traced out by the common *points* of all the lines which are paired by the projection. When the projective construction is pointwise, that is to say, when *any three points* of a line are in projection with *any three points* of another line, the curve will arise enveloped by the common *lines* of all the points which are paired by the projection (Figure 13).

All we have to do is to place the points with their raying lines and the lines with their streaming points in various dispositions on the page and allow them to interrelate with one another according to the projective process we have just learned. It is a *process*. We need only be active, setting the process in motion and thinking clearly in order to perceive how the guiding thought leads from one step to the next, until at last, when we have "come full circle" and have run our course, the curve is born—*a form arising out of movement*! As ripple and wave arise in water or in air, the curve of the harebell flower in the sunlight, the form of man upon the earth, so in the new geometry thought becomes form. In the rhythmically moving interplay of opposites, which brings about the projective relationship, the thought creates a form. A thought itself becomes a form.

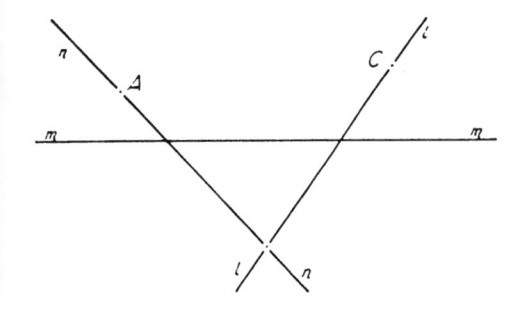

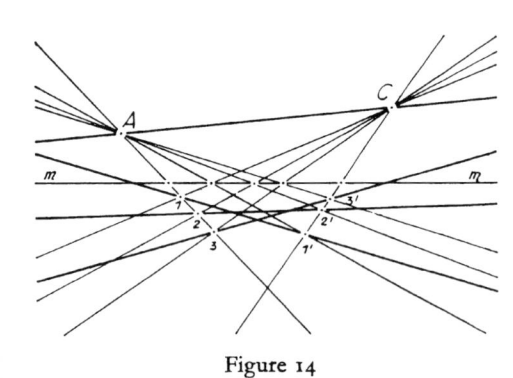

To construct the linewise curve: We must first consider how best to arrange the conditions whereby the curve will arise so that it comes on the page before us. In projective geometry it can so easily happen that necessary elements of a drawing arise far beyond the limits of the paper.

We need the familiar elements: two projecting points, two lines to be related projectively and an axis of projection. Let us call them A and C, n and l, and m respectively.

We note that in carrying out this projective construction, we are not concerned with any measurements which will ultimately belong to the actual curve, once we have it on the page. Neither are we taking our start from any central point or focus; on the contrary, one would rather say that in creating the curve as we are now about to do, we shall be working from the outside. Alone the relative situations of the five elements working together in relation to one another will govern the shape of the curve. This interplay is the *sole governing factor*. According to the infinite possibilities of variation in the situation of these basic entities, there will be many shapes and sizes of ellipse and hyperbola. It is, however, characteristic of the parabola, as also of the circle, that they always have the same shape, due to their closer connection with the infinite.

Let us set the scene for the creation of an ellipse. Having positioned A and C somewhere on the page, we draw in them n and l, the two lines whose points are to be related projectively. We will allow the common point of n and l to be on the page, for the sake of greater ease of construction, and we will put m, the axis of projection, somewhere "between" this common point and the two points A and C (Figure 14).

Having freely set out A, C, n, l and m, we begin by choosing freely, one after the other, points of n and projecting them into points of l (or vice versa). Our procedure might take the form of a kind of refrain, as follows: "Choose a point of n. A line of C takes this point of n into a point of m. A line of A takes it into a point of l. The point of n and the point of l are in projection, and *their common line will be an enveloping line of the curve.*"

Then choose a next point and do the same, and so on and on. We choose as many points as possible (the more the better) all along n, out into the infinite in one direction and back again from the other, *including* the infinitely distant point.

To find and use points which are off the page presents difficulties; one must simply work with an extra sheet of paper and lengthen lines where necessary. Where, however, the points to be used are actually in the infinite there will be no difficulty at all, for it is easy to draw parallels accurately.

There are three lines in the picture which will play a dominant part, and they will also help us to predict where the curve will lie, namely, the lines n and l themselves and the line common to A and C. These three will all turn out to be lines enveloping the curve. Thus:

1. The line n, which is a line of A, takes the common point of l and n (as a point of l) into the point of m which is common to itself and m. The ray from C leaves this point unchanged. The line joining the common point of l and n with the common point of m and n is the line n itself; therefore it envelopes the curve.

2. Similarly for the corresponding line l of C.

3. A line of A, common to A and C takes the point C (as a point of l) into a point of m. The line of C which projects this point of m into a point of n is the line common to A and C. So A and C project into one another and the line joining them envelopes the curve.

As we carry out the construction, passing from step to step, we come in turn to these three lines and find them already included among the enveloping lines. Moreover, we find that infinitely distant points play their part, just as do all the other points; in fact, if we were to leave them out, we should find that our curve would not be whole.

To construct a pointwise curve: The Principle of Duality will of course allow us to reverse the roles of point and line and carry out the dual construction. (Figure 15).

We set the scene with a and c, the two projecting lines, which contain, each in the other, the two points N and L, the lines of which are to be projectively related, with the help of M as centre of projection.

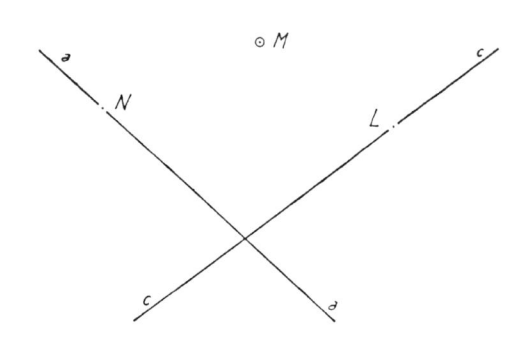

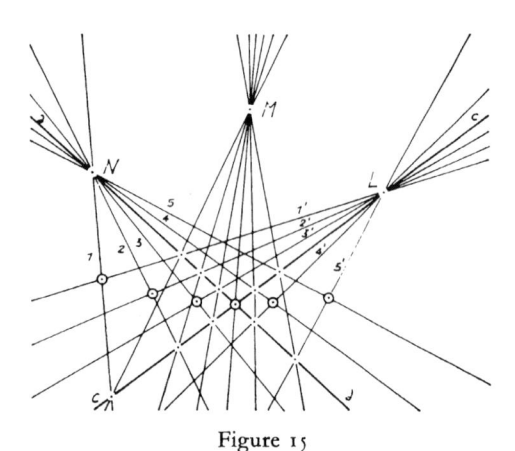

Now we choose freely, one after another, lines of N and project them into lines of L (or vice versa). Our refrain will now be: "Choose a line of N and find its common point with c. This point of c will relate the line of N with a line of M, which also contains a point of a. This point of a will relate the line of M with a line of L. Thus the line of N and the line of L are in projection, and *their common point is a point of the curve.*"

Once the principle of these constructions has been mastered, there will be endless ways in which the creative elements may be arranged in relation to one another, to create different variations of the shape of the circle-curve. We need not be restricted to keeping the two projecting points *in* their respective lines. The five elements may be set anywhere in the plane, as the examples show, and some may be in the infinite. It will be seen that all that is required is the possibility of freedom of interplay between the creative elements. When this is in some way hampered, should certain of the elements coincide, an interesting phenomenon may arise, namely, that the curve will degenerate; it may die into a point-of-lines or a line-of-points. Where, however, the conditions allow us to work freely and accurately, the circling movements of lines and points will bring forth a beautiful curve (30).

Once the curve is there, an outcome of all this movement, we shall want to know more about it; how comes the measured harmony of its form?

Once created, the curve surely has its own characteristic measurements from the Euclidean aspect. How then does the fixed measure of the created form arise? We must look for an incipient quality already inherent in the projective process—the moving interplay of the formless entities, points and lines—which must bring it about that the form, with its actual measurements, takes shape before our eyes. Moreover, we must approach the deeper meaning of the dual aspect of the curves.

Let us turn to another of the fundamental figures upon which projective geometry rests. It is one with which we might well have begun all our considerations.

Figure 15

The Harmonic Forms: Four-point and Four-line

Take a pencil, put down any four points, no three of them in line, and then draw any four lines, no three of them in a point. The lines by which it is possible to combine all four points will be six in number, no more and no less. In the dual case, just six points are determined by the four lines; any four lines, no three of them in a point, will determine six points (Figure 16).

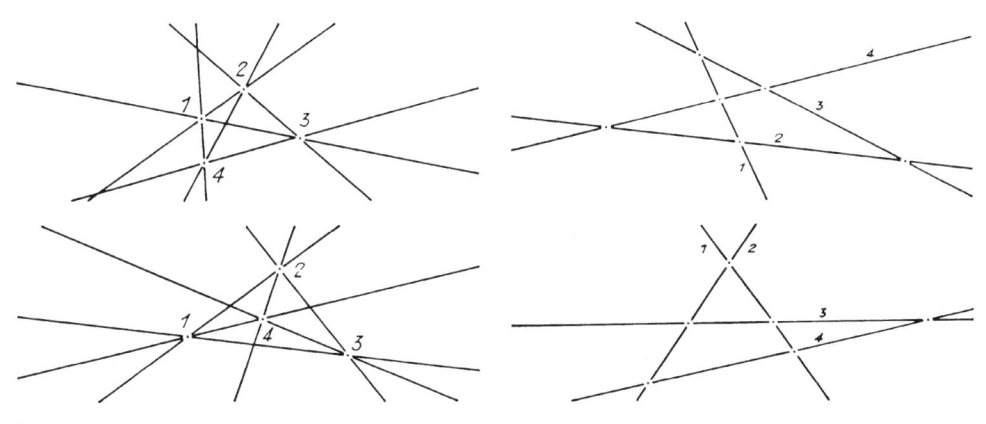

Figure 16

These are the figures called in projective geometry the complete quadrangle and the complete quadrilateral. We shall also call them Four-point and Four-line, and shall see why they are also called the harmonic forms.

We have already made the acquaintance of the complete four-point (p. 56), when we saw that given any three points on a line, the four-point construction will pick out an unique fourth point, which will be paired with one of the original three in such a way that one of the pairs will contain two lines each, and alternating with them a second pair will each contain only one (Figure 17). The pairs may be interchanged, as their function is mutual. Moreover, if two point-pairs on the line have this characteristic relationship with one four-point, they will have it with infinitely many four-points (p. 58).

The theorem, which derives from the work of von Staudt (1798–1867) in its full dual aspect, may be stated as follows (25):

Given a point-pair, and a third point on a line o, any number of quadrangles may be drawn so that five of their six lines pass two by two through each point of the pair and one through the third point. Their sixth line will always pass through one and the same fourth point of the originally given line.	Given a line-pair, and a third line in a point O, any number of quadrilaterals may be drawn so that five of their six points lie two by two on each line of the pair and one on the third line. Their sixth point will always lie on one and the same fourth line of the originally given point.

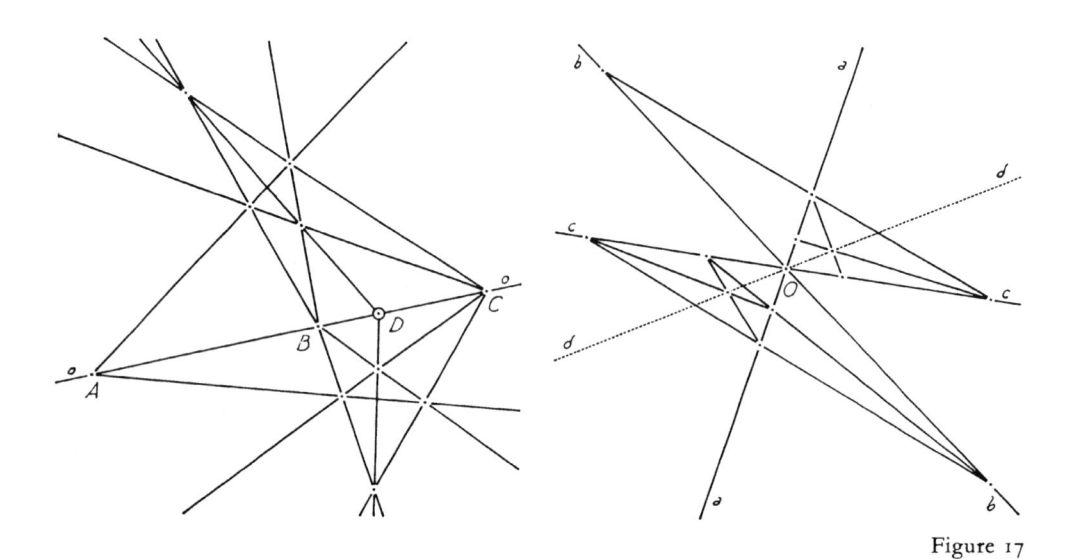

Figure 17

It is important to realise that just as the line-pairs in O separate one another, so do the point-pairs in the line o. A and C are really next to one another on the line o; we have only to let C move out to the right, pass through infinity and return again from the left, and it will approach A without encountering either B or D.

We must pause a moment and consider the ordering of points along a line. Given a number of lines, all meeting in a point and in a plane, they obviously follow one another in "cyclic" order, like points along a circle. The same is true of planes that are in line, but also of points along a line. Thinking of three points on a circle, we cannot say unambiguously that A is between B and C, or that B or C are between the other two respectively. As soon, however, as we take four points, we can pair them in three different ways, say:

DA, BC;   DB, CA;   DC, AB.

With four points along a circle there will always be one of these three pairings, of which we can say that the pairs *separate each other,* while of the other two this will not be true. This is the kind of order which belongs also to four lines meeting in a single point and plane, to the planes of a line, *and* to the points of a line, in our wider, and fundamentally truer conception of geometry. We can pass freely along the line o from A to B by going out to the infinite in one direction and returning from the other, the infinitely distant point of the line in each direction being one and the same. Setting a third point C in the outer section between A and B, and a fourth point D in the inner section, we can truly say that the pairs AB and CD separate one another. (The same applies in the case of the other pairs. Figure 18.)

Figure 18

Harmonic Fours and the Anharmonic Ratio

These pairs—two point-pairs of a line, or two line-pairs of a point—always determined by a four-point in the one instance and a four-line in the other, are called Harmonic Conjugates, or a Harmonic Range. We speak of the "harmonic fourth" in relation to a given three. We note once more that this determination of a harmonic fourth was independent of any measurement of distance or angles. However, hidden in this relation of harmonic fours are beautiful and exact proportionalities of distances or angles, as the case may be. Moreover, these hidden properties are indestructible when the forms in which they are expressed undergo changes by perspective or projection. The harmonic property is typical and fundamental and persists throughout all possible projective metamorphoses. It is expressed by a certain so-called cross-ratio, which is a property of any harmonic range.

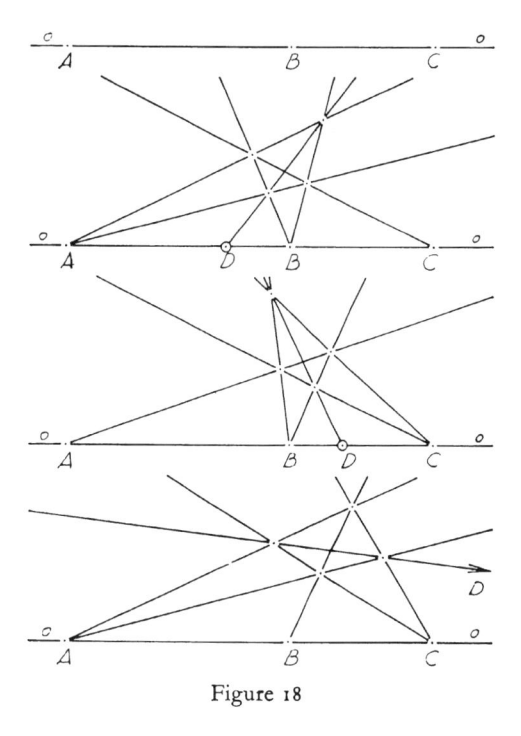

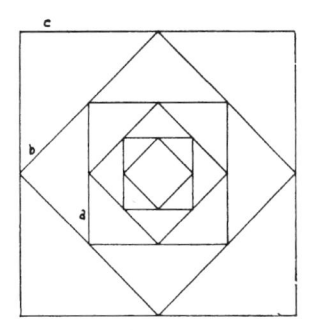

Figure 19

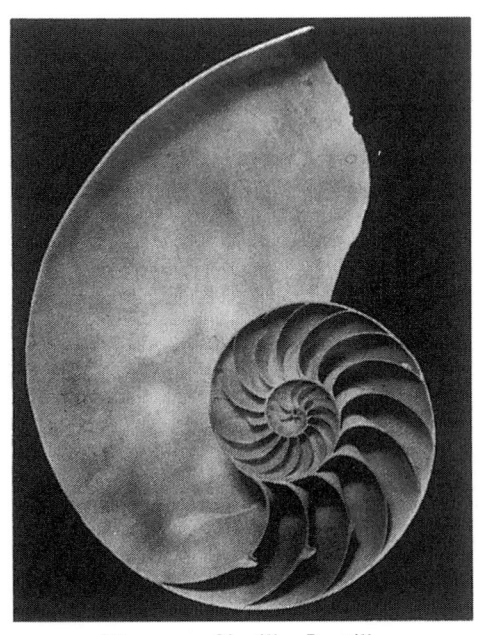

Figure 20: *Nautilius Pompilius* (Vertical Section)

What kind of measure is it which comes to expression in these proportionalities?

While the Euclidean and other metrical geometries take their start from measurement, in projective geometry a measure comes about as the end-result of a mobile process. For example, in the quadrangle net, we saw how step-measure (arithmetical progression) arose in its projective form when the four points of the "horizon-line" were in the finite, and in its normal Euclidean form (equal steps) when one or all of the four points were infinitely distant.

Suppose, however, that instead of setting the forms side-by-side we were to put them *one inside the other*. Then, as we know also from Euclid, a different measure will arise (Figure 19). Whereas with step-measure the actual steps remain constant, in the kind of measure we get when forms are one inside the other there will be a *constant proportion*. It is the ratio or proportion between any two consecutive lengths which remains the same if, for example, squares are inscribed one within the other. The lengths a and b are in a definite proportion to one another, the lengths b and c in the same proportion, and so on. This type of measure is called geometrical progression; we will also call it *Growth Measure*; imitations of this measure are to be found throughout the living kingdom of nature. It is, as we have seen, the type of measure of the equiangular or logarithmic spiral, an approximation of which occurs in many shells, as for instance in the Nautilus (Figure 20).

Growth measure is intrinsically different from step-measure, depending as it does not simply upon repeated steps, but upon a relationship—a proportion—which is maintained between the steps (26).

When, as we shall be doing in Chapter V, we draw the quadrangle net putting the quadrangles one inside (or outside) the other, we shall be working in a field where this kind of measure prevails. The measure along all the lines in this type of net is a growth measure seen in perspective. It appears as an actual geometrical progression when the quadrangles are derived from the infinitely distant line (pp. 125, 127).

There is a third type of measure which is related both to step-measure and growth measure; we will call it *Circling*

Measure (Figure 21). A harmonic range of points in a line (or lines in a point) with the anharmonic ratio therein preserved, is a special case of such a measure, as we shall see in more detail as we go on. The harmonic property maintained among the harmonic pairs, although it is a kind of measure, is still further removed from the actual measurements of a form, directly ascertainable by means of a tape-measure or ruler. It is not even expressed simply by a proportion, but by a constancy between two proportions; it is the *constant relationship between two different proportions—a proportion of proportions*.

For the harmonic quadrangle in Figure 22 we might express it thus: the lengths AB and BC establish a proportion, so too the lengths AD and DC; both may be expressed by a number. The ratio of these two ratios is the Anharmonic Ratio (Cross Ratio); it is the expression of a relationship between relationships, and it holds for every harmonic range.

The projective construction of circle-curves is permeated through and through by the anharmonic ratio, the measure arising in the interplay of relationships. These harmonic proportions give the curves their delicate beauty of form and their musical quality.

In describing these three types of measure, we have touched upon three fundamental aspects of all geometry. We have come by stages away from the type of measure which is typical of the earth. From given concrete measures, we have progressed to *proportions* and then to *proportions of proportions*. The third stage is the one which belongs to projective geometry; it involves the transition in the study of forms from the physical and concrete to the sphere of pure metamorphic thinking.

Many textbooks of projective geometry today still take their start from metrical considerations, although it is in the very nature of this archetypal geometry that it comes to expression in the much freer realm of relationships between relationships. The different measures crystallize out and are end-products of its mobile projective processes, wherever some limiting factor or special spatial situation is brought to bear upon the interplay between these creative elements.

Given a harmonic relation among four points of a line (or four lines of a point), they will follow one another in cyclic

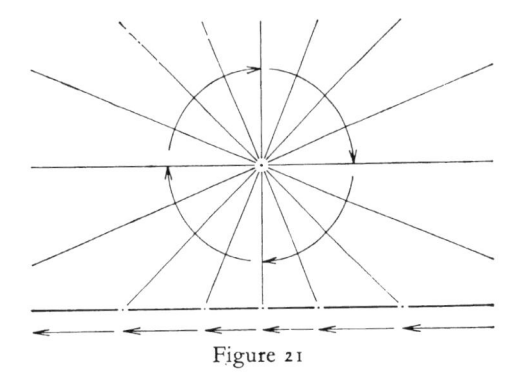

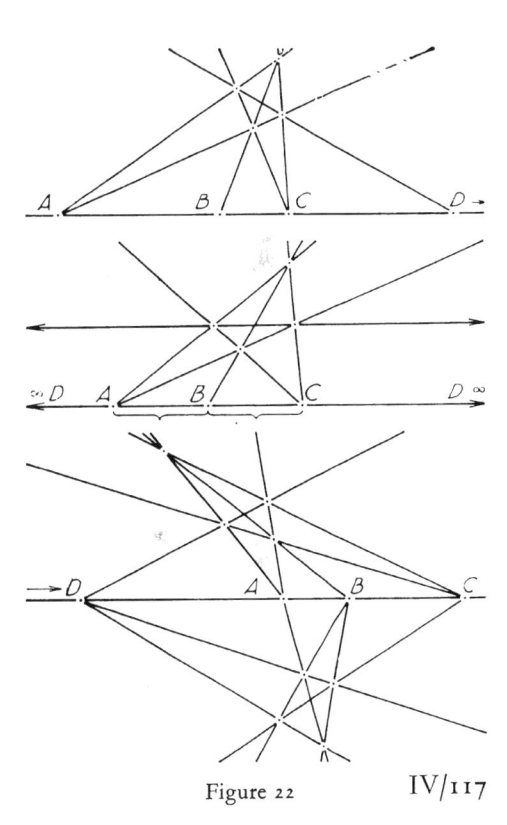

Figure 21

Figure 22

IV/117

order, like points along a circle. As we already know (p. 59), if a point of one of the pairs has moved to the infinite, its fellow will come into a central position, while the paired points harmonic to them assume a symmetrical position (Figure 22). This is the simplest instance of the harmonic relationship, and it exemplifies one of the archetypal phenomena of modern geometry, namely, that *the middle is the harmonic counterpart of the infinitely distant.*

However, not only in this, the simplest instance, is there an exact relation between the distances of four harmonic points along a line. *Any* two such pairs will be so situated on the line that the one pair will divide the distance between the other two internally and externally in the same proportion. We call them harmonic pairs; the relation between them proves to be mutual.

Let us take first the simple case, when one point of the harmonic range is infinitely distant.

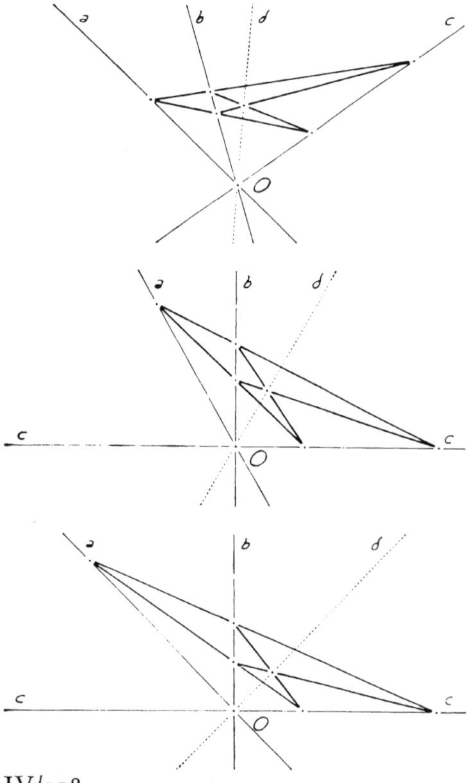

We could travel from A to C via B and compare the two distances AB and BC. They are equal; B divides the distance from A to C internally in equal proportions, 1:1. The ratio of the distances AD and CD is also 1, when D is in the infinite. *Thus the point B divides the length AC internally in the same proportion as D divides it externally.* (Only in the special case when one of the points is infinitely distant does this ratio of ratios equal 1.) Equality of the two proportions is characteristic of the harmonic range (27).

In order to come to the concept of cross-ratio, or anharmonic ratio, we started from concrete measurements and progressed not only to the concept of proportions but to that of the ratio of these proportions—in fact, to a potentised proportion. Starting from the concrete measures, we reach with our concept of measure into the realm of pure thought.

Only at this level is it possible to reach the archetypal, underlying idea of space, where space is freed of all particular metrical attributes. In this sense it is justifiable to speak of a free, archetypal space (p. 248). By means of this threefold potentising process we reach to the fundamental phenomena of all space-formation (28).

Anharmonic ratios are in fact characteristic for that geometry which relates to our eyes and our activity of seeing, to the extent to which this is of spatial nature, while the concrete

Figure 23

measures play their important part in the realm of touch, among material things. The development of a spiritualised natural science depends upon such steps in thought.

Having described the harmonic pairs in connection with the quadrangle we must not forget that the principle of duality allows us to do the same for the quadrilateral. Four lines form a harmonic range of lines in a point (often called a "pencil"), if the points they have in common with any given line form a harmonic range of points on a line. It is clear that in the symmetrical position the lines of one conjugate pair will bisect the internal and external angles of the other pair. In this special case, the bisecting pairs of lines are at right-angles and form a cross (Figures 23 and 24).

The Invariance of the Harmonic Property

The wonderful thing about the anharmonic property, as we have already said, is that it is indestructible by any number of perspectives. It is truly "projective", a so-called projective invariant. Stated simply: a property which is preserved by every projection is a projective property, while a property which is preserved by all rigid motions is a metric property.

Thus the harmonic quality of four points along a line, or of four lines in a point, is always transferable by perspective (Figure 24). If we draw any line to cut across four harmonic rays, the four resulting points along this line will also be harmonic. If we confront a range of four harmonic points with an eye-point, then the four rays of this point will be harmonic, and so in turn will be the points in which the four rays are met by any other line that crosses them. The harmonic quality among points of a line will be demonstrated, if it is possible to draw a four-point construction upon them, while the harmonic quality among lines of a point will show up, if in a given set of four lines, a four-line can be spanned within them. Thus, beginning with a four-point construction, in order to determine a range of four harmonic points, we can go on projecting at will and see that the harmonic property remains invariant.

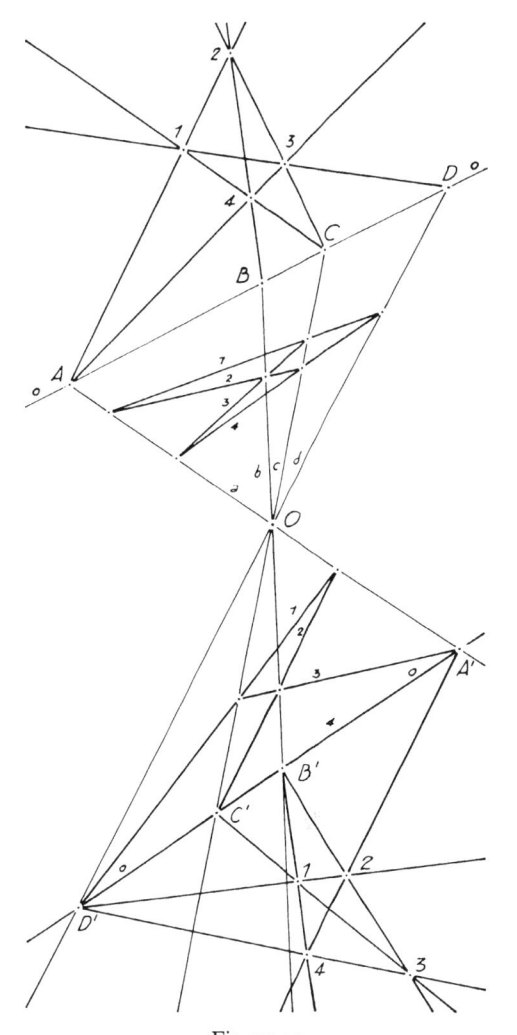

Figure 24

The Thirteen Configuration and the Diagonal Triangle

As we practise the constructions we find them pervaded through and through by the harmonic property. Once a range has been established, either by a four-line or a four-point, it reappears everywhere, wherever four points are on a line or four lines in a point. The four-point and four-line, with their harmonic conjugates, sustain one another mutually.

It will therefore not surprise us to find that they may be fused together in one configuration, and that they share a common bond in the form of a triangle, linewise for the four-line, pointwise for the four-point. This triangle is called the Diagonal Triangle, or the Polar Triangle. (It might well be called the Dual Triangle.)

All six lines of the four-point and all six points of the four-line are really equivalent, and to distinguish between "sides" and "diagonals" is to lean on Euclidean terminology, which for the linewise case especially is awkward and actually inapplicable. We should therefore not be led astray in our geometrical thinking, when the mathematician uses the word "diagonal" in this connection.

Each complete four-point (four-line) gives rise to a particular triangle. The six lines of the four-point (points of the four-line) not only lie in threes in the four given points (lines); they also lie in pairs in three further points (lines). Call the points of the four-point (lines of the four-line) 1,2,3,4; the six lines (points) give rise to "opposite" pairs: 41, 23; 42, 31; 43, 12. These pairs lie in the three points (lines) of the Diagonal Triangle of points (lines) (Figure 25).

Thus:

The triangle determined by the meeting points of opposite sides of a quadrangle is called its diagonal triangle, its points being the "diagonal points" of the quadrangle.	*The triangle determined by the lines joining opposite points of a quadrilateral is called its diagonal triangle, its lines being the "diagonal lines" of the quadrilateral.*

In the quadrangle construction we look for the one point which we have hitherto disregarded (the point S), while in the quadrilateral we find a line which has not hitherto been drawn in (the line s). In the drawings, QRS is the diagonal triangle of the quadrangle and qrs of the quadrilateral (Figure 25).

Figure 25

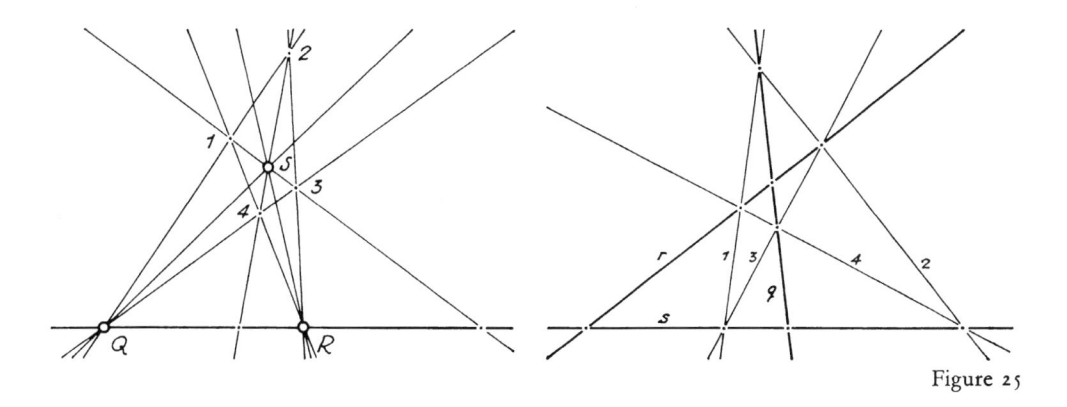

Drawing the two figures together (Figure 26) demonstrates the wonderful interdependency and harmony of this fundamental construction. It consists of thirteen lines and thirteen points in groups of Three, Four and Six, which are reciprocally related to one another; the common bond between the four-point and four-line is the Triangle in its dual aspect. We call this construction the Thirteen Configuration.

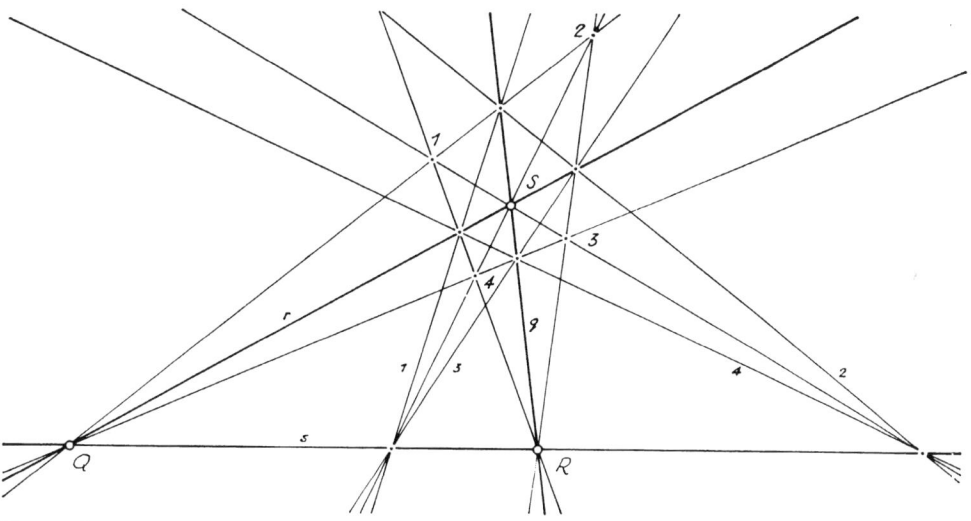

Figure 26

Each diagonal triangle point contains four lines in which lie the six points of the four-line. Two of these lines are diagonal lines, while the other two are two of the six lines of the four-point.

Each diagonal triangle line contains four points in which lie the six lines of the four-point. Two of these points are diagonal points, while the other two are two of the six points of the four-line.

Looking at this configuration with the theorem of Desargues in mind, it is clear that the Desarguean construction is also basic here. (In fact, by means of Desargues applied four times over, we can prove the existence of the harmonic quadrangle related to the harmonic quadrilateral.)

Thus to every given harmonic quadrangle there belongs a certain harmonic quadrilateral, and vice versa. Each diagonal point with its four lines can also be regarded as the point of four harmonic lines upon which the four-line is based; each diagonal line with its four points can be regarded as the line of four harmonic points upon which the four-point is based. We see that the harmonic pairs of points and lines are fully interchangeable. The Thirteen Configuration is an expression of complete reciprocation.

It is a good exercise to construct the Thirteen Configuration for the special case, so that the following may be experienced: starting from projective laws and processes, it is easy to make the transition to the metrical aspect involving special cases, by sending certain elements to the infinite. On the other hand, it is often difficult to begin with a figure already fixed by metrical laws—a figure determined, as it were, by the laws of the earth—and to return to the much freer, projective, more cosmic figure, from which the specialised form arose in the first place.

We can now proceed to drawings in which we shall bring together the Line-woven Net, Growth Measure and the Thirteen Configuration. We shall see how our constructions, without regard for any restrictive measurements or formal premises, will lead to the free creation of a whole family of circle-curves, each of which will express the harmonious beauty of a hidden metric.

Before continuing, we should however not leave unmentioned another most interesting construction based on the principle of duality as it relates to the harmonic forms.

Looking back to Figure 17 (p. 114), it will be clear that just as we started with the range of four harmonic points (left) on a horizon-line in order to draw the harmonic net in the plane, so there must be a corresponding dual construction taking its start from four harmonic lines (right) in an "all-relating" point. Let us compare the process of constructing the net with that of constructing this kind of web.

To draw the harmonic net we began with a four-point drawn from a range of four points on a line, and by drawing in further lines, those common to points of the harmonic range and points of the four-point, we saw the net begin to spread out into the plane. The quadrangles set themselves side by side, becoming smaller and smaller in a kind of "échelle fuyante" as they reached towards the "vanishing line", which functioned as an outer infinitude.

To draw a harmonic web we begin with a four-line drawn into a range of harmonic lines in a point. We then seek further points, those common to lines of the harmonic range and lines of the four-line, and we see that these points will determine new lines of the web. Continuing, we see that the web will begin to weave in the plane all around the point which contains the four harmonic lines, and which functions as an *inner* infinitude.

The exercise is not easy, because of the natural tendency to think of forms spreading out spatially into a plane. Here the consciousness of a spider might come in handy, to spin a web around an innermost point! The spreading out of the quadrangles in the plane took its course in a way that is familiar in our normal experience of space; not so the interweaving of the web of quadrilaterals. This is an exercise leading to an appreciation of the kind of "space" we shall be concerned with in Chapter XI. Figure 27 gives an indication of the way to begin constructing a harmonic web of four-lines in a range of four harmonic lines; this would be the dual of the net in Figure 6, Chapter III (29).

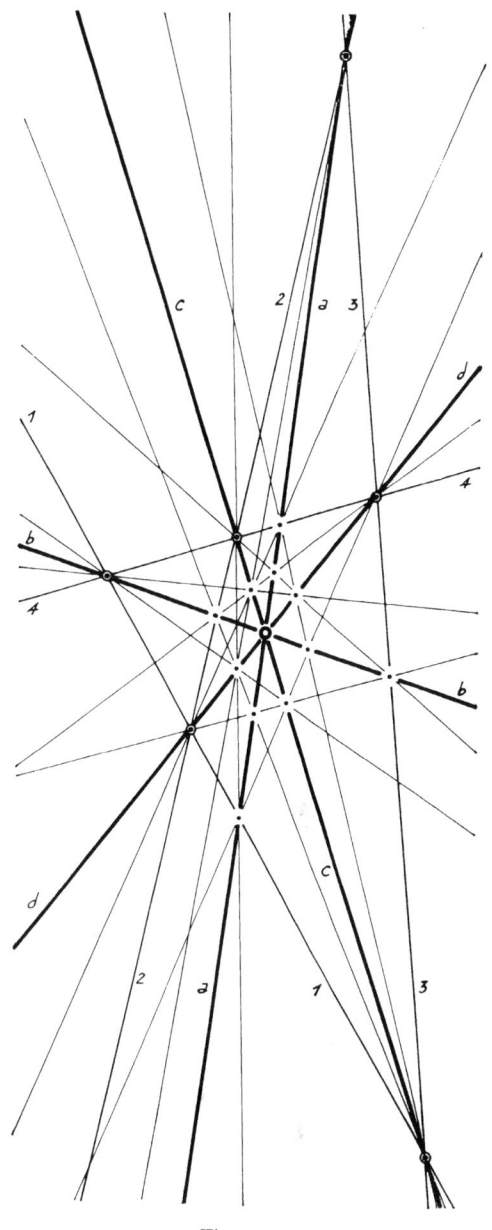

Figure 27

V Projective Laws of Curves

There is a world of difference between the nature of forms, which appear side by side and those which come one inside the other. As little children we played happily with bricks, setting them side by side, or one upon the other, until the exciting moment when they fell. But the fascination of the boxes, the eggs, the little Russian dolls one inside the other is of quite another order, and long outlasts the age of childhood.

It is in the nature of the lifeless realm that things lie side by side like pebbles on the beach, or stand like soldiers on parade, each one a repetition of the last. Here step-measure prevails.

The occurrence of forms one within another is characteristic of life. Plants above all, and many lower animals, reveal this phenomenon most simply; their shapes often approximate very closely to the equiangular spiral, or are arranged in other ways according to some growth measure. Here the inner and the outer take on a different character, for the forms are arranged around a special centre.

Line-woven Net in Growth Measure

In the Thirteen Configuration, four-point and four-line are so arranged that the points of the one are in the lines of the other, and that both are related on the one hand to the same "horizon-line" (s) and to the same special point within (S). On the basis of this arrangement of points and lines, it is possible to continue the construction by putting a smaller four-point inside the four-line and a larger four-line around the four-point, and then repeating these operations to create a network as far as the drawing will allow (Figure 1). This construction shows growth measure in perspective.

The figures converge towards the point within and widen out, flattening towards the line, without ever reaching either.

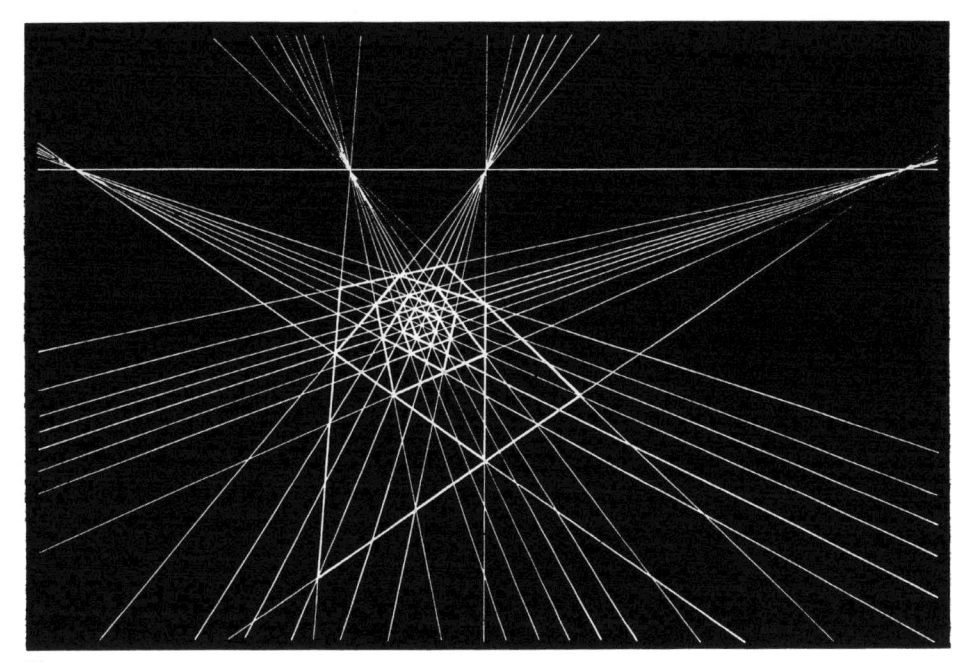

Figure 1

The step-measure net filled the whole plane with its figures, the horizon-line alone functioning as infinitude, towards which the forms tended on both sides of it. The growth-measure net is spanned between two infinitudes, the one concentrated in the point within (S), the other is the expanse of the horizon-line (s). The forms of this type of net are related at one and the same time to a centre and to a periphery, both of which function as infinitudes. Figure 2 also shows a growth measure in perspective, this time in the form of a net of hexagons.

To draw the complete harmonic quadrangle-quadrilateral net in growth measure (also the hexagon net) is a test of accuracy and perseverance which might well be used in the classroom long before the time comes to bring any deeper thoughts to bear on the construction. It is not easy to find the correct continuity of the individual forms as the net opens out below and returns through the infinite from above. The overcoming of this very difficulty, however, gives a practical experience of the infinite continuity of line and plane.

Figure 2 ▶

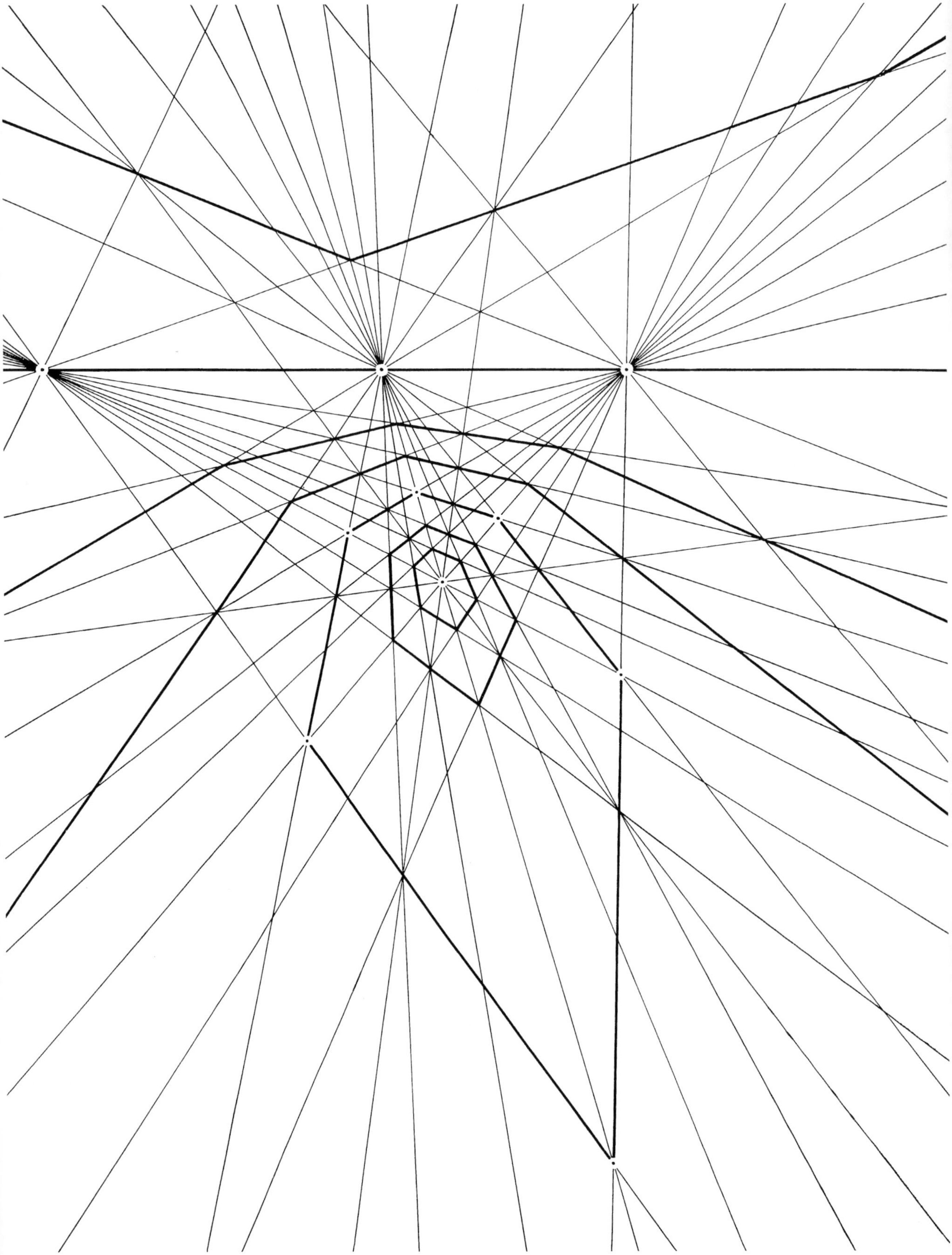

Projective Concentric Circles

With the growth measure net, as with step-measure, we can send the horizon line to the true infinite and then the forms will take on a relatively regular, concentric character (Figure 3). They would become completely regular, forming squares, if the point-pairs were to be evenly arranged on the line at infinity which would, of course, mean that the line-pairs would form equal angles in the central point.

A glance at the concentric patterns will remind us of concentric circles and also of the spirals which might be drawn into such a matrix. Each square or hexagon is inscribed in a circle, while at the same time it circumscribes a smaller one. Pointwise and linewise the forms are related to a family of concentric circles or ellipses.

So are the forms drawn in perspective from a horizon-line related to a family of curves. With a little skill it is possible to sketch them in, either directly on the drawing or by using transparent paper. These curves are what we may call a *family of projective concentric circles*. This family of curves consists of ellipses which close in around the inner point, and of hyperbolae which flatten from two sides into the outer line. Among them there might occur a parabola and a circle, though these will not appear unless we take care to catch them in the setting out of the drawing (p. 174). We shall return to this family of curves later on and their relation to the points and the lines of the construction arising from the Thirteen Configuration. (It is interesting to note that spirals too, transformed projectively, are hidden in this drawing.) Compare figures 1, 2 and 3 of this chapter with figures 41 to 47 in Chapter VI.

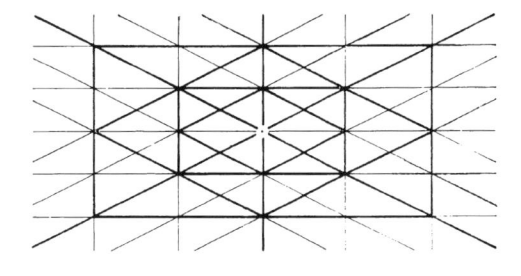

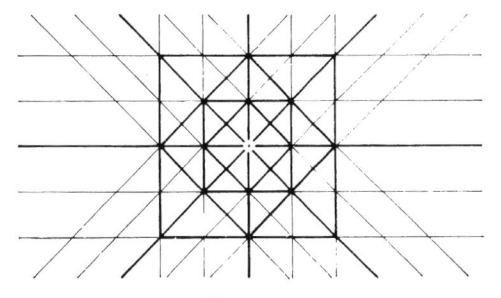

Figure 3

The Theorem of Brianchon

We remember the theorem of Pascal, concerned with six points of a conic. Now we are able to realise something which Pascal, in the year 1640, was not able to see. It was not until 1806 that the dual of Pascal's theorem was discovered by yet another Frenchman, Charles Brianchon (1785–1864). As the dates show, Brianchon was at that time twenty-one.

Stated briefly the two theorems are as follows:

Pascal's Theorem
The points common to corresponding lines of a hexagon inscribed in a conic are collinear.

Brianchon's Theorem
The lines common to corresponding points of a hexagon circumscribed about a conic are concurrent.

All that we found in the case of Pascal's Hexagrammum Mysticum we find again in reverse if, guided by the principle of duality we draw, instead of any six points of the conic, any six of its lines and proceed to make the dual of the Pascal construction, where, instead of a "Pascal line" of three points, we find a "Brianchon point" of three lines (Figure 4).

(A line of a conic is its *tangent line*. In the case of the circle, the tangent is at right-angles to the radius, and is easily drawn. To draw a tangent to the other curves without a more elaborate construction, it will be sufficient to place it by eye. The reader will be aided in making accurate drawings from now on, if he avails himself of a large transparent right-angled set-square, upon which a line has been marked which is exactly at right-angles to the longest side. To draw a tangent to a circle from a point outside it then becomes a simple operation. Allow the line marked on the set-square to pass through the centre of the circle, while the long side passes over the point from which the tangent is to be drawn, and at the same time over one *single* point of the circle. The line drawn at the long side of the set-square will then be an accurate tangent to the circle.)

As with Pascal's hexagon, there will be sixty possible orders arising from the given six lines of the curve (though some may coincide with one another). Whereas the Pascal Line is determined by three points in a line, the Brianchon Point arises in the meeting of three lines in a point. There is a natural tendency, if the hexagon grows fairly regular in shape, for the Brianchon Point to come towards the middle. Indeed, with a regular hexagon circumscribed about a circle, the Brianchon Point will be the very centre of the circle. This corresponds exactly with the tendency of the Pascal Line to go out into the infinite when the inscribed hexagon grows regular.

Taking the six lines of the curve in any order gives an

unexpected variety of shapes, and involves an exercise requiring considerable concentration. Each line of the hexagon may of course only be counted once, but the so-called "corners" may be at any part of it (two out of a possible five on each line). The hexagon may take on a very bizarre appearance. Whatever the order and the shape, the three lines joining "opposite corners" will always meet in a point! This is the necessary and sufficient condition for the six lines to touch a conic. If they fulfil it in one order, they will do so in all; if they fail in one they will fail in all.

It is wonderful to see how Brianchon's theorem complements that of Pascal. The Principle of Duality reveals the interdependence of the pointwise and the linewise curve. Pascal still thought in terms of the point and the pointwise curve, whereas it now becomes not only possible but *necessary* to include the linewise aspect of a curve—its envelope—without which we do not truly apprehend the whole curve. To consider a curve only as a pathway of points and not to conceive of its mantle of lines is like looking at the cellular structure of a petal without perceiving the living beauty of the whole form. The pointwise, radial formation, deriving from centres of contraction, and the tangential, plastic formative process, originating from the "encircling round", weave together in all forms.

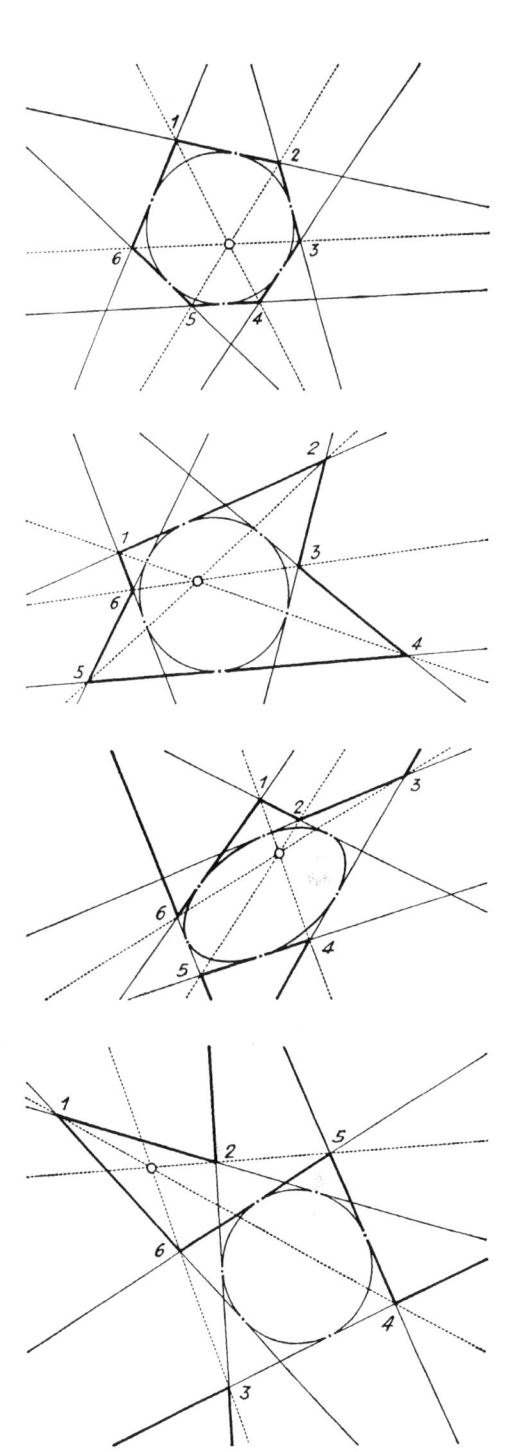

Figure 4

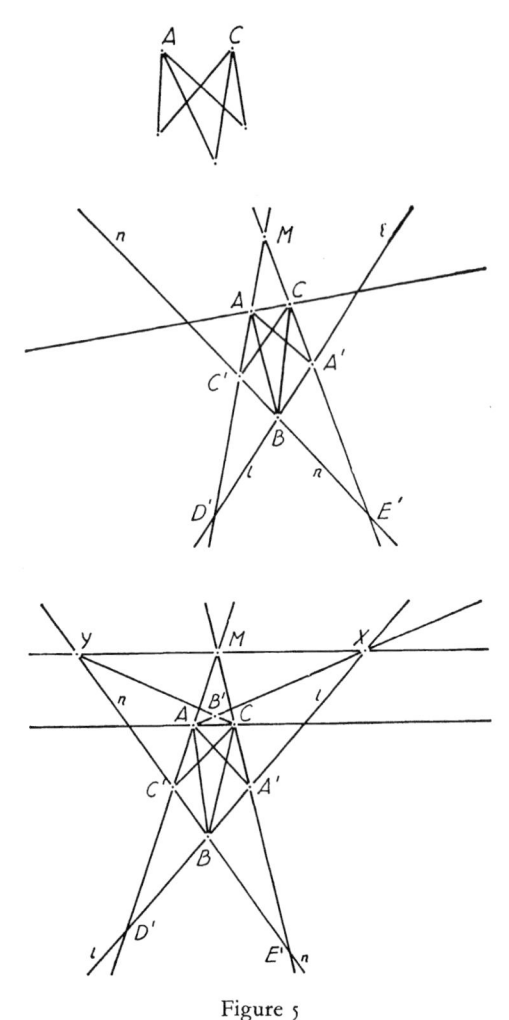

Curves Through Five Points and Touching Five Lines; Pascal and Brianchon

A conic is determined by five points or five lines (24); that is to say, we may choose at will five points or five lines from among all the points or lines of the plane, but as soon as we have chosen all five, the conic is fixed. (No three points may be in line and no three lines in a point.)

Pascal's theorem has reference to six points of a conic and Brianchon's theorem to six lines. These theorems express the condition which *six* points or lines must satisfy, if they lie on any circle-curve (18).

Let us now look again at the projective formation of curves arising as the result of projectively paired points or lines, from the Pappos construction (the projection of Three into Three). We used five creative elements, and these naturally allow of a pentagonal arrangement.

1. A and C, l and n, and M.
2. a and c, L and N, and m.

1. Three lines of A and three lines of C meet in pairs in three points, call them C′, B, A′. These three points together with A and C make a pentagonal figure, which we may complete (its head at M and its feet at D′E′). Make l the line BA′ and n the line BC′; M is the intermediate point (Figure 5).

Now take any other ray of A and find the point in which it meets l, say X; then find the point of n, which is in perspective with it as from the centre of perspective M, say Y, and finally the ray from C to the point Y. Thus two further rays have been established, which are projective with one another; their common point B′ is therefore *a sixth point* of the curve determined by the other five.

We only have to bring the line YMX into movement, pivoting it about the fixed point M, with X and Y running up and down the lines l and n respectively, and by this means we find for every line of A the corresponding line of C. As all the rays from A sweep round, each one having a common point with the corresponding ray from C, this common point too will have to make some kind of circling movement. As it moves round it will pass through A and C themselves.

The moving point B′ will in fact describe a circle-curve, which passes through the five points of the originally given pentagon.

Moreover, at every situation the six points together make a hexagon—A B′ C A′ B C′—of which the three opposite pairs of sides, namely BC′, CB′; CA′, AC′; AB′, BA′ meet in three points Y, M, X respectively, which are in line.

This then is the condition, both necessary and sufficient, that the sixth point lies on the curve determined by the given five: the hexagon made by the *five given points and any sixth point* that is to be on the curve, must be such that the meeting-points of its opposite pairs of sides are in line (Pascal's theorem). The ellipse in Figure 6 passes through the five points A, C, P_1, P_2, P_3.

We must ask: Is there only one such curve passing through the five points? Of the given five, A and C fulfilled a different function from the other three. With A and C as the raying points, there is obviously only one curve. But of the five points, we might also have chosen two others, and we should again have obtained a curve through the given five. The complete functional symmetry, both of the hexagon-picture and of the statement shews that the condition for any sixth point B′ to belong to a curve engendered by any other pair of raying points is precisely the same. Through five given points there is only one such curve; it matters not which two of them we choose as the mutually raying points.

2. By the Principle of Duality: if instead of relating two points projectively, we relate two lines and carry out the dual of the construction we have just made, we shall be able to determine the movement of a *sixth line,* which in a continuous circling movement will engender a line-conic among whose lines will be included the original five lines of the pentagon (Figure 7). Instead of the Pascal Line moving round in M, we shall see the Brianchon Point moving along m. It will be evident that at each moment this point is in the situation of a Brianchon Point, and that the condition required by the Theorem of Brianchon is the necessary and sufficient condition for the sixth line to be a line of the curve. The ellipse in Figure 8 touches the five lines a, c, p_1, p_2, p_3.

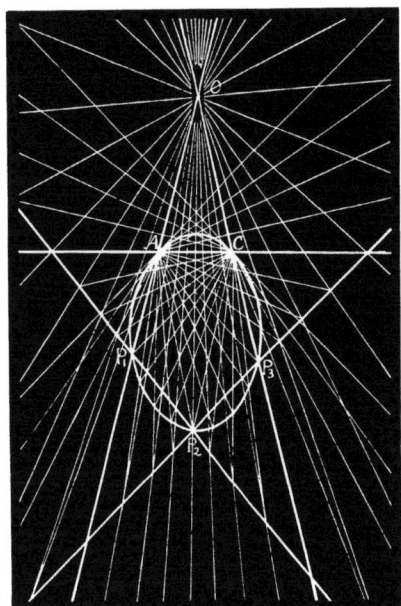

Figure 6

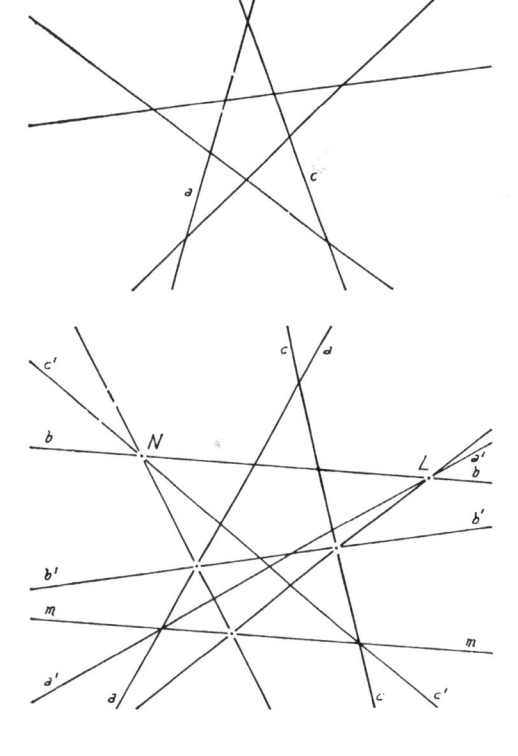

Figure 7

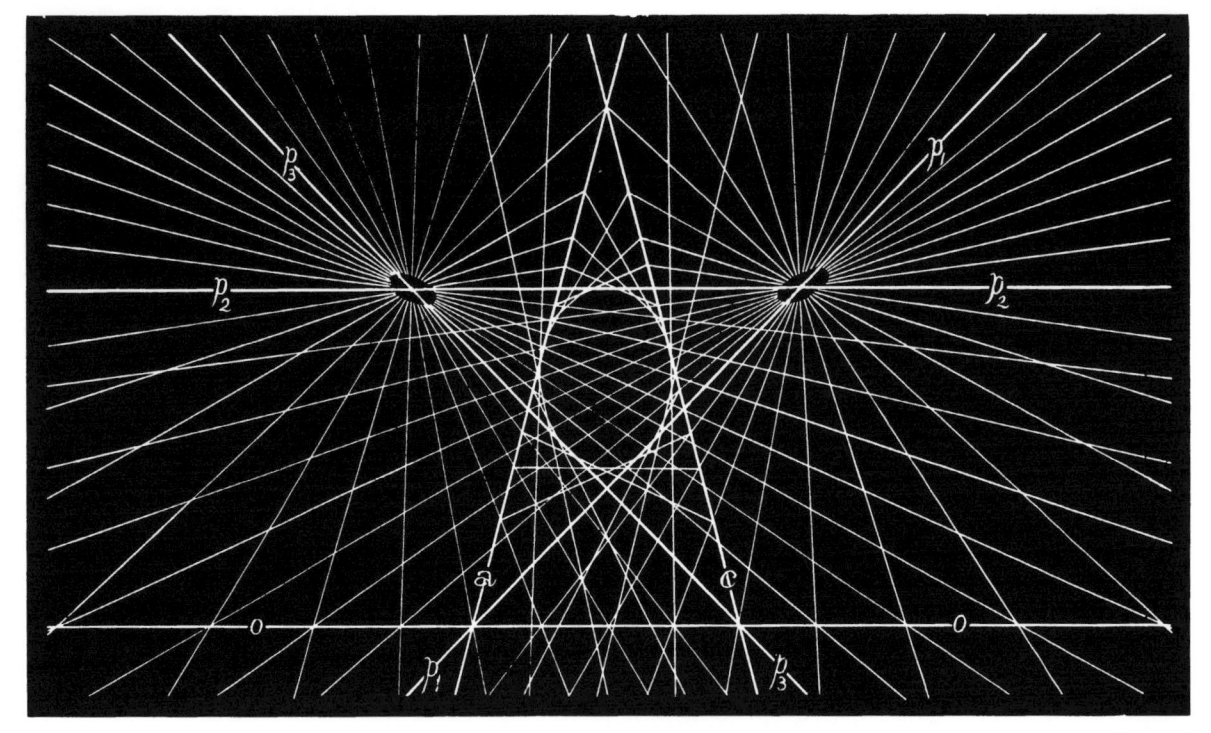

Figure 8

Thus, we may formulate the dual statement:

Point-conic
Through any five points of the plane, no three of which are in line, there is one and only one conic. A moving ray in the centre of perspective M plays the part of Pascal Line, giving rise in each position to a further point of the curve.

It sweeps right round, pivoting on the fixed point M, for the whole curve to be engendered by the meeting-point of corresponding lines of A and C.

Line-conic
Touching any five lines of the plane, no three of which meet in a point, there is one and only one conic. A moving point in the horizon of perspective m plays the part of Brianchon Point and gives rise in each position to a further line of the curve.

It runs the whole length of the fixed line m, completing the cycle through the infinite and back again, for the whole curve to be engendered, by the common line of corresponding points of a and c.

Steiner's Theorem

We have met here with a fundamental property of the circle-curve. We first determined the curve by the projective relation of the lines of two fixed points or the points of two fixed lines. From this relation arose infinitely many other points or lines, making the curve. And now we find that *any* two of them—*any* pair—are in the same kind of relation by virtue of the curve itself. The two we first envisaged—although it was by thinking of them in such relation that we first came upon the curve—can claim no special privilege; the curve once given, they melt into it and are like any other two. In modern geometry the entities again and again reveal just this quality—determining yet self-effacing, creative and yet selfless, we might say. In their co-operation to form larger entities between them, they seem to indicate the archetypes of truly social relationships.

This projective property of conics is known as Steiner's theorem, after the Swiss geometrician, Jakob Steiner (1796–1863).

The curve engenders a projective relation between any given two of its own points. Namely, if from the two points lines be drawn to all the other points of the curve, then those lines which meet in points of the curve, are corresponding pairs of a projectivity.

The curve engenders a projective relation between any given two of its own lines. Namely, if in the two lines points be sought each containing a pair of lines from among all the other lines of the curve, then those points containing lines of the curve are corresponding pairs of a projectivity (Figure 9).

We shall return to this in the following chapter.

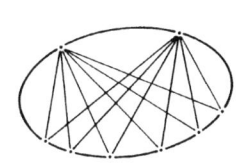

Figure 9

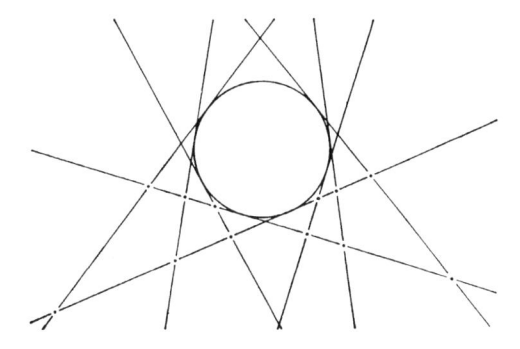

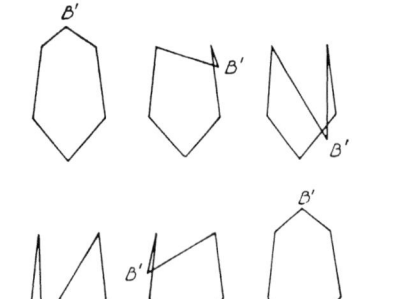

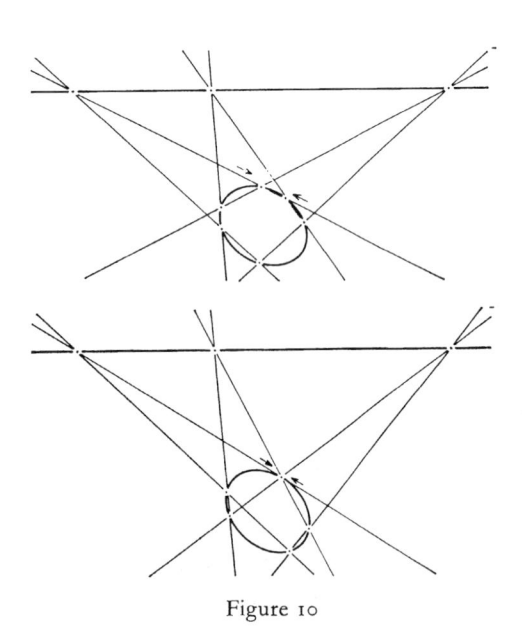

The Tangent and its Point of Contact

With these considerations we have come right into the heart of the questions concerning the *tangent*. It is a surprising and a significant fact that problems concerning the tangent to a given curve, which have their roots in Greek mathematics, were only satisfactorily solved at such a comparatively recent time in history. It was the recognition of the fundamental principle of duality—a true child of projective geometry—which was the main reason for the great advances in geometry during the last hundred years.

Let us now follow the movement of the point B' as it traces out the curve. As it passes through one of the five given points, the hexagon changes its type (Figure 10).

In the moments of transition, when the sixth point merges with one of the five, a line of the Pascal hexagon, from being a "*chord*", changes momentarily into a *tangent*, touching the curve. Though the hexagon has changed into a pentagram, as two of its points have merged into one and become the point of contact of the tangent, we still have a Pascal line with its three points, enabling us in effect to construct the exact tangent of the curve at the given point of the pentagon, which now counts as two.

Similarly, in the Brianchon construction (Figure 11), we watch the movement of the sixth line and see it merge into one of the five other lines. At the moment of transition, two sides of the hexagon merge and become a single tangent. What was their common point then becomes the *point-of-contact* of the tangent line. By the Brianchon construction, which still holds for the pentagon (in which one side counts twice), we are led to the point of contact of the tangent.

Note that the divergence of two points from one another is generally experienced as the *length* of line between them; when this grows infinitely short the two points merge into one. On the other hand the divergence of two lines is experienced as the *angle* which they make with one another; when the acute angle becomes infinitely small the two lines merge at last into one. (Only if the points are infinitely distant or if the lines are parallel the types of measure are reversed; the divergence of two infinitely distant points, as of two stars in the heavens, is experienced as an angular

Figure 10

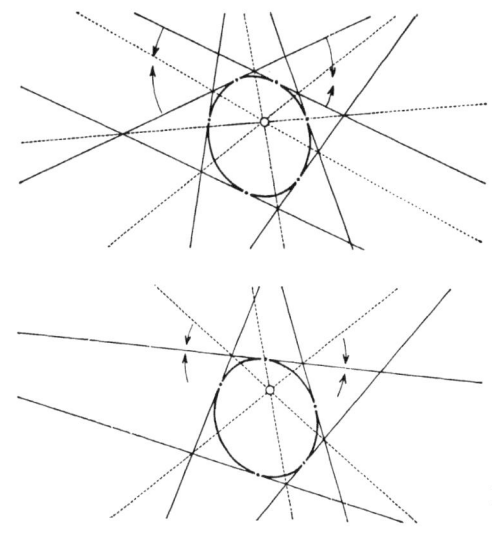

Figure 11

magnitude; that of two parallel lines as the perpendicular distance between them).

Thus we may say, in effect:

The tangent line at any point of a curve (conceived originally pointwise) is what becomes of the line joining two distinct points of the curve in the moment when these two points—remaining always points of the curve—move up to one another and at last melt into one point.	*The point-of-contact* of any line of a curve (conceived originally linewise) is what becomes of the common point of two distinct lines of the curve in the moment when these two lines—remaining always lines of the curve—move nearer to one another and at last melt into one line.

It should be mentioned that we have intentionally used a certain similarity of construction throughout these descriptions, in order to make it easier for the reader; yet this does not accord with the spirit of projective geometry, which asks for change and metamorphosis. The reader will do well to use his initiative, setting the basic elements of the constructions differently and experiencing to what a great extent the result may vary. This way of learning to see a concept or idea manifesting itself as a phenomenon in a drawing in a variety of ways is a truly Goethean approach to geometry. It is of greater value than the mere memorizing of the theorem and its proof.

Identity of Pointwise and Linewise Circle-curve

We have so far been holding apart the conceptions of pointwise and linewise curve, discovering and defining each in turn by a projective process. The two processes are as yet distinct. We have only compared and contrasted the two aspects of the curve, passing from a truth about the one to a corresponding truth about the other. Since to each point of the one curve we can find the tangent, and to each line of the other the point-of-contact, we can say:

The tangents of a pointwise curve constitute a linewise curve.	*The point-of-contact of a linewise curve constitute a pointwise curve.*

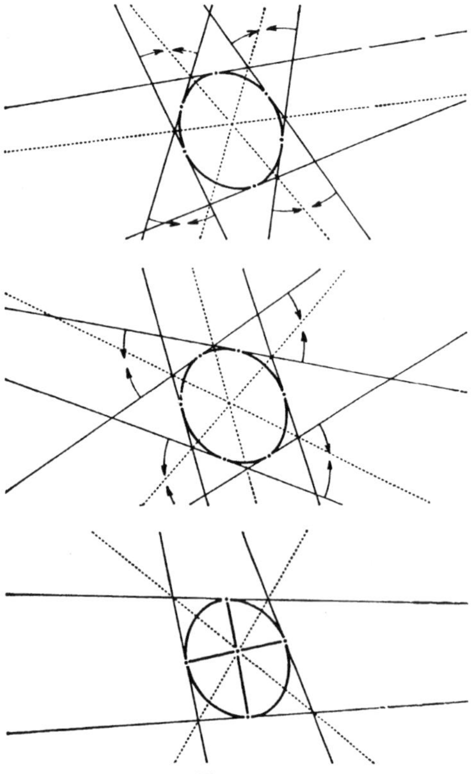

The very idea of a continuous curve being formed by an infinite sequence of points or of lines leads to this intimate synthesis of point and line. The one begets the other; the one cannot be without the other.

The question, so seemingly obvious, must, however, be asked: given a pointwise conic, projectively defined, do its tangents form the same identical curve linewise? The answer, which is in the affirmative, must needs and can of course be proved. It is only from the projective aspect that this question arises, due as it is to the principle of duality. The ordinary metrical definitions of a conic all define it as a pointwise curve. Projective geometry must define it in both aspects and prove that the two are equivalent.

We have so far seen the relation of the conic to the Six and to the Five, and the possibility of the metamorphosis of the Pascal and Brianchon constructions from the hexagon to the pentagon. If now we apply the same process—the merging of two tangents—this time twice over, we shall be led to an important truth concerning the relation of the conic to the Four, namely, to any four lines that touch the conic and their points-of-contact (Figure 12).

By the merging of two neighbouring tangents into one, the Brianchon hexagon becomes a pentagon. Apply this process a second time, and the pentagon becomes a quadrilateral. The two remaining corner-points of the hexagon have changed into points-of-contact, which may now be

Figure 12

included in the Brianchon statement. As such a point-of-contact may always be regarded as the meeting-point of two tangents which have merged into one, the Brianchon statement is valid also for the quadrilateral circumscribing the circle-curve.

Thus: the lines joining opposite points of contact and the lines joining opposite corners of the quadrilateral lie in a point.

Conversely, in metamorphosing a Pascal hexagon into a pentagon and then into a four-point, two sides of the four-point become tangents of the curve and may be included in the Pascal statement. We find that the Pascal line still holds for any four points of the curve, two of which are points-of-contact of tangent lines (Figure 13).

Looking now at a quadrilateral arising from a metamorphosed Brianchon figure, we see that we have in fact two lines joining opposite corners of the quadrilateral, which are two of the sides of the diagonal triangle, and two lines joining opposite points-of-contact of the curve with the quadrilateral which meet in a point of the diagonal triangle (Figure 14). If we complete the figure, we shall have pairs of lines joining the points-of-contact of the quadrilateral with the curve, which will meet in the other two points of the diagonal triangle (Figure 15).

Figure 13

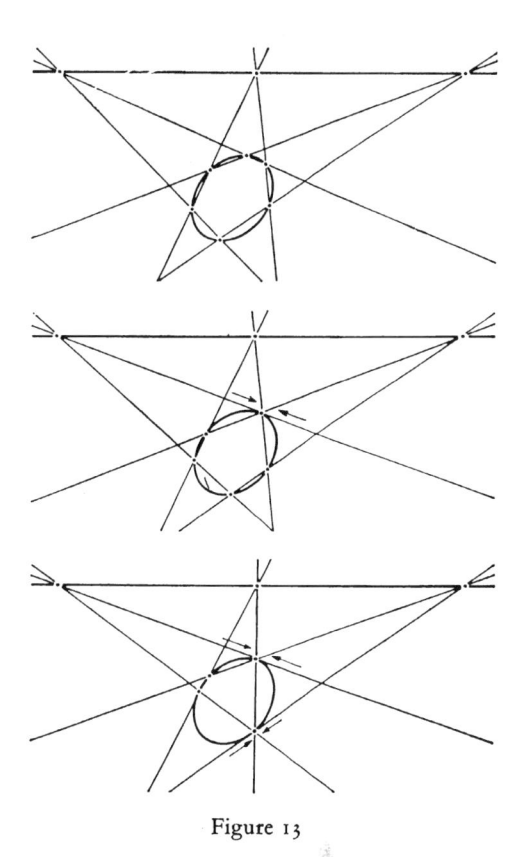

Figure 14

Figure 15

Correspondingly, in the case of the Pascal metamorphosis, the opposite sides of the quadrangle meet in the points of the diagonal triangle, while the meeting-point of the two pairs of tangents determine lines of the diagonal triangle. These lines are metamorphosed Pascal lines and contain four meeting points.

Thus the curve, as a synthesis of line and point, leads naturally and organically from the four given tangent lines to the four points-of-contact, i.e. from the four-line to the four-point.

The four-line has a diagonal triangle which we arrive at linewise; the four-point has a diagonal triangle which we arrive at pointwise. We know that the two triangles are identical.

Thus we may say:

Given four lines of a conic and their four points-of-contact, the three lines joining opposite point-pairs of the four-line will meet in pairs in the three points of its diagonal triangle.	*Given four points of a conic and their four tangent lines, the three meeting-points of pairs of opposite lines of the four-point will be joined in pairs by the three lines of its diagonal triangle.*

We have come to this balanced result; the harmony is perfect. The synthesis of the two aspects shows that the pointwise and linewise conic are in fact the same.

We must not forget that any of these constructions may be made in relation to *all* the different forms of the circle-curve. Naturally, it is easiest to use the circle itself, to which tangents can be most accurately drawn. At this, as at numerous other points in our considerations, it will be instructive to remember the regular situation, namely, the special Euclidean case (Figure 16). Here we see that the diagonal triangle becomes the cross in the centre of the circle. Considered pointwise, it has one point in the centre and two in the infinite; considered linewise, it has one line in the infinite and two in the centre.

There is here revealed the perfect harmony of point and line in weaving interplay in and around the circle-curves. Four-point, four-line, Pascal and Brianchon constructions are intimately interwoven with the circle-curve. Mediating between them all is the diagonal triangle, pointwise and linewise. The picture represents the most harmonic relationship of a curve to quadrangle and quadrilateral.

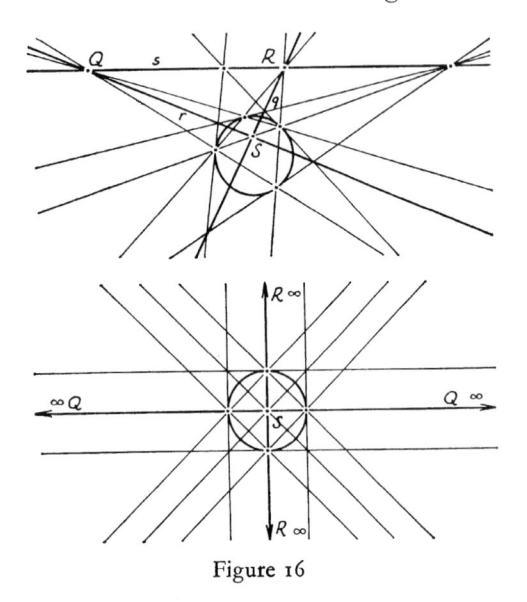

Figure 16

This manner of experiencing the metamorphic quality of the theorems and facts of projective geometry is indeed in accordance with Goethe's scientific method. Goethe sought to understand the laws of transition from one organic form to another, not in the study of the parts alone, but by reaching out to apprehend the encompassing whole, the archetypal Idea, which can but express itself in the ever-changing external phenomena through the interplay of contrasts and of polarities.

VI Projective Transformations; Collineations

Projectivity and Involution within the Line

The complete harmony and interdependence of the Thirteen Configuration, and the simple, basic structure of the Euclidean case will lead us to think—and rightly so—that we are here concerned with a construction which is fundamental. We shall therefore not be surprised to find ourselves led back again to this construction along the various paths we may choose to follow.

We have so far been concerned with the projection of the points of one line into the points of another line, or the lines of a point into the lines of another point, and have been able to see how and why this mobile process, freely applied, leads to the creation of the circle-curves, in which point and line are fused.

Now we shall consider the projective transformation of the line or the curve *into itself,* whereby we shall become acquainted with further properties of the circle-curve.

We remember our original definition of a *Line*. It can have one of three aspects:

1. It is a simple, undivided entity—a line pure and simple.
2. It is a line of points.
3. It is a line of planes.

Now when a line is transformed into itself one must think of it as absolute identity. The line as line does not move at all, it is utterly at rest; it only slides within itself. That is to say, its points move within it, or its planes move about it. We will consider first the pointwise aspect of the line.

In projective geometry we are led to think more deeply than in ordinary, physically biased geometry. We should be thinking only from a physical point of view if we were to say that the line moves like a straight ruler along a straight groove. The movement is really *the streaming of the points* within or upon it. There is good precedent for thinking thus;

the English mathematician W. K. Clifford mentions that St. Thomas Aquinas distinguished, as two different ideas, a line and the sum-total of its points. A better illustration for this type of process than the ruler moving in a groove or the circling of a spinning wheel would be the streaming of water through the standing wave or the flow of cells through the developing organism.

Thus, a line is projected into itself by means of two projecting points and an intermediary line. This may happen in either of two ways (Figure 1):

1. A given sequence of points is projected into another sequence. A B C D $\overline{\wedge}$ A' B' C' D'

2. A point is projected into a second, the second into a third and so on indefinitely, in a kind of rhythmic potentising process.

In the projection of a line into itself it will usually happen that some points return into themselves, that is to say that they act as *Double Points,* being self-corresponding in the sequence of perspectives. Here there are three possibilities. There may be no double points, one double point or two double points. (We will leave aside for the moment the case when there are no double points, and we will show the projection taking place *in the potentising process.*)

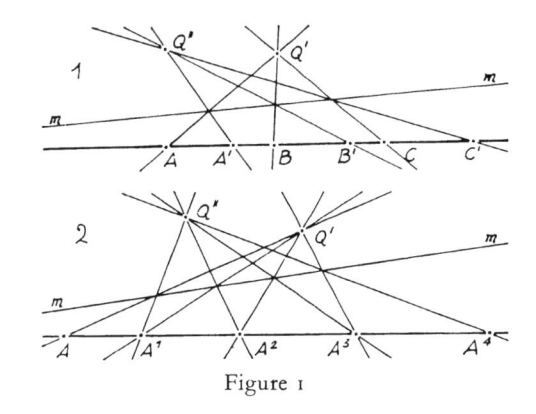

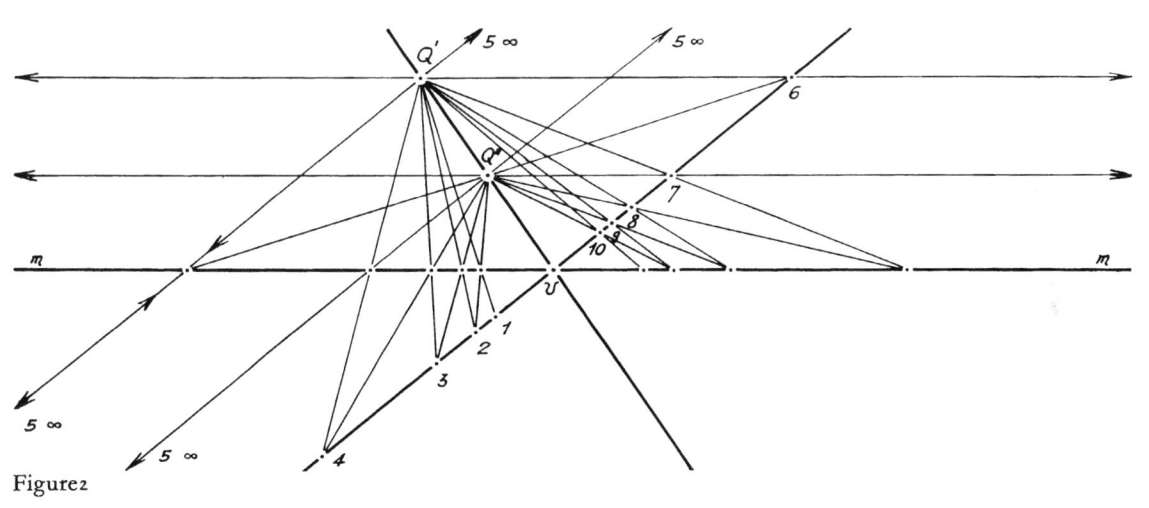

Figure 1

Figure 2

Projection with One Double Point; Step-measure

This occurs when the line joining the projecting points meets the line of projection at the point where the intermediary line meets it (Figure 2). This point, call it U, will obviously be self-corresponding; the points as they are projected one into the other will crowd infinitely in towards U on either side and race away from it on either hand. This creates a projective step-measure on the line, with U as its functional infinitude. Send U to the infinite and the measure will assume its Euclidean aspect (Figure 3).

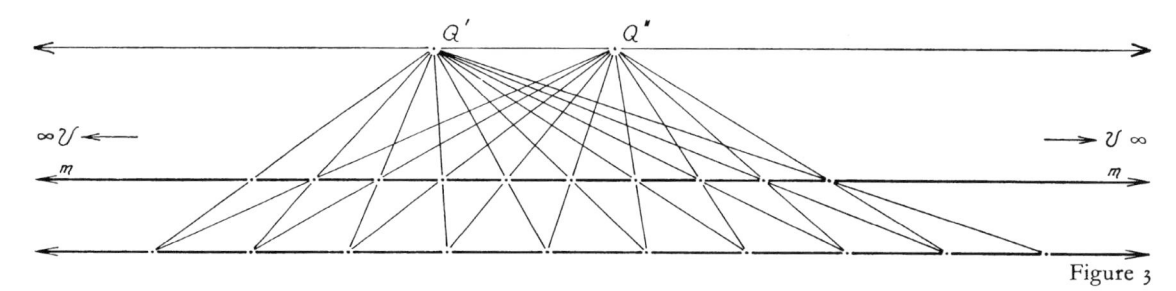

Figure 3

Projection with Two Double Points; Growth Measure

More often it will happen that there will be two double points, namely, the point in which the intermediary line meets the line to be projected and also the point in which the line joining the projecting points meets it. The whole line will then be divided in two parts by the two double points,

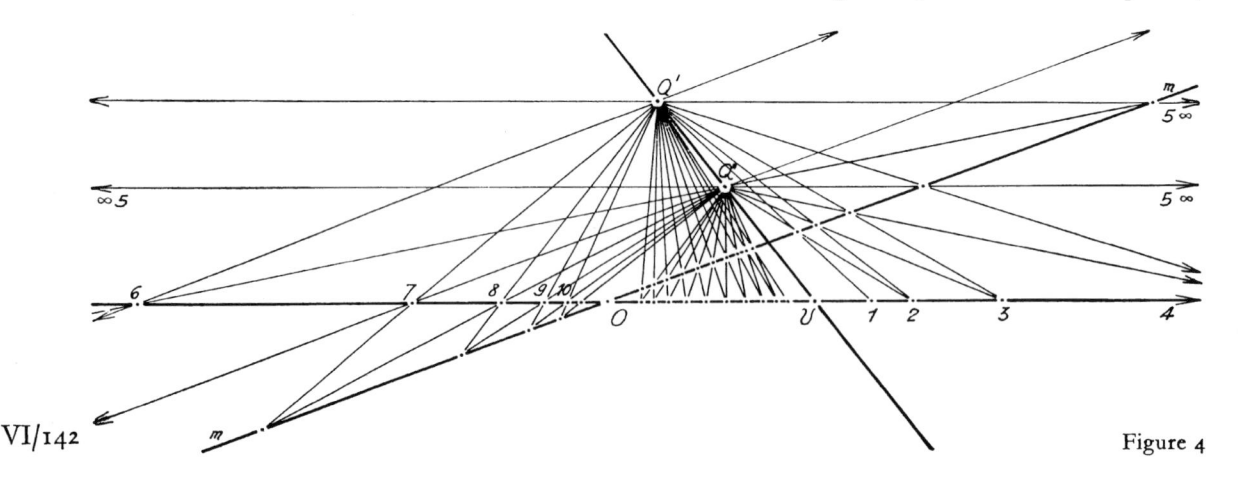

Figure 4

and its points, as they are projected, will crowd in towards these double points and spread out into the spaces between them. Here there are two possibilities:

1. The projecting points are not divided by the intermediary line and the line OU (Figure 4).
2. The projecting points are divided by the intermediary line and by the line OU (Figure 5).

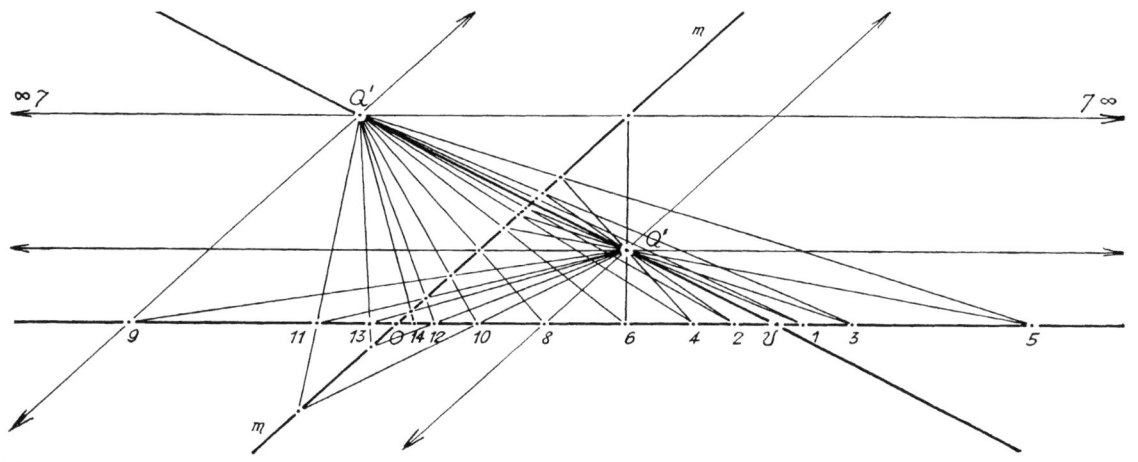

Figure 5

When two double points occur in the projection, the sequence of points is in growth measure. In the first case (Figure 4) the points stream in and out, towards or away from O and U, which are infinitudes and separate the line into two distinct parts; they are like an infinite *source* and an infinite *sink*, to borrow the expressions from hydrodynamics.

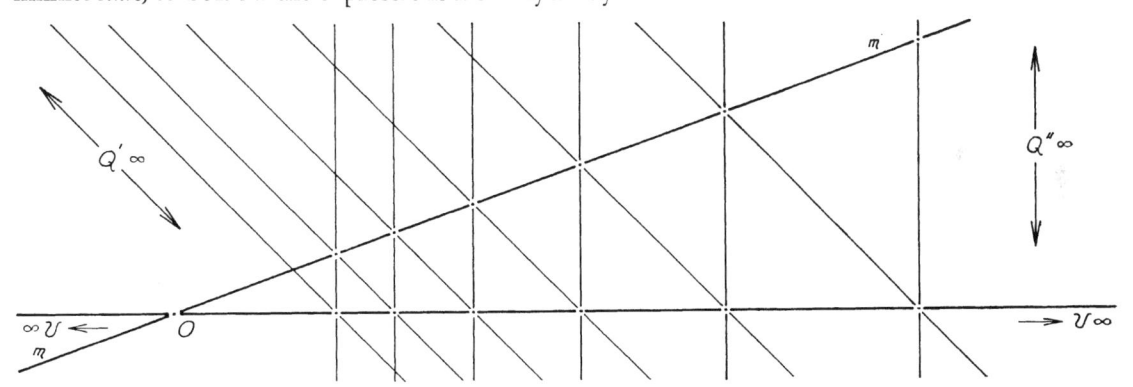

Figure 6

In the second case (Figure 5), in the process of projection the points leap to and fro alternately between the two spaces of the line which are bounded by the double points towards which they converge from either side. This is a kind of potentised process dominated by a negative number; just as two negatives make a positive, so at every second step the point returns to the side from which it started.

Here too, of course, the Euclidean case occurs, if, for instance, we put U and the projecting points in the infinite, when we shall recognise the familiar picture of growth measure as it is in the normal aspect (Figure 6). The projective process brings the realisation that in its inner character a growth measure may be present just the same, even when none of the determining entities is in the infinite—or what *appears* to us as the infinite from a physical point of view. It may take place between points which function as infinitudes.

Involution

This latter instance of the projection of a line into itself with two double points in a negative potentising process takes on a very particular character when it so happens that the projected point returns at the second step to its original starting-place. The projection then resolves itself into a rhythmic process which changes these two points continually back and forth into one another. The projection then becomes what is called an *Involution* and the two points are said to be "in involution". It will easily be seen that this can only happen when the projecting points and lines form a harmonic quadrilateral (Figure 7).

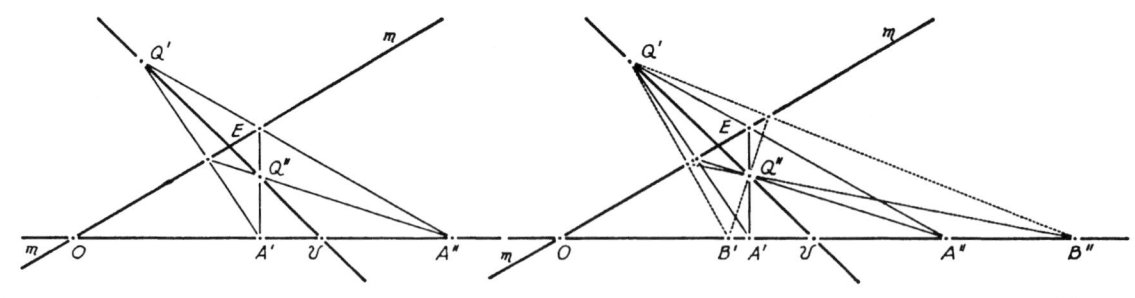

Figure 7

The existence of the harmonic quadrilateral in this process shows that for a projection to become an involution it is necessary for the various points and lines involved to take up positions in which pairs of points (and pairs of lines) throughout the figure are harmonic. For example, Q' and Q" are a pair harmonic with U and the point, say E, at which the line Q'Q" meets m; so, too, O, U and A'A" are harmonic pairs. This means that *when, in a projectivity, two points are in involution, then this is so for all other point-pairs of the projectivity.*

Distributing the lines differently in the plane we will make two more constructions, which in an interesting way show the significant difference between a projectivity and an involution (Figure 8). In the first, the point pairs Q'Q" and EU are not a harmonic range; in the second, as the presence of the harmonic quadrilaterals show, they are.

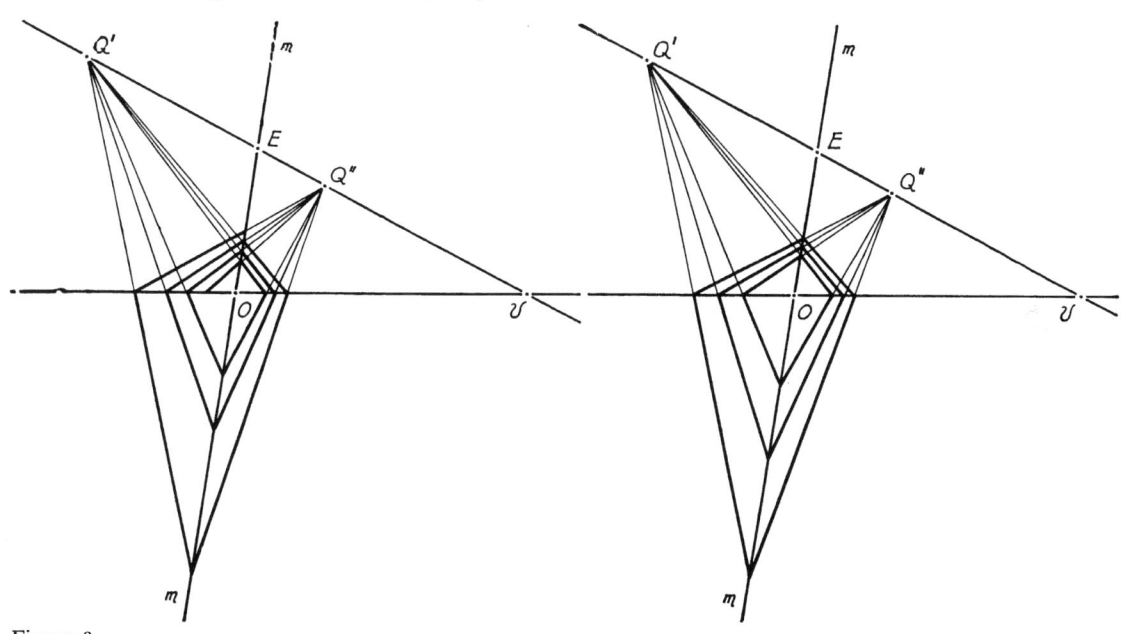

Figure 8

Thus we may say:

Given two pairs of points on a line, the necessary and sufficient condition for them to be in involution is that a quadrilateral may be drawn in them.

Furthermore, it will also be clear that:

Among the six points of a harmonic quadrilateral determining an involution of points on a line, each of the three pairs of points may function as the projecting pair.

We have considered the projective transformation of the points of a line; we could also do the same thing with the lines of a point and carry out the whole of this section from the dual aspect. Instead of sequences of points in projection, we should be thinking of the lines of a point, and we should come at last to the situation when two pairs of lines are in involution. There would then be a harmonic quadrangle which would demonstrate this fact. It is a valuable exercise to carry out this dual process.

Projection and Involution on a Circle-curve

All that we have seen taking place when a line is projected into itself, we can see happening when a circle-curve is projected into itself. (Where, in our considerations here, the one or other dual construction is not included, the reader will have the opportunity to supply this lack. He will also find that constructions which we bring here based only on the circle itself, will be valid for all the other circle-curves!)

The projection of the curve into itself is still a one-dimensional transformation, for although a circle lies in a plane, and is as such a two-dimensional form, yet the process we are concerned with takes place *within the one dimension of the curve of the circle-curve*.

There will be a direct perspective between the lines of a point of the curve and any other points of the curve; also

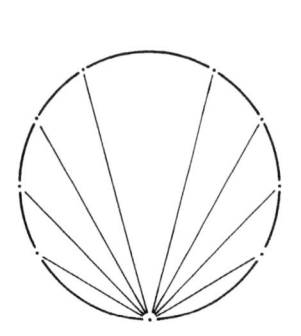

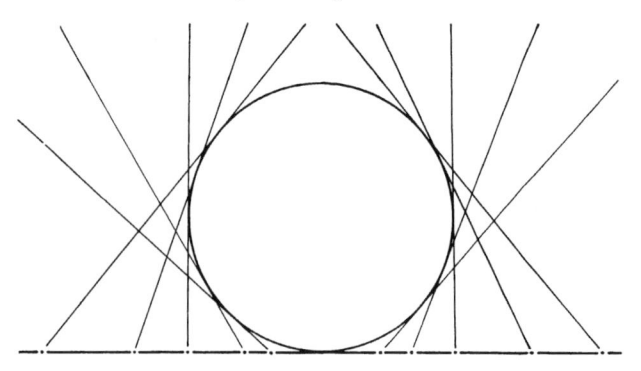

Figure 9

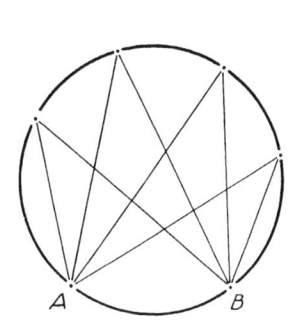

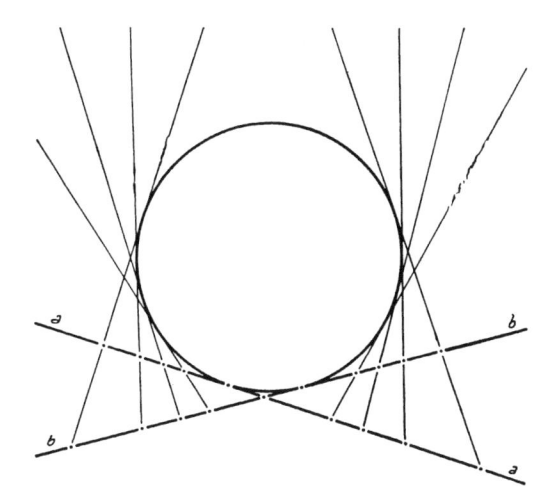

Figure 10

(the dual) between the points of a line of the curve and any other lines of the curve (Figure 9).

As we have already seen, the lines of two points of the curve are in projection with one another by means of all the points of the curve. Conversely, the points of two lines of the curve are in projection with one another by means of all the lines of the curve (Figure 10).

As with the projection of a line into itself, a circle-curve may be projected into itself by means of two projecting points and an intermediary line. (It may also be projected into itself by means of two projecting lines and an intermediary point.)

As with the projection of a line into itself, this may take place in the two ways already described (Figure 11):

1. One sequence of points is projected into another sequence, thus $ABCD \barwedge A'B'C'D'$

2. The projection takes place in a rhythmic potentizing process.

Given three points projected into three other points, any pair of corresponding points can rank as the projecting points, when all other points in the sequence will be fixed (Figure 12). We see that the line m is none other than the Pascal Line for the hexagon of six points on the curve.

Moreover, if we imagine the conic to be a very long ellipse, which then degenerates into two straight lines, then Pascal's

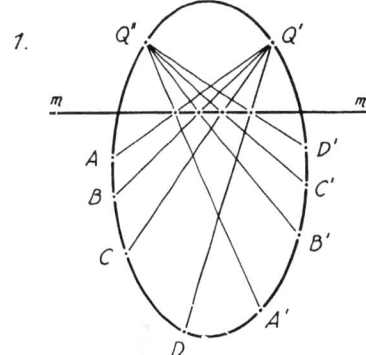

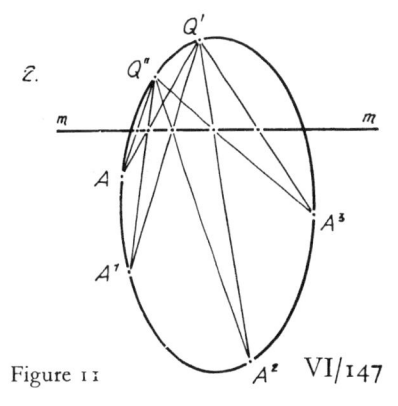

Figure 11

Figure 12

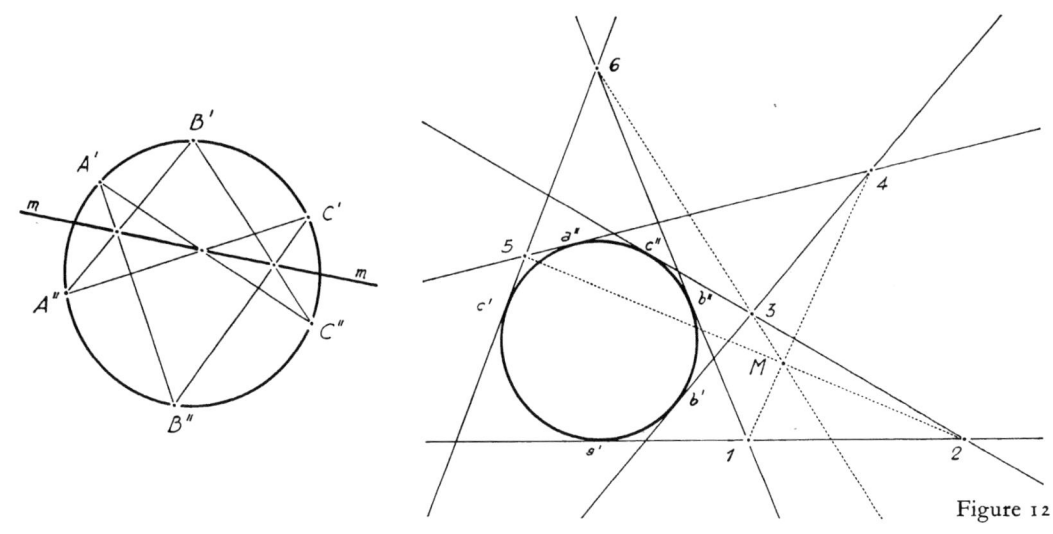

theorem becomes Pappos's theorem. The two points of the curve in common with the line m are the double points of the projection (30).

Conversely, we think of any three lines of the curve being projected into any other three through the mediation of a centre of perspective M. This point then turns out to be the Brianchon point of a hexagon which circumscribes the conic. Degeneration of the conic into two points relates the theorem of Brianchon in turn with the converse of the theorem of

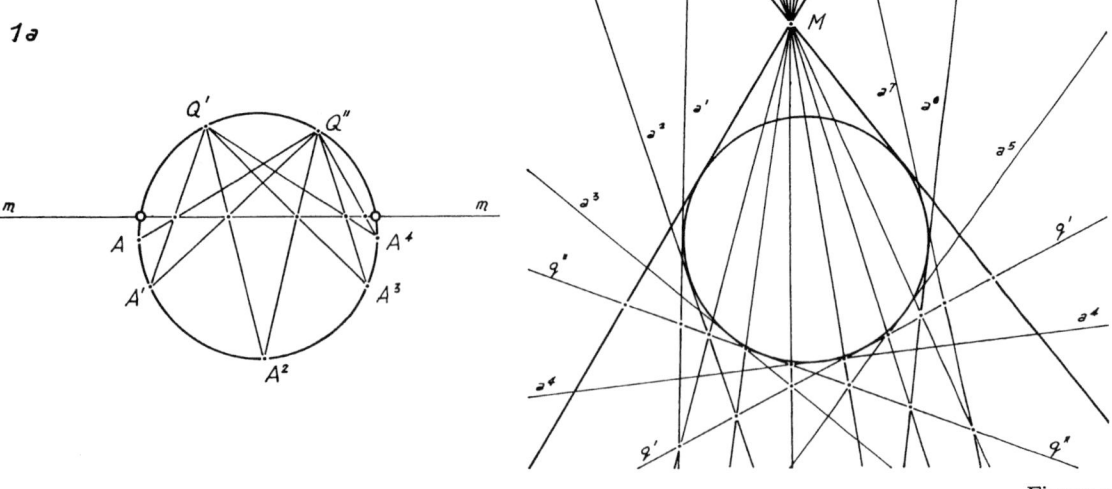

Figure 13

Figure 14

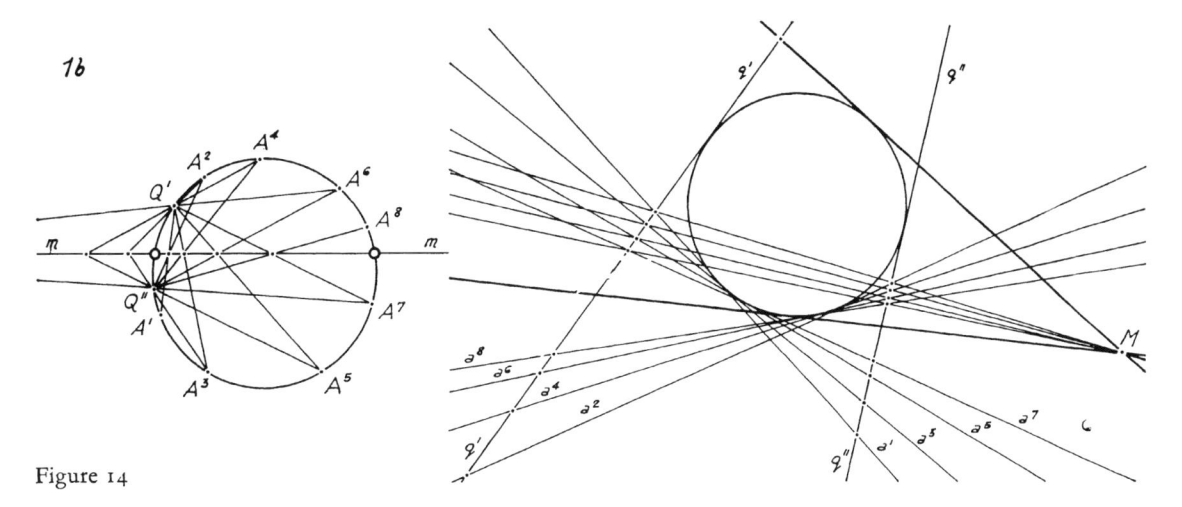

Pappos. The two lines of the curve which are in common with the point M are the double lines of the projection.

From the foregoing figures it is easy to see that, as with the line, so with the curve; there are three possibilities of the projective transformation:

1. When the Pascal line passes through the curve. It then contains two real points of the curve, and the transformation, having two double points, will be in growth measure. Conversely, the Brianchon point will be outside the curve, and contain two of its lines (Figures 13 and 14).

2. When the Pascal line is a line of the curve. Its point of contact will give one double point and the transformation

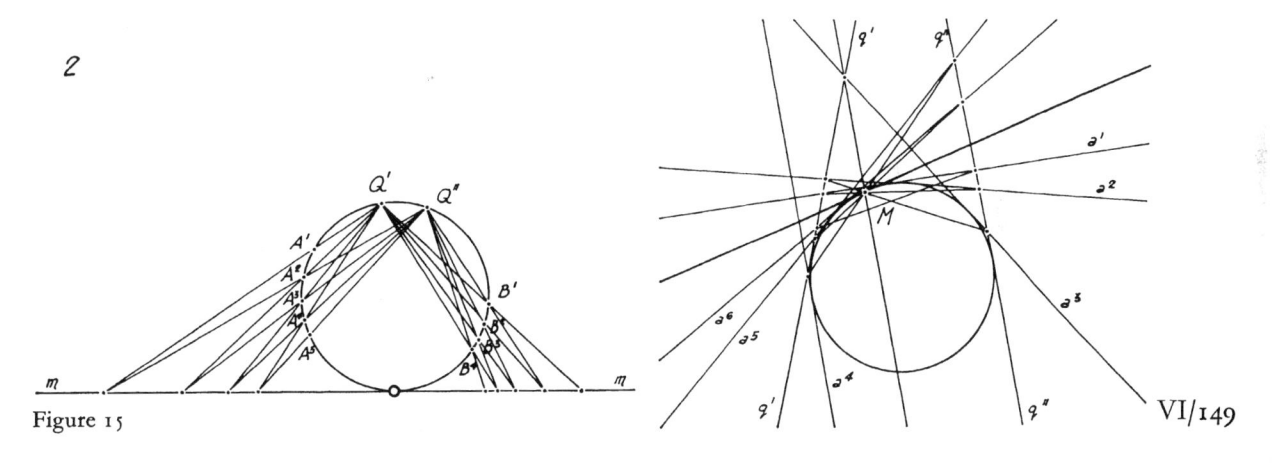

Figure 15

will be in step measure. Conversely, the Brianchon point will be a point of the curve and will therefore contain only one of its lines (Figure 15).

3. When the Pascal line passes outside the curve. It then contains no points of the curve, and the transformation, having no double points, will be in circling measure. Conversely, the Brianchon point will be inside the curve, and will therefore contain no real lines of the curve (it is not possible to draw tangents from it to the curve) (Figure 16).

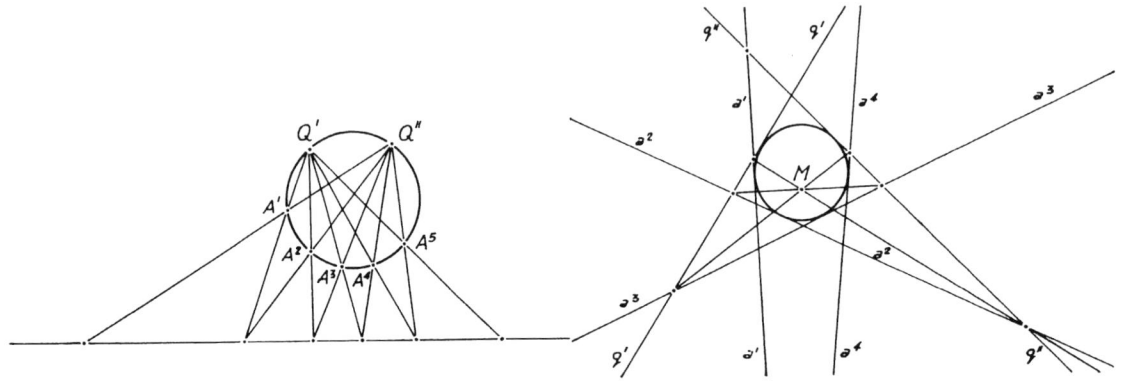

Figure 16

The illustrations picture the three possibilities. The points Q' and Q" (the lines q' and q") serve as fixed projecting points (lines). In the case of the projection with two double points (double lines) there are again two possibilities; the sequence is either continuous or the movement leaps to and fro, using the two parts of the curve alternately. Figures 13 and 14 illustrate these two possibilities.

Projection without Fixed Projecting Elements; Potentizing Process

Now it is possible, in this process of projecting a curve into itself, instead of using two special points (or two special lines), as we have been doing, to use *each successive point* (*or line*) as the projecting elements. The result is a kind of potentizing process (see Figure 12).

In a projectivity, one of the two special points (lines) is projected into the other (Q' into Q" or A' into A"). Our method of procedure might have been described as follows: A' projects into A''; to find where a third point B' goes, join it to A'' and then join the point determined by this line on m to A', which line will determine the new point B'' of the curve.

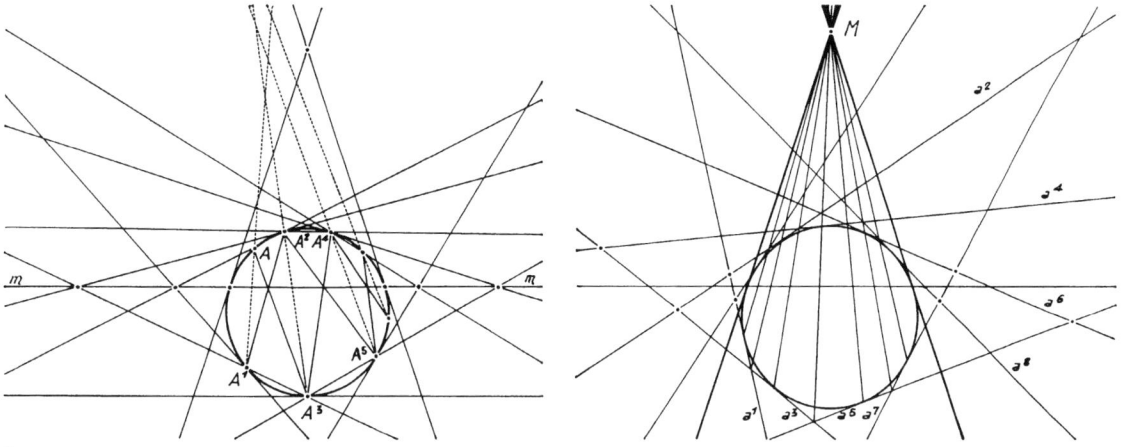

Figure 17

Thus we can say (Figure 17): A projects into A_1; where does A_1 project to? *Join A_1 to itself* (it being the second projecting point); that is to say, *draw the tangent at A_1* and then join the point given by this tangent on m to A (the first projecting point). This will give the point to which A_1 moves. Then, to find where the new point A_2 goes, repeat the process, using it and A_1 as projecting points, and so on. Figure 17 is an example for the projective transformation of a curve into itself with two double points (double-lines).

Breathing Involution

In the previous example, for the pointwise case (Figure 17), we have drawn in lightly the lines joining pairs of points projected in sequence one into the other, and we see that they all come near the common point of the tangents at the double-points. These tangents are the double-lines of the linewise projection. Correspondingly, the meeting-points of pairs of lines related by the projection come near to the line joining the double-points (Figure 17 right).

When the projection resolves itself into an involution, and each point returns into itself again at the second step, the lines joining the points paired in the involution will meet in a point (M), namely, the common point of the tangents (the double lines) at the double points. Moreover, the meeting-points of the lines which are paired in the involution all lie on the common line (m) of the double points (Figure 18).

Figure 18

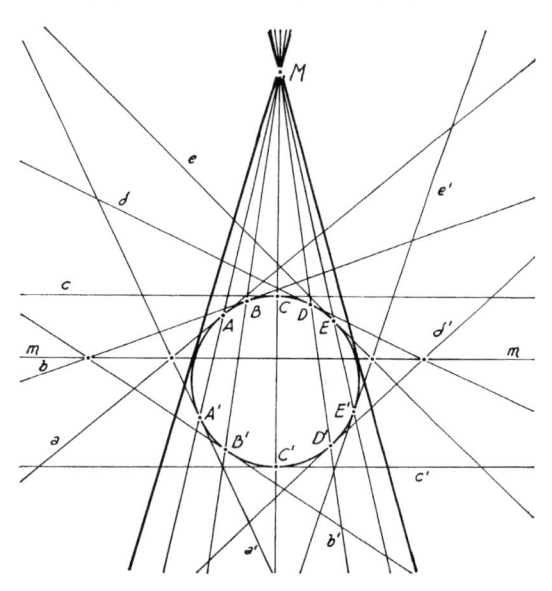

Here the linewise and the pointwise projections fuse together in involution and the whole is represented by a far simpler form. If we have called the line joining the double points m or the Pascal line, so we must call the point joining the double lines M or the Brianchon point. The Pascal line is called the Axis of Involution and the Brianchon point the Pole.

In Figures 17 and 18 we have considered the case of a projection with two double-points (lines), which takes place in growth measure and is then resolved into an involution, when the pairs project back and forth into one another (Figure 18). The projected points stream to and fro (the lines swing back and forth) in the spaces bounded by the double-points (or lines). The pairs in involution are always separated by the double entities, which act like boundaries or guardians of the whole process. The pairs either crowd in towards the one or the other guardian, pressing in towards a double element, or they spread out as far away from them as the space allows. It is like an in-and out-breathing movement. Using a rather imaginative terminology, we will call this a "Breathing Involution", distinguishing it from an involution which comes about when there are no double elements in the projection (18).

Cyclic Projectivity

Considering now the case of a pointwise projection with *no real double elements*, we find that such a process might result in the elements circling indefinitely around the curve as they project one into the other. Circling measure results from this process. Here, too, the points of the sequence in turn take on the function of projecting points, so that we have no fixed Q' and Q'', as in Figure 16. On the other hand, it is possible that the projectivity might be a cyclic one, that is to say that after a number of steps it might return to the point at which it began. There would then be a cyclic projectivity of the number three, four, five and so on along the curve (Figure 19).

Circling Involution

If, however, at the *second* step the point or line circles back to its starting-point, the projectivity resolves itself into an involution once again—this time a Circling Involution. Here, the involutionary movement of pairs circles round, either clockwise or anti-clockwise, with no double elements to act as guardians or boundaries to dam the flow (Figure 20).

While in the Breathing Involution, each of the point-pairs which are in involution has a line of M in common and each line-pair a point of m, in the Circling Involution the same applies, but this time the axis (m) of the involution is outside the curve and the pole (M) is within (18).

As in the breathing involution, so too here; there will be no limit to the number of pairs which may circle round in the involution. There may, however, in both cases, be pairs which

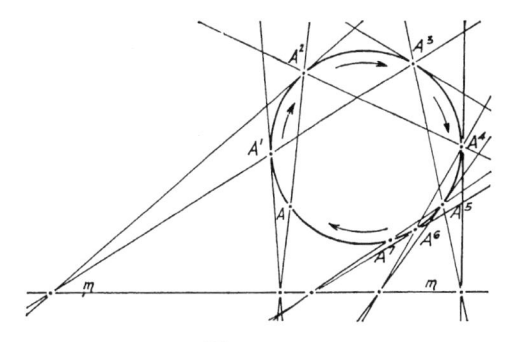

Figure 19

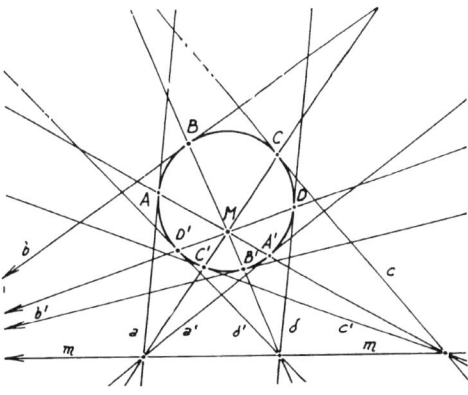

Figure 20

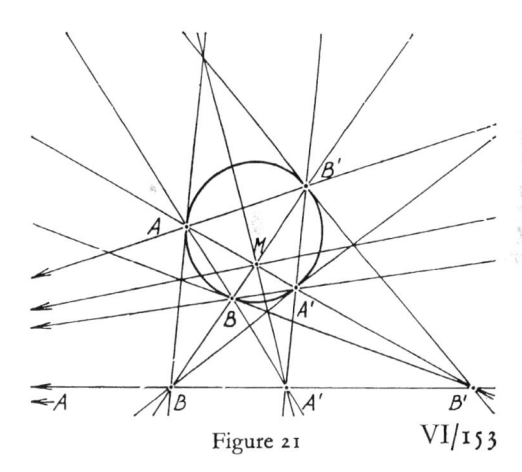

Figure 21

are in a special relationship with one another, namely, the harmonic relationship; under certain circumstances it may happen that a familiar construction arises, where two pairs of points or lines in involution form a harmonic quadrangle or quadrilateral (Figure 21).

Furthermore, from this construction, it will be seen that the pairs in involution can be projected from any one of the four points on the curve to the line outside the curve, the axis of the involution. The points on this line are also in involution, and clearly the pairs are harmonic in relation to one another (cf. Figure 7).

The conditions of involution of point- or line-pairs of a circle-curve are thus consistent with what we know about four-point and four-line and the diagonal triangle. Not only may an involution be brought about when m and M function as axis and pole, when it will be a circling involution. Axis and pole might also be the other two point-line pairings respectively in the diagonal triangle! The involution may in fact be brought about by using either of the three lines or points of the triangle as the axes or poles. In one case, the involution will be circling, in the other two it will be breathing. Thus we have returned again to the beautiful simplicity of a well-known figure, approaching it from yet another aspect.

If now we move the pole of the involution into the centre of the circle, so that the line (axis) is in the infinite, we shall see that if the pairs in involution are projected from any one point on the curve, the circle's right-angled circling relationship to the infinite is at once revealed. The rays projecting corresponding pairs are at right angles to each other. (The angle in a semi-circle) (Figure 22).

This right-angled circling property of the circle may in fact be used to find the centre of a given involution. Given two pairs of points in involution on a line, a semi-circle is described on each of the pairs, and their point of intersection will give the point from which the right-angled circling rays will project on to the line, giving point-pairs which will be in the same involution as the given two pairs. Harmonic pairs of lines in the involution will here be determined by the fact that these pairs mutually divide one another's right angle.

The paired elements in an involution are said to be con-

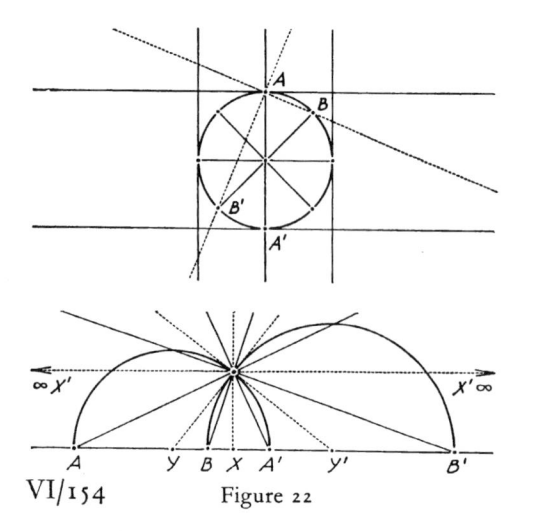

Figure 22

jugate to one another; each element has its partner. An involution is determined by two such pairs; it is an arrangement of elements (points, lines or planes), which are conjugate in pairs. A harmonic range consists also of conjugate pairs. As we shall see in greater detail in Chapter 7, every circle-curve can call forth such a relationship between point-pairs of any line or line-pairs of any point of the plane to which it belongs.

The conjugate points of a line in the plane of a circle-curve are the related elements of an involution.	The conjugate lines of a point in the plane of a circle-curve are the related elements of an involution.

Indications concerning the Imaginary Double Elements

The type of involution which we have called "breathing" is also called "hyperbolic", while the "circling" involution is called "elliptic".

In the first instance, there are two double elements (Figure 18). The pairs are separated by the double elements, into which they merge or from which they emerge. The members of each individual pair in the involution do not separate one another, but they are always separated by the double elements. There are an infinite number of conjugate pairs, all of which are harmonic with respect to the double elements.

In the second instance, circling involution (Figure 20), there are no visible double elements; the pairs in involution are not separated by double elements, but the members of each individual pair always separate one another. There are an infinite number of conjugate pairs in involution, which are harmonic with respect to one another, though not *all* pairs are harmonic pairs.

In the case of a circling involution, the mathematicians speak of *"imaginary"* double elements, in contrast to the "real" double elements in a breathing involution. In both, they recognise the self-corresponding elements, whether real or imaginary.

In a circling involution of points on a line, two imaginary points play the part of double points.	In a circling involution of lines in a point, two imaginary lines play the part of double lines.

It was Christian von Staudt's great contribution to mathematics that he introduced the theory of imaginary elements into pure geometry and developed it to a high degree of perfection (25). Although the conception of the so-called "imaginary" numbers resulted directly from the analytical method, v. Staudt was able to relate the imaginary to geometry, without the help of analytical calculation. This brought the science of pure geometry a great step forward. It goes beyond the scope of the present work to embark upon this field, beyond explaining to some extent what is meant in the present instance. It should, however, be said that to pass in geometry from the real to the imaginary in no way presupposes a lesser degree of clarity and accuracy in thought. On the contrary, it becomes perfectly possible to picture processes taking place between real and imaginary elements. An intensification of the pictorial activity is required, and a readiness to picture moving processes accurately. It is yet a further step in the direction of spiritual activity and away from mere dependence upon external, physical pictures. In trying to make pictures in geometry of the physical traces of the imaginary, we enter in clear thought into a realm which, while entirely free from the limitations of physical matter, is nevertheless fundamental to all created form.

In the example before us we think as follows:

Every line in the plane of a circle-curve has just two points in common with it. If the line passes through the curve, the two points are real. If the line is a tangent to the curve, the two real points have become one. If the line passes outside the curve it has two imaginary points in common with it.	Every point in the plane of a circle-curve has just two lines in common with it. If the point lies outside the curve, the two lines are real. If the point is a point of the curve, the two real tangents have become one. If the point is inside the curve (when no tangents can be drawn) it has two imaginary lines in common with it.

The Imaginary can only be grasped through dynamic movement. Any such dynamic movement may, however, be carried or receive a form, by taking into account the properties typical to it. An imaginary point-pair is always carried by a real line (a line-pair by a real point) and can be represented either as two pairs of a circling involution or as a harmonic four.

In a breathing involution the two real double-points (or lines) are like the product of a damming up process in the reciprocal inward and outward streaming of the points or swinging of the lines. In the circling involution the points (or lines) are in a continuous cyclic flow.

The self-corresponding elements in the involution of points along the axis or lines in the pole are, for the breathing involution, the elements into and out of which the movement streams, in either direction. These elements are fixed, and they dam up the flow.

For the circling involution, these fixed elements no longer exist, but the flowing movement is still there, lines swinging round and round in the pole of the involution, points streaming round and round on the axis in either direction. It is this *movement of point-pairs (line-pairs)* which represents the two double- or self-corresponding points (or lines) of a circling involution. These are the imaginary double elements of a circling involution.

Imaginary points glide, as it were, like comets along the line which bears them, and it is akin to their nature to depict them in the form of a harmonic range. One imaginary point is pictured by the movement of the four points along the line in one direction, and its conjugate is the movement in the opposite direction (Figure 23).

Correspondingly in the case of an imaginary line we picture the rotational movement of four harmonic lines held in a real point; one imaginary line is depicted by the movement in the one direction, while its conjugate imaginary line is depicted by the movement in the other direction (Figure 23).

If, now we consider the case when the pole of a circling involution on a circle has moved into the centre of the circle, the axis of the involution being the infinitely distant line of the plane, we must think as follows: the right-angled circling property of the circle is due to the two conjugate imaginary

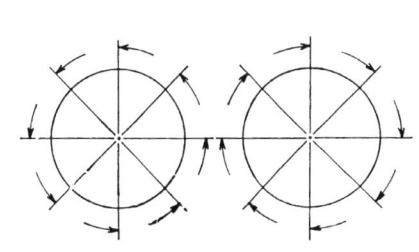

Figure 23

points which every circle has in the infinite. They are called the Circular Points, or I and J; one is the clockwise right-angled circling movement, the other the anti-clockwise movement. These two imaginary points are carried by a line which, *though infinitely distant is nevertheless real*; the pole of the circling involution is then the centre of the circle. It is a *real* point bearing two imaginary lines—the clockwise and the anti-clockwise movement of the four harmonic lines (31).

Plane Path-curves in Breathing and Circling Involution; One-Dimensional Transformations

These result from breathing and circling involutions and projectivities. *The curve is transformed into itself*; it slides, as it were, pointwise or linewise within itself. Each point and line, as it changes into the next one, traces the history of its transformation into the plane in the form of the curve. As, however, the curve lies in the plane, *all* the points and lines of the plane relate themselves to the transformation, and, as we shall see, there arise not only single curves, but a whole sheaf or family of curves.

It is interesting to reflect on the philosophical implications of such a geometrical concept. The change within the one-dimensional individuality—the curve within itself—has a potential effect upon the whole environment. The points and lines of the plane as a whole take note, as it were, of the transformation within the single curve. The streaming movement of points and lines in and around the single curve is echoed throughout the whole plane, like ripples upon the surface of water.

There will be a pointwise path curve (or a whole family) when a projectivity of rays from two given points is determined by perspective with an involution (breathing or circling) of points on a line.

There will be a linewise path curve (or a whole family) when a projectivity of points of two given lines is determined by perspective with an involution (breathing or circling) of lines in a point.

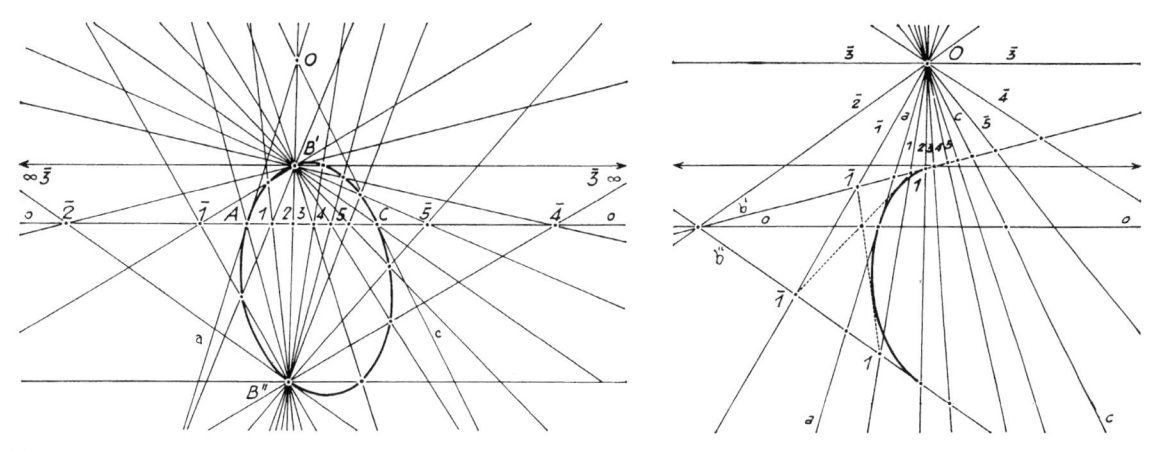

Figure 24

Construction for a Pointwise Ellipse: Breathing Involution (Figure 24)

First establish a breathing involution on a line, using a construction as in Figure 18, and transfer the points to a line o. Choose a point O to be the common point of two lines a and c, which pass through the double points of the involution. On some line of O, choose two perspectiving points B' and B'', harmonic with respect to O and o.

Now draw in the rays from B' and B'' to the points of the breathing involution, thus establishing a projectivity. The common point of a ray from B' to a point of the involution with a ray from B'' to its harmonic pair in the involution will be a point of the path-curve. As the rays of B' and B'' move round, thus paired by the involution, their common point will trace the path of the curve. Naturally, a different curve will arise if we choose different situations for O, o, B' and B''. In Figure 24 (right) is shown the way to construct the ellipse linewise through the movement of lines common to pairs of points in involution on the lines b' and b''. We are reminded of the way we first drew curves in a free projection in Chapter IV.

Figure 25

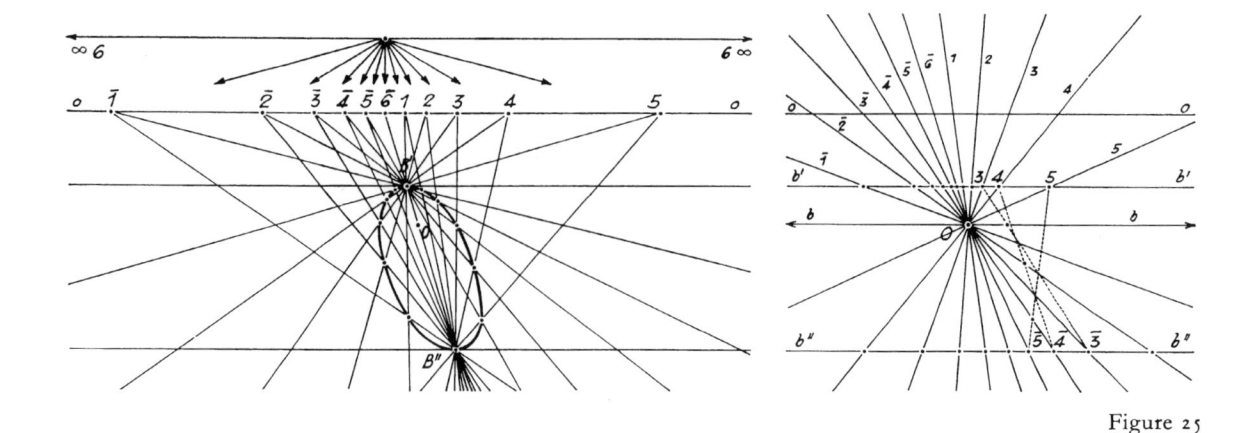

Construction of a Pointwise Ellipse: Circling Involution (Figure 25)

The ellipse in the next figure is determined by a similar construction, except that here we have used a circling instead of a breathing involution on the line o. This involution is established in the simplest possible way, by perspective from the right-angled circling involution on the line at infinity of the plane, which results of course in equal angles among the lines of the perspectiving point.

The more points we use in the involution along o, the more guidance we shall have in drawing the curves. Among the most useful of the circling measures to take is the twelve-cycle. The canonical Four quarter the entire 12 into 4 equal circling steps. If $\bar{6}$ is the central point of the involution on o and $\bar{3}$, 3 the harmonic pair on either side, then these two points give the "amplitude" or measure of the involution. Setting O somewhere outside o, were we to draw all the rays which these points have in common with O, we would be free to choose on any one of these rays the two points B' and B'', remembering, however, that they must be harmonic with o and O.

Now we begin the actual construction of the curve; we look for the meeting-points of projectively related rays. These give twelve points of the curve in cyclic order. The curve is determined by these twelve points and can now be drawn in freehand.

In these one-dimensional transformations we can choose B' and B" anywhere on any ray of o, but they must be harmonic with O and o. As we know—and this is particularly evident in the case of the circling involution—each new pair arising through the projection may be used as the projecting pair. In Figure 26, where an ellipse is being created through a breathing involution in which three pairs are already determined, we may continue the construction, either by using the original point-pair B' and B", or we can pass on and use the following pair. Each pair, once found, may be used in finding the following pair. The same may be carried out linewise, beginning with b' and b".

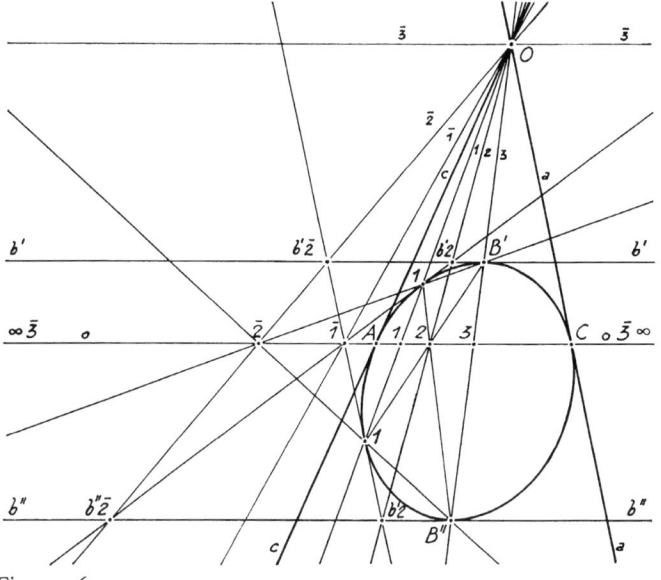

Figure 26

In Figures 24 and 25, for reasons of clarity, we have shown the dual aspects separately; it should not, however, be forgotten that they always belong together. We should really always work both pointwise and linewise at the same time. The lines (tangents) which in Figure 26 touch the curve in B' and B" are the projecting lines. We look for the common points of b' and b" with corresponding lines in O (for example, the line-pair O2 and O$\bar{2}$) and join them, thus finding lines of the curve. If later on we want to draw curves without using

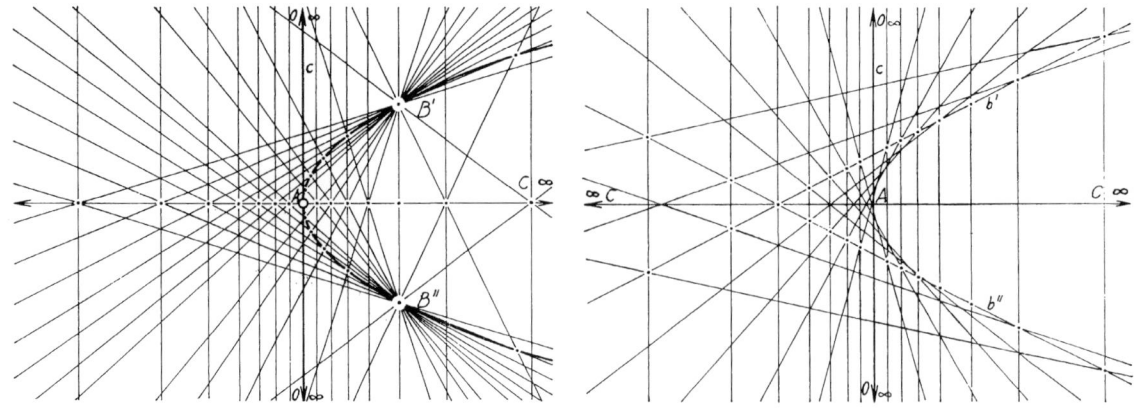

Figure 27

too many construction lines, it is important both for the end-result and also in order to experience the true shape of the curve while drawing it, to use the tangents. If we know where a point moves to, we can then find the tangent at this point which moves with it and gives the exact direction of the curve.

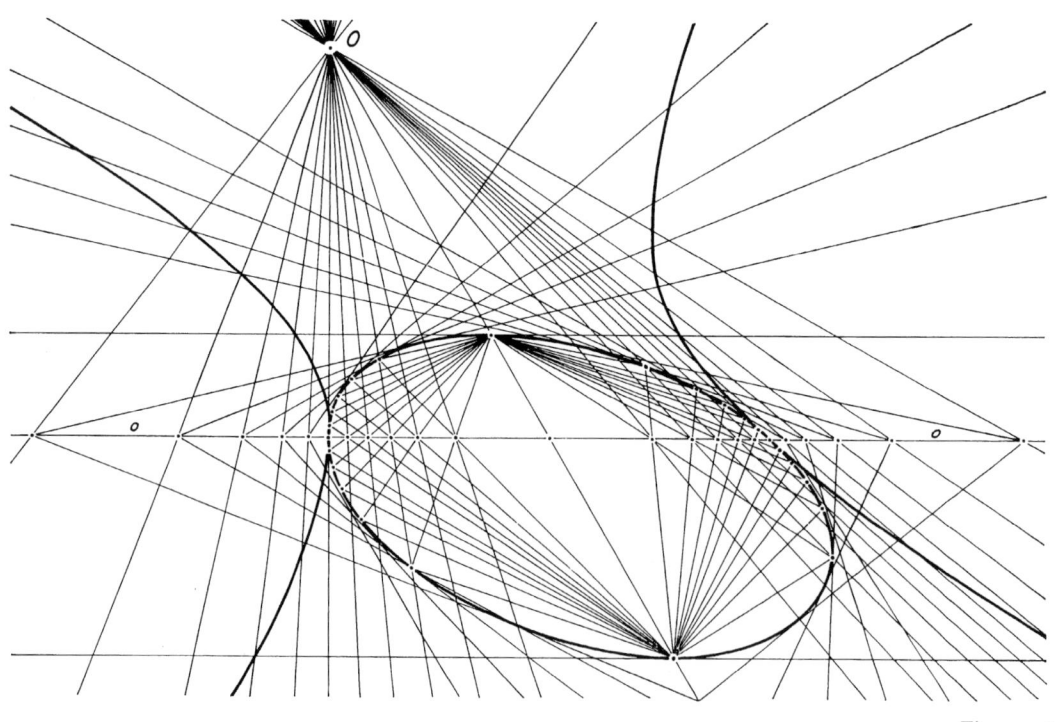

Figure 28

It is possible, with this construction, to create all kinds of circle-curves in different situations. In Figure 27, O and C are in the infinite and the resulting curve is a parabola. Figure 28 shows an ellipse and also a hyperbola as the outcome of a breathing involution; in Figure 29 they appear in a circling involution.

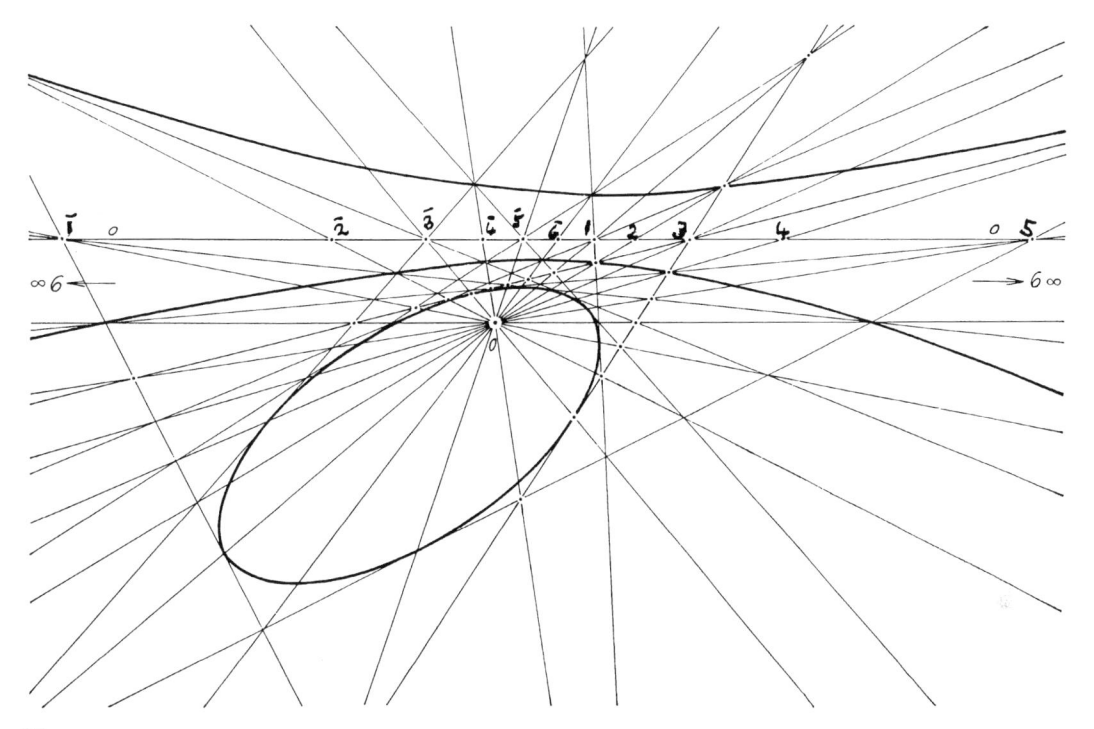

Figure 29

According to the way we position O, o, B', and B'', we shall arrive at different curves. If, however, we wish with certainty to determine either a circle or a parabola, we must put B' and B'' on a ray of O, which passes through the central point of the involution on o. The parabola will come about (when o and O are in the finite) if B'' is infinitely distant and B'o = B'O (p. 175). The circle comes about (when O is on a ray passing through the central point of the involution and at right-angles to o) only when two rays such as, for example, B'1 and B''$\bar{1}$ are at right-angles to one another (Figure 30). Figure 31 shows the linewise aspect.

VI/163

Figure 30

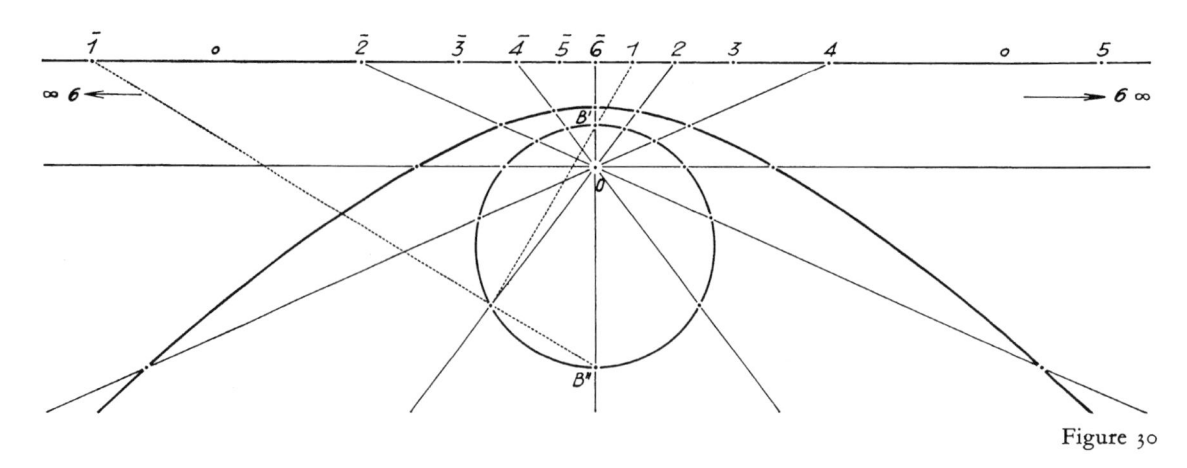

Having created one curve, one may then choose to relate a second in some harmonious way to the first and so on in sequence to bring about a family of curves; that is, a sequence of curves all related to one another. This has been done in one way or another in the various examples (Figures 32, 33). For instance, in the breathing involution with O in the infinite (Figure 33), the points of the curves where they pass through a line in their midst parallel to the two double-lines, will be found to be in a circling measure along that line; we chose to start the new curve at one of the points of this

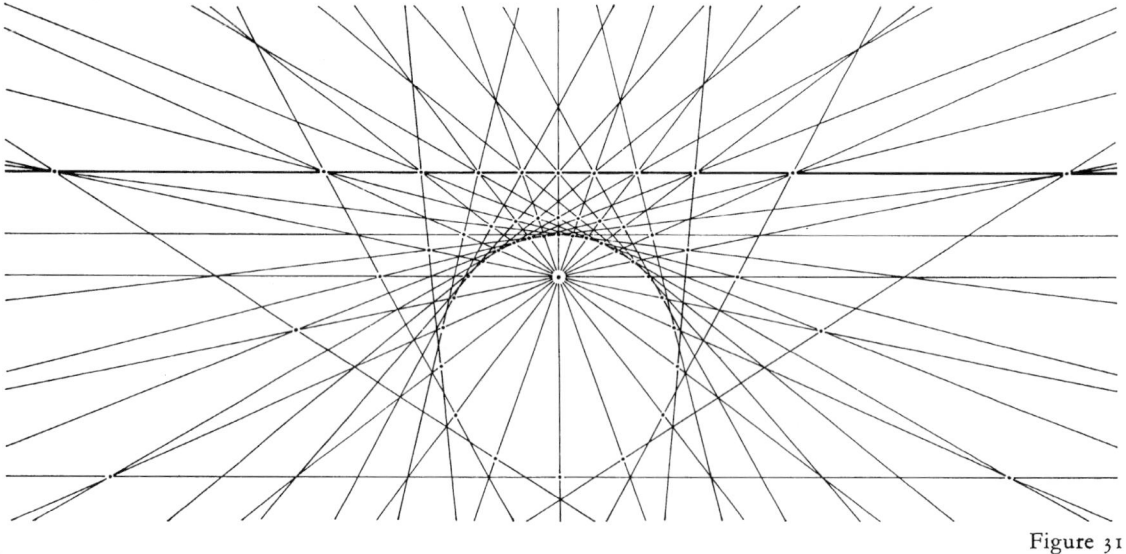

Figure 31

◀ Figure 32

Figure 33 ▶

circling measure. So, too, growth measure or step measure may be used in the transition from one curve to the next.

To bring about the family of curves in circling measure, as in Figure 41 we must fix O and o for all the curves and then allow the situation of B″ to arise as a result of our choice of the situation of B′. Here there is a growth measure between o and O.

The family of curves in a breathing involution give a very different impression from the family in a circling involution. In the breathing involution the two "guardian" points or lines dam up the flow of curves, while in the circling involution there is no such congestion at two places in the picture. In the breathing involution picture, not only do the two points dominate the process, but also the meeting point of the double tangents. In fact, there is a "guardian triangle" of the transformation of curves in a breathing involution. It is a real triangle; its three lines and its three points are real (18).

A circling involution also has its "guardian triangle", but it is less evident, because it is partly *imaginary*. The point O and the line o are the real members of a triangle, and they are the bearers of the other points and lines of the triangle, which are imaginary. The members of this semi-imaginary triangle are: the real point O and the conjugate imaginary pair of points determined by the involution on o, and the real line o and the conjugate imaginary pair of lines determined by the involution in O (25).

Two-dimensional Projective Transformations; Homology and Elation

The other aspect of the projective transformation of curves is the two-dimensional one; that is to say, there may be a transformation taking place throughout the whole plane in which the curve lies. Such a transformation changes curves one into another over the entire plane. We remember that it is characteristic of all projective transformations that without exception they leave unchanged such qualities as harmonic ones, so that the curves in such a transformation all have the same fundamental qualities and belong together like members of a family. We can speak of a family of curves.

We must clearly distinguish in thought between the one-

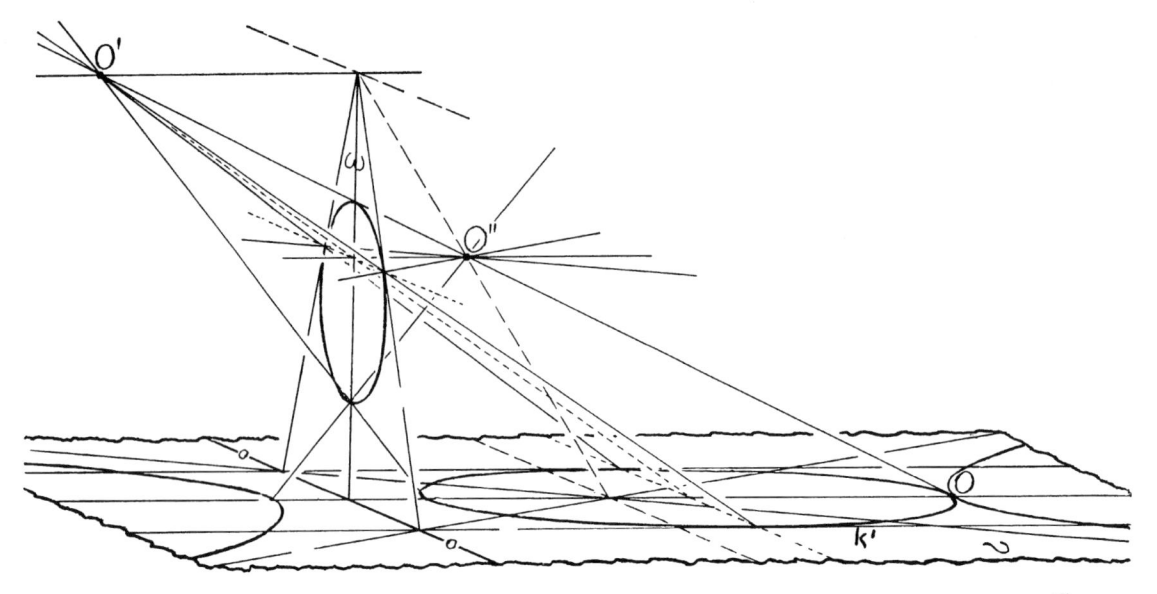

Figure 34

dimensional and the two-dimensional projective transformations, although, as we shall soon see, the two fit in with one another and assist each other. The projective transformation of a plane which we shall be considering here is called, after Poncelet, a Homology (a particular type of which is an Elation).

Just as in the process of transforming a line into itself we needed to go out of its one-dimensionality and used other points and lines of the plane, so in the projective transformation of a plane we will go out of the plane itself and use other entities in space. Thus, for example, in Figure 34 we see a transformation of a plane being brought about by means of an intermediary plane ω and two perspecting points O' and O''. (32).

As an aid to the understanding of this figure, let us first consider the simpler process of perspective taking place from one plane to another by means of the rays of a single point. The illustrations 35, 36 and 37 show how a circle in a plane (the slanting one) changes by perspective into one or another of the circle-curves, according to the relative situation of the perspecting point. When the circle becomes a parabola,

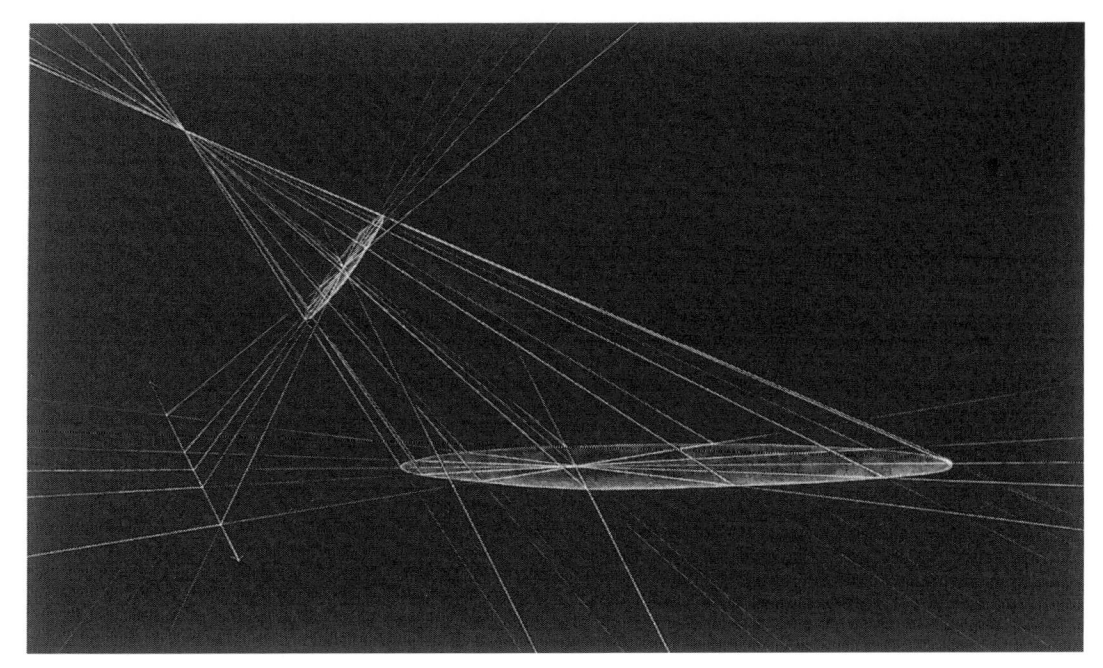

Figure 35

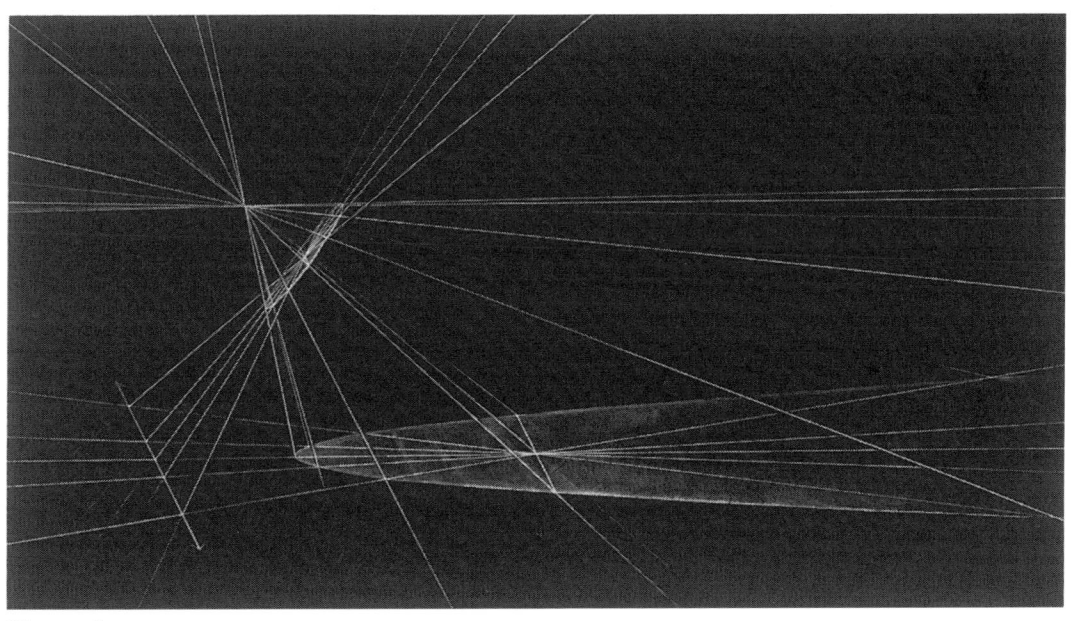

Figure 36

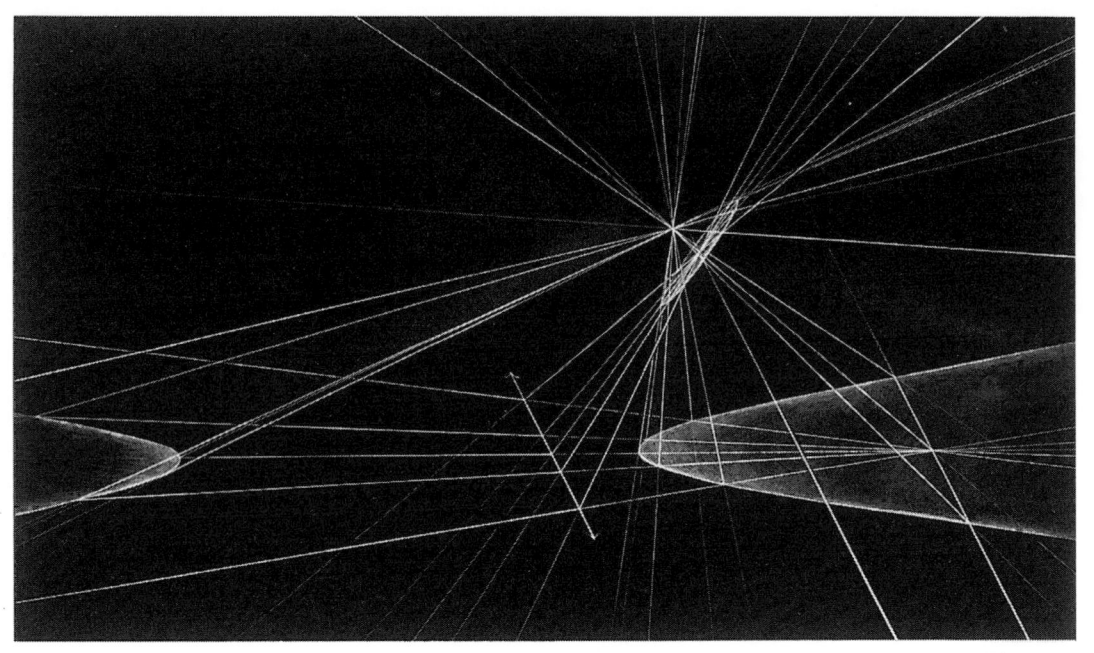

Figure 37

this "eye" point will be so situated that one of its rays is parallel to the horizontal plane, thus taking a point of the circle to the infinite on this plane (Figure 36). And to understand how the circle becomes a hyperbola in perspective, we must be able in thought to consider the line *in its entirety*, and to see that some of the rays send some of the points of the circle away out *through* the infinite, bringing them back from the other side as they come in from below the horizontal plane (Figure 37). We are thus enabled to follow the perspective change *through* the infinite, from the closed form of the circle and then the ellipse to the "open" form of the parabola, which has just one infinitely distant point, and then to the two-branched form of the hyperbola. In this process as such, with all its continuity, the infinite plays no special part; it is only from a one-sidedly physical point of view that it contains problems.

Returning now to the projective process as pictured in Figure 34 we see that the point O' perspectives the curve k' (a circle), which lies in the horizontal plane v , into the ellipse in the intermediary (vertical) plane ω. The second point O'' then perspectives this ellipse back into the horizontal plane, where it appears as a hyperbola. Thus the circle k' in the horizontal plane is transformed by projection into another curve *in the same plane*.

By means of such a projective process, any form in any plane may be transformed into another form in the same plane, the actual result of the transformation depending on the relative situations in space of the elements concerned in the process.

It is important to distinguish between the *method* and the *result*. The homology transformation in the plane is the result of a sequence of two perspectives (from which fact it deserves the name "projective"). Just as in the one-dimensional transformation of a line or curve into itself it was necessary to go out of the one-dimensionality of the line into the plane, so now the *method* of transforming the plane into itself by projection necessitates a construction in three dimensions. Having become clear about the method, we contemplate the result in the plane v, and discover that as a necessary consequence, all points of the common line o of the two planes must remain at rest in the transformation, while all lines of the point O, situated as it is on the common line of the two perspecting points, must also remain at rest; that is to say, they are transformed into themselves. We may call all those points and lines of the plane which remain at rest "latent"; they are fixed, and do not take part in the transformation. Yet without them it could not take place, for in the process all points which move do so along lines of the fixed point O, while all lines which move do so about points of the fixed line o.

This describes how a homology may be carried out within the plane itself; it is in fact how a homology is defined. Quite apart from the method by which we were led to this, we can say:

A Homology is a transformation in the plane, such that all points move along lines of a fixed point, while all lines move about points of a fixed line.

An Elation is the special case of a Homology, when the fixed point is in the fixed line; the fixed line is in the fixed point.

Figure 38

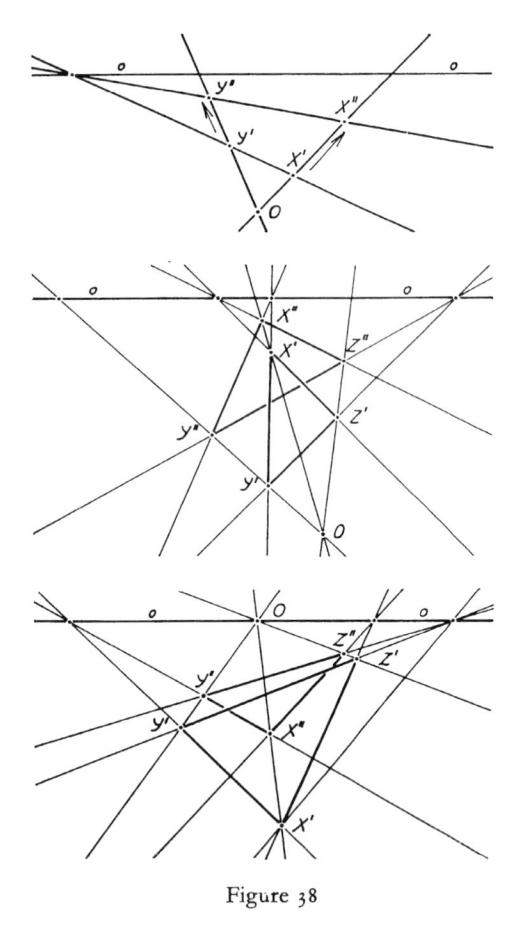

Bearing this formulation in mind, we are enabled in effect to carry out a given transformation in the plane quite simply by homology, by-passing, as it were, the cumbersome and difficult method of a three-dimensional construction. In order, however, fully to grasp this process, we must not forget that the underlying projective method of its construction involves the third dimension. All this becomes the more understandable when, as we shall see, the dependence of a homology on the two triangle Theorem of Desargues becomes evident (Figure 38). Without the theorem of Desargues, no homology would be possible. (As we already know, this theorem only becomes self-evident when seen in three dimensions.)

Comparing now the one-dimensional and the two-dimensional aspects of projective transformations, we can say: In transforming a line into itself we have an intermediary line and two perspecting points. This gives two "latent" *points*—the double-points O and U. They will be one, if the common line of the perspecting points meets the line to be transformed in the same point as the intermediary line does. (See Figure VI/2). The presence of two double-points, as we saw, gives rise to a growth-measure transformation (in potentized projection), while, when the two fall together in one, a step-measure transformation results.

We must note the difference between the two aspects of a perspective transformation. Changing the plane into itself we have, not two double-points, but a double-point and a double-line. The line joining the two centres of perspective in this case still leads to a double-point, but the intermediary entity is now a plane (ω in the figure) having a whole line-of-points in common with the plane we are transforming. In the one-dimensional case, the intermediary entity was a line, having only a single point in common with the original line.

Having seen how single curves and then whole families may arise as the result of a one-dimensional transformation, we will now see how, given one curve, another curve and then a whole series of curves arises, as the result of a two-dimensional transformation called homology or elation.

To carry out a homology transformation: Choose a centre O and a periphery o. Having decided that X′ goes to X″; if Y′ is any other point, then (Figure 38):

1. Y′ must move along a ray of O (OY′),
2. X′Y′, X″Y″ must have a common point in o.

Hence Y″ can only be the point where OY′ meets the line from X″ to the common point of X′Y′ and o. While the points of the plane move along lines of O the lines of the plane swing round in points of o.

The refrain for this construction might be: "When X′ moves to X″, where does Y′ move to?" Answer: "Find the common line of X′ with Y′; see where it has a common point with o; swing it round so that it takes X′ into X″ along a line of O (OX′), in so doing it will take Y′ into Y″ along *its* line (OY′) of O."

If Z′ is a third point, Z″ can be found *either* from the movement of X′ to X″, *or* from the movement of Y′ to Y″, already found. From this it may be seen that Desargues' theorem shows the condition that the two results are the same. Thus, once the homology has been set going we are free to use *any* already determined movement of points or of lines to determine any other.

It is a simple matter to transform a given curve through homology or elation. Beginning, for example, with a circle, we choose o and O somewhere in the plane and relate them with one another. Then we choose any point (or any line) of the circle and determine freely where it shall move to in the transformation; for example, in Figure 39, 1 moves to 1. This movement of 1 into 1 determines the movement of all other points and lines of the circle, for all points of the plane move along rays of O and all lines turn in points of o. The same happens in an elation, but in an elation O and o are coincident.

In Figure 40 the sequence of homologies of a circle reveal the possibilities of this type of transformation. Here the line o is somewhere below the curve and the point O a little above this line, but far outside the picture to the left.

We see that a transformation by homology is a perspective transformation, in which the forms either contract inward towards O or spread out towards o. If now we make a chain

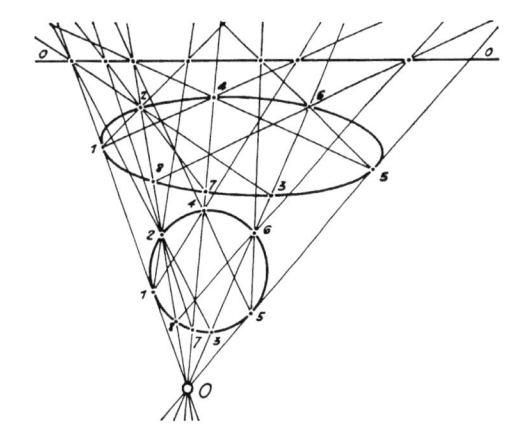

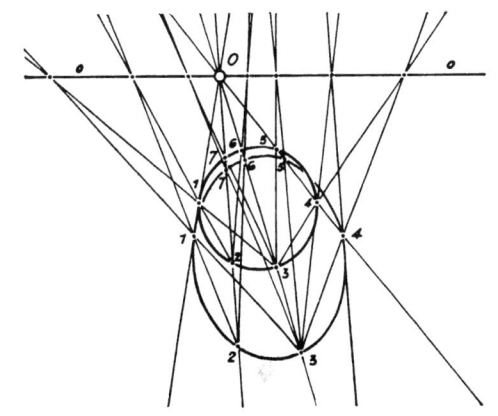

Figure 39

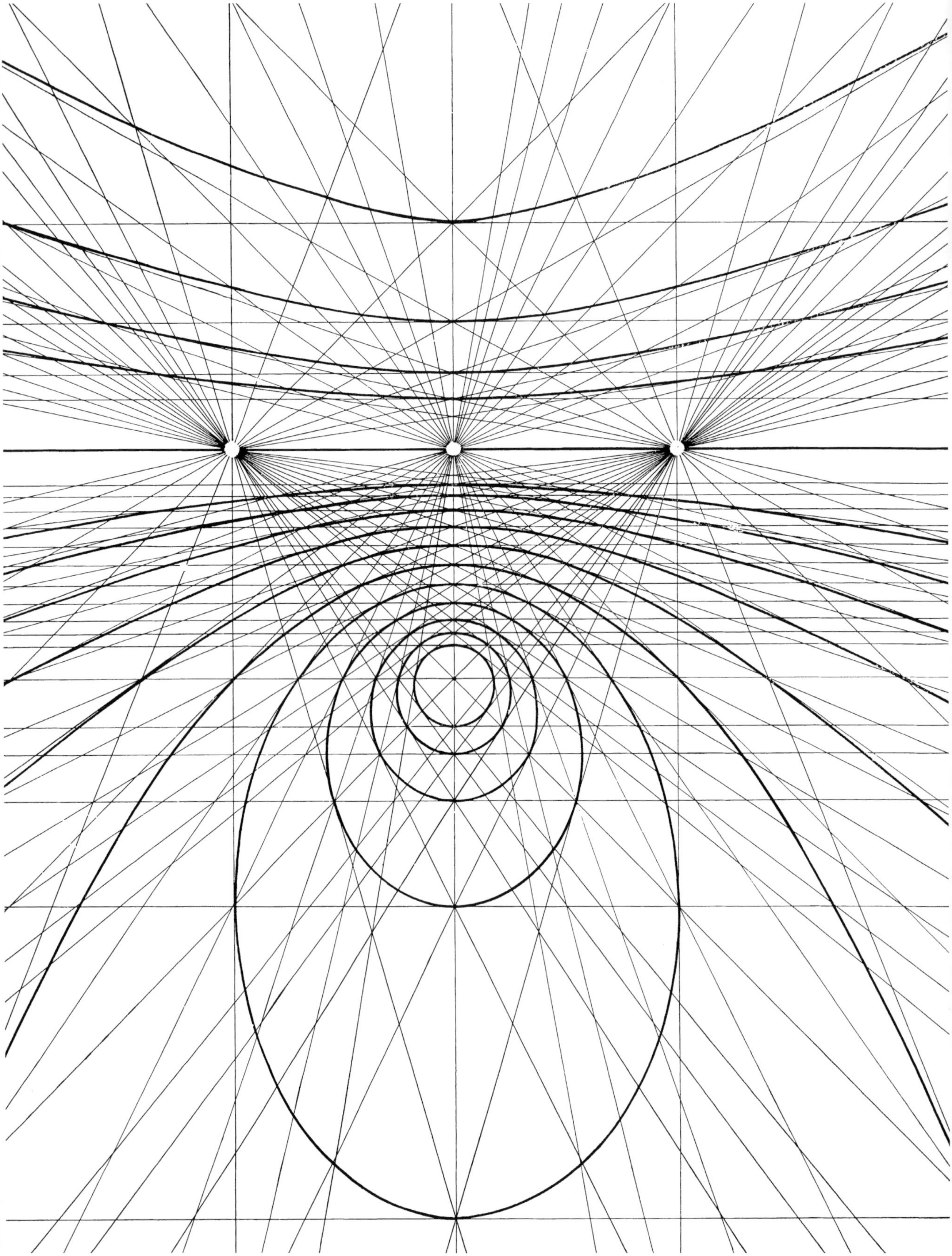

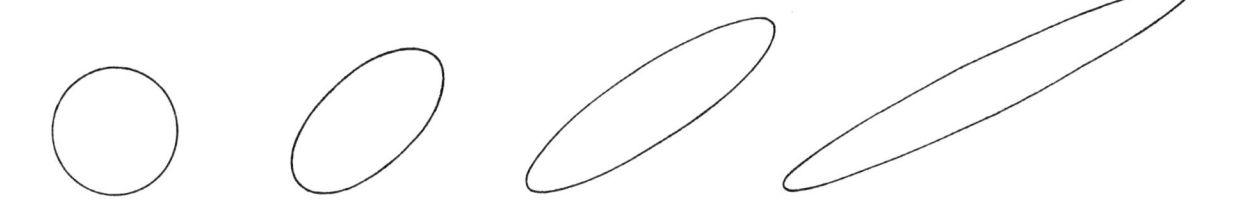

◀ Figure 41

Figure 40

of homology transformations, allowing one form to arise from the previous one, in the continuous or potentising process with which we are already familiar, we shall see how one form in the plane begets the next in a growth-measure sequence of expansion and contraction between the two extremes, point and line (Figure 41). It is in the very nature of this type of homology transformation that it gives rise to a growth measure between its fixed point and its fixed line, while the special case—the elation, when the fixed point is in the fixed line—results in step measure.

Supposing then we start with a circle in the plane and put O inside the circle and o somewhere outside, we can construct our family of circle-curves by means of a continuous homology process in growth-measure, when the curves will close in around the point and flatten out towards the line. The ellipses within the circle will at last degenerate into O, while the hyperbolae outside it, passing through the infinite, will die away into the line o, their two branches coming together, back to back, as it were.

Construction to include the Parabola. In order to be sure of "catching" the parabola we must take special steps to do so, as described for the path-curve transformations (page 163). It is often, however, best to begin with it. A point on the parabola will always be equidistant from O and o, and therefore we must proceed as follows: If Q is the central point of the involution on o, and $\bar{D}$, D the harmonic pair on either side giving the amplitude, then (whether or not OQ is perpendicular to o) the apex of the parabola, the harmonic pair of which with respect to O and Q is in the infinite, must be halfway between O and Q. Moreover, the two points of the parabola on the line of O which is parallel to o must have a distance from O which is determined by the amplitude $\bar{D}$, D (Figure 42).

Thus:

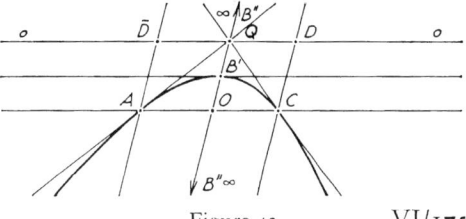

Figure 42

(i) The infinitely distant point B″ of the parabola, giving the direction of its axis, lies along the line OB′Q.

(ii) Beginning the construction from A and C, along the line parallel to o, the distance AC must equal the amplitude $\overline{DD}$. If OC = QD, the tangents from Q will touch the parabola at A and C.

(iii) Beginning the construction from points on the line OQ, one of the two points is already given, namely B″ in the infinite, while the other, being harmonic to the infinite, is the point B′, half-way between O and Q. Given the parabola, it is easy to draw the rest of the curves by homology.

Starting from a given curve, some of the points of the next curve will already be determined by the meeting points of tangents (Figure 41). If the circle is to be included as well as the parabola, the instructions on page 163 must be taken into account. In the beautiful, slanting family of curves (Figure 43) there can be no circle. The fact that these curves are not dominated by a rigid right-angle symmetry and measure gives them their freedom of form and movement, in contrast to the family of curves with a central axis of symmetry.

In the symmetrical picture (Figure 41), among all possible circling, involutary movements of the lines of O and the points of o, it was the right-angled one which was chosen. As

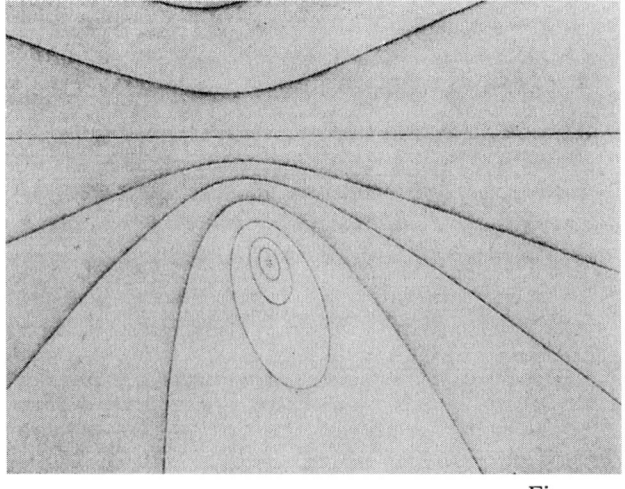

Figure 43

a result, O is a focus common to all the curves and o is the common directrix (p. 40). This arrangement gives the picture its more metrical character. Both pictures illustrate, each in its own way, how in the regular, circling measure one curve after another is born, resulting in a homology sequence which moves inward and outward over the entire plane.

Families of Curves and the Harmonic Net

We have already seen that a family of circle-curves may be drawn into a harmonic net construction in growth measure (p. 127). If now we start with a circle, set O inside it and o outside, then we draw two tangents to the circle from *any* point on o (Figure 44). The line joining the points-of-contact of these tangents will give rise to a second point on o, from which it will then be possible to draw two more tangents to the curve. We shall now find—surprisingly—that the line joining the second pair of points-of-contact will lead back to the point of o from which we started! The common point of the two chords will be O.

We shall shortly be concerned in more detail with this construction (p. 193). For the moment we note that continuing the growth-measure net, beginning from the quadrangle of tangents to the circle, we shall in fact be carrying out a homology transformation as the result of which a family of projectively concentric circles will arise. Once more, we have been led back from a new aspect to a now very familiar construction (Figure 44). Although we began with a circle, the Thirteen Configuration appears once more as the basic structure.

Figure 44

Spiral Matrix

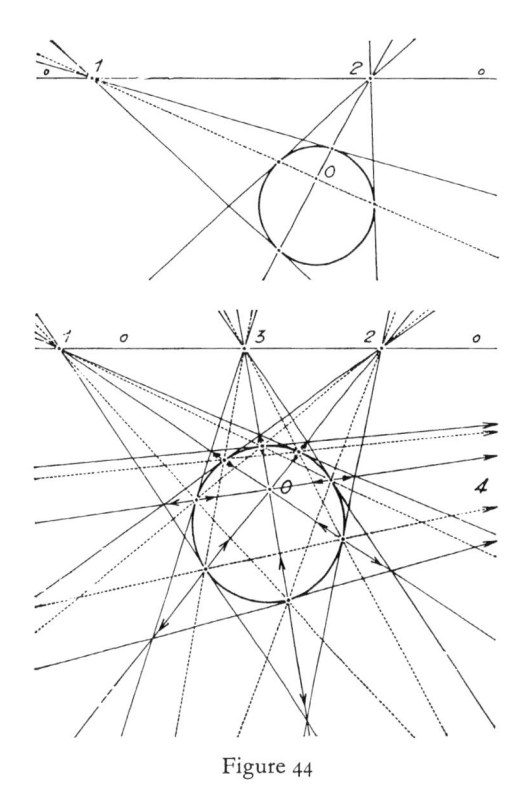

Having thus considered these two types of projective transformation which can take place in a plane—the one-dimensional and the two-dimensional—we will now consider them in mutual interplay. In the one type, the points and lines of the plane circle round the curves on which they lie; each curve being transformed within itself in either a breathing or a circling involution. These are "path curves" of a transformation. In the other type, all points of the plane move along rays of a point and all lines turn about points of a line, so that the curves of a family change into one another in an inward and outward movement. This is homology.

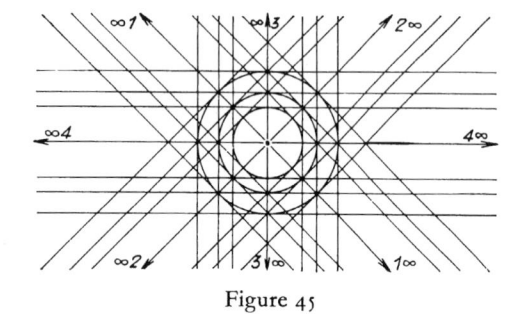

In both instances, the curves involved are usually circle-curves—the circle and all its variations. They are called curves of the second "Order" and the second "Class". It means that *any* line in the plane of the curve will have two points in common with the curve and *any* point two lines (tangents). It matters not at all, whether the points or lines in consideration are in the finite portions of the plane, or whether they are in the infinite. Moreover, as we have seen, the two points or lines may melt into one or be imaginary.

If, now, we combine the two types of transformation, we get more complicated curves, namely spirals. These are path curves, which combine in varied measure the inward and outward movement with the movement circling round. Of these, the archetypal examples are the equiangular spirals of Bernoulli (logarithmic) obtained from concentric circles in growth measure and their radii.

We will move the line o, bearing the circling involution into the infinite, thus bringing the right-angled circling law of the circle into our picture (Figure 45). As O is the pole of the circling involution on the line o, it will appear as the normal centre of a family of concentric circles. The right-angled circling of the imaginary points I and J on the line at infinity of the plane induces the constancy of angle in the lines of O (25). Upon the lines of O, a constant growth measure appears in the rhythm inward or outward in which the circles move from circle to circle.

It is a valuable exercise to carry out the construction using the twelve-cycle along the infinitely distant line o (Figure 46). The result will be to construct circles but without using compasses.

First choose twelve points in the infinite with corresponding ones between, so that each right-angled pair of directions has the same number (with or without bar). Then number the lines in O to correspond. As with the ellipse in Figure 25, we begin with one pair of points and tangents which are harmonic with respect to O and o. But as the line o is now in the

Figure 45

infinite, these pairs which are to create the circle are equidistant from O. The lines $\bar{x}$ and x, meeting as they do in a point of o (the line at infinity), will be parallel. Figure 46 shows the scene set for the construction of the circle, analogous to the construction of the ellipse as in Figure 25.

If then we carry out a homology, in which we shall immediately be guided by already existent interweavings, we shall see that the first circle construction will beget further circles, larger and smaller ones, spanned in growth measure between centre and infinite periphery.

Such a construction (cf. also Figure 45) is the matrix or "mother-form" of ever so many spirals (Figure 47). According to the way we combine the circling and radial components of the network, so the spirals will be more radial or more circular (Figure 48). The network has its linewise as well as its pointwise aspect. The *points* of the network lie in growth-measure along the lines of O and in circling measure around the circles. The *lines* lie in (parallel) growth measure families in the twelve infinitely distant points of o and in circling measure around the circles. The spirals may be drawn through the points or enveloped by the lines, which is a fine exercise in freehand drawing. It is useful to make tracings based on the matrix; every kind and degree of circling measure can be united with every kind and degree of growth measure in the creation of spiral families.

In carrying out the drawings and especially when drawing curves, it is good to work on a fairly large scale and possibly to use colours. The drawings should give practice in accuracy, but they should also give the soul the opportunity to expand and to enjoy the beauty of the forms.

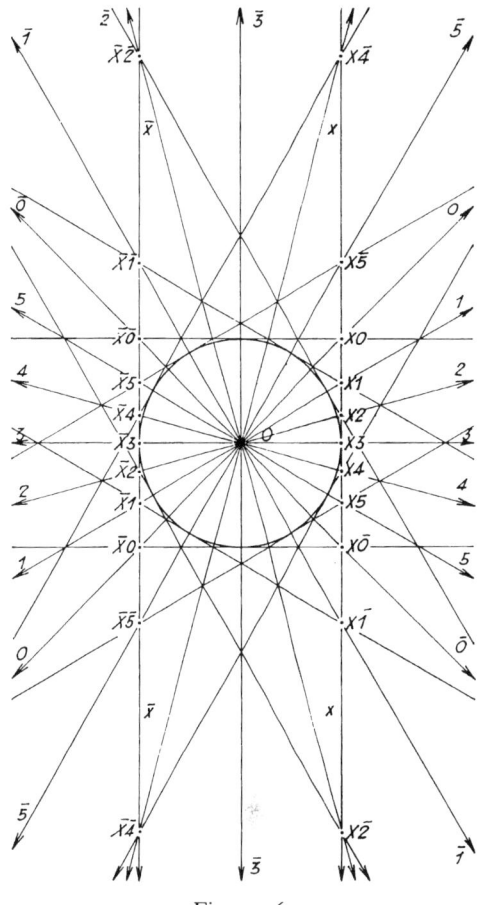

Figure 46

Figure 47

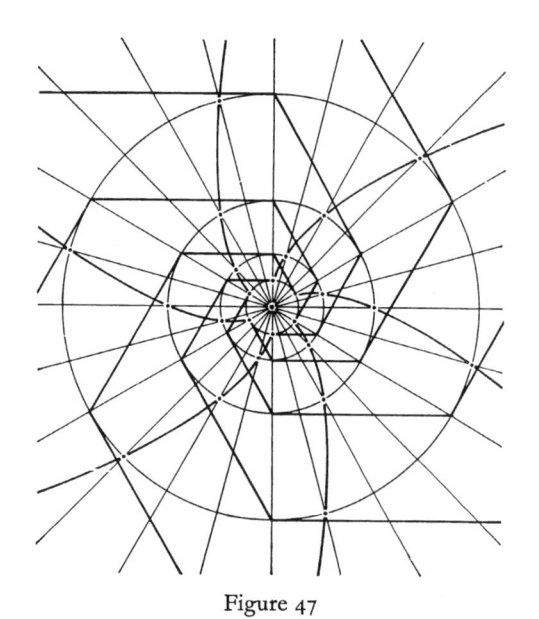

Figure 48

Projective Transformations in Space—Plastic Perspective

For the sake of completeness in the line of thought we have been following, but without going into detail and giving practical instructions, which would lead us too far in one particular direction, we will ask the question: Having considered transformations in the line and the plane, may we not do so in space itself?

It would seem that we should have to go outside three-dimensional space, into a *four*-dimensional space which contained it, and in this higher space, choose a *different* three-dimensional space as "intermediary space" and, once again, two centres of perspective.

An interesting reflection! There is nothing logically inconsistent in it; as a *formal* process of thought, it is perfectly possible, *but*—our imagination deserts us. True, in thinking projectively, and even in conceiving of imaginary elements in our geometrical constructions, we have become less bound to the rigid three-dimensional framework of space as such; this was our intention. If, however, we begin to think formally and abstractly of spaces with more than three dimensions, we have entered quite another realm and we cannot escape the question: Is there any real value in going on thinking like this? (33).

So that, for space as a whole, the *method* of projective transformation becomes—to say the least—problematical. However, having learned to know in one and two dimensions, the kind of ultimate *result* which this method leads to, we can quite well extend to three dimensions the characterisation of projective transformation in terms of ultimate result.

First, let us recall the fact that to deserve the name "projective" this kind of transformation of space into itself must change point into point, line into line and plane into plane; and, moreover, it must preserve unchanged the harmonic quality and all harmonic ratios. This is what is called *linear* transformation, which means that the dimensionless quality of the point and the formless, uncurved quality of line and plane remains unchanged. A single line, for instance, never changes into a curve, nor a single plane into a plastic surface.

Remembering that a homology in the plane, seen as the result of a projective transformation, takes place between a resting point and a resting line—they are the double elements, which we may also call *invariants*—it would be logical to expect that the result of a projective transformation in space will take place between *an invariant point and an invariant plane*. This is true. We can say:

As in the transformation of the plane into itself, all lines of a given point and all points of a given line remain invariant,   *so in the corresponding tranformation of space into itself, all planes and lines of a given point and all points and lines of a given plane remain invariant.* (34).

The illustration (Figure 49) shows a plastic perspective transformation of a pentagon-dodecahedron. As we saw in Chapter III, the regular form involving the right angle of the three dimensions is *based*, as it were, on the infinitely distant plane of space. The cube has a triangle as its archetypal pattern in the Absolute, the dodecahedron has a pentagram. In Chapter III we were concerned with *perspective* transformations; we are now considering a *projective* transformation of a three-dimensional form, which, as we can see, is thereby changed far more plastically.

Contemplating the picture, it is not difficult to see that the movement of any one part of the form would result in the movement of all other parts in such a way that *all points*

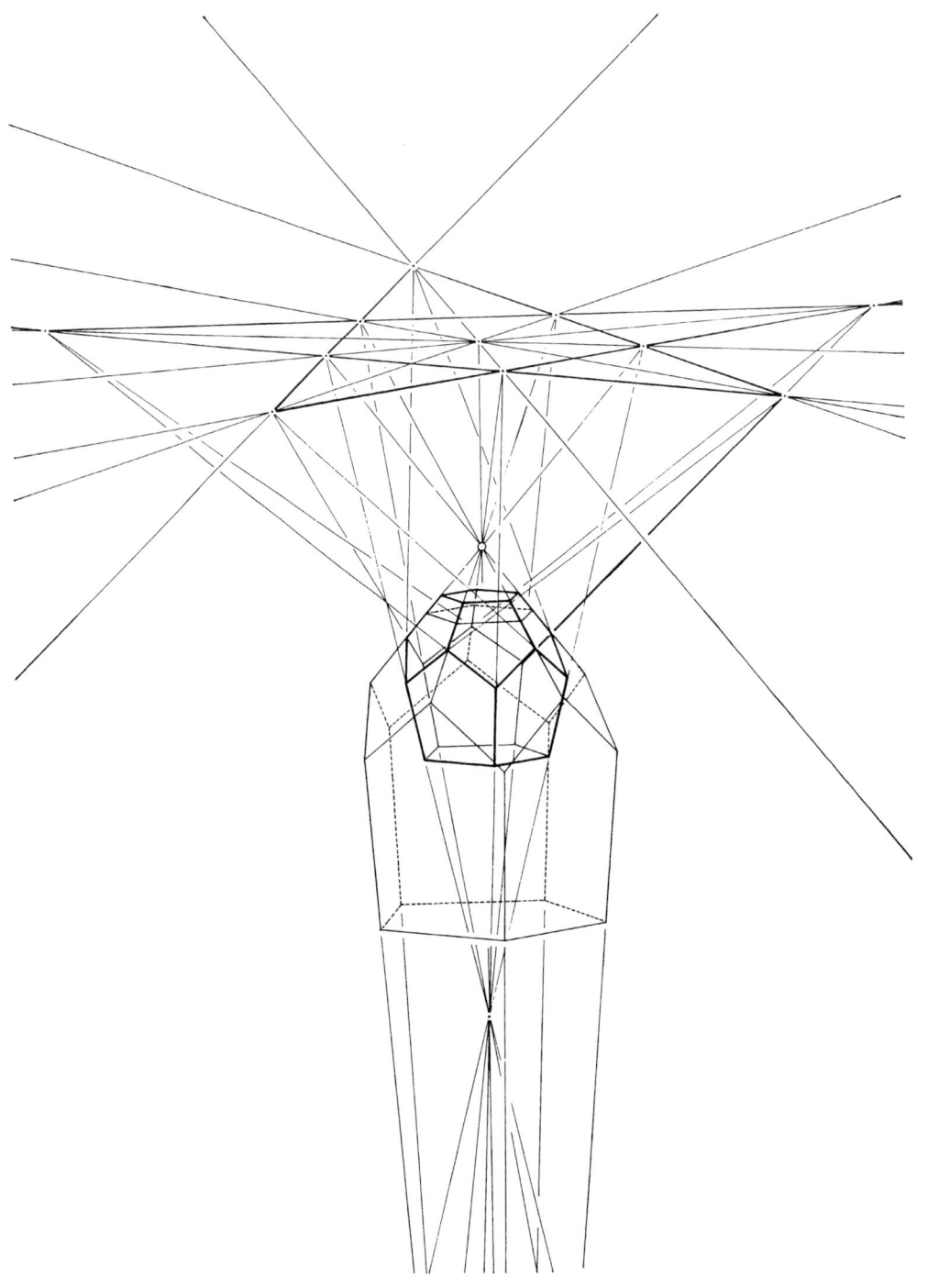

VI/182

Figure 49

would move along lines of the guardian point, while all lines and planes of the form would turn about the points and lines of the invariant or guardian pentagram in the plane.

A beautiful, organic transformation of the pentagon dodecahedron comes about when the fixed point of the homology is set within the form and harmonic with respect to the upper and lower planes of the form and the plane of the pentagram.

Thus, it is perfectly practical to carry out a homology transformation of any three-dimensional form in space, such, for example, as the regular Platonic forms. Having mastered the construction, which requires considerable spatial imagination, there is a manifold possibility of change and variation of an individual form, whereby it might under some circumstances become almost unrecognisable, without ever becoming untrue to its regular archetype (Figure 50). One might say of such a transformed form, that if mathematically true, it will sound forth as a modulation of its dominant chord, but it will never be out of tune. (The arbitrary setting together of planes to create some more or less regular form may, however, very well give a discordant impression.)

The type of projective linear transformation we have hitherto been considering (in one, two and three dimensions) is called *"Collineation"*. It transforms like into like. This type

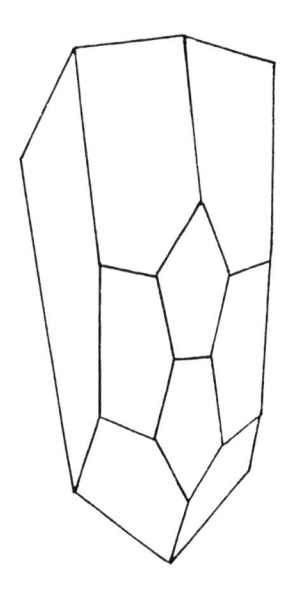

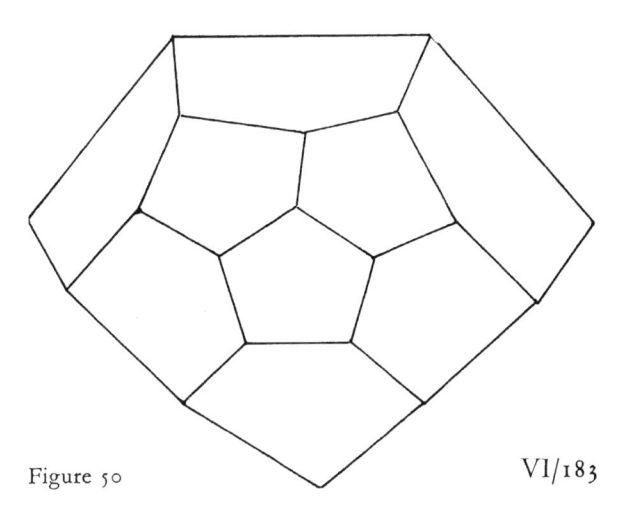

Figure 50

of transformation does not, however, change one form into its opposite, but is non-polar; *it keeps the intensive intensive and the extensive extensive,* for point remains point and plane remains plane.

In contemplating the transition from one three-dimensional form into the next in a homology transformation, it is important to experience the *movement of the planes* as they open out or close inward, modelling the forms. We are not concerned with an arbitrary alteration in the measurements of a shape, but with an organic change from one related form to another. It is in experiencing these planar movements that we shall be led over from that aspect of space in which measurement is the all-important factor, to a much subtler aspect of space in which the principle of duality—polarity, in the true sense of the word—plays a fundamental part.

In the next chapter we shall be considering the type of linear transformation called "*Correlation*", the polar kind, which interchanges the intensive and extensive qualities of forms and, as we shall see later on, leads to the conception of the polar transformation of space itself.

VII Polar Transformations on Circle-curves; Correlations

It is a remarkable fact that the simplest and yet most mysterious of forms, the circle and the sphere (with their projective variations) may so govern the plane, or space in the case of the sphere, that all elements and all forms are changed into their polar opposites.

When we think of the almost magical qualities of the circle and the sphere, we might divine that they have an individual part to play in geometrical transformations; in fact, the law which is called Pole and Polar is of deep significance.

This law rests upon the concepts we have already considered, basically upon the Pascal-Brianchon unity. We might, however, very well have begun our considerations with Pole and Polar and then have passed on to all the other laws, because projective geometry is based simply upon relationships—relationships between elements, but also between theorems—and is not dependent on the setting up of a logical pathway from one particular theorem to the next, as though in a continuous sequence of cause and effect. There are many pathways along which one may pass in thought, linking one part, one theorem, logically with another. In learning to know the whole, we may pass along these pathways at will, provided we see clearly the interdependence of the various parts and their validity.

Imagine taking a walk for the first time along a track through a wood; step by step, from tree to tree, you arrive at the other end at last, learning to know the track but not the whole wood. It is rather like this with ancient geometry; its metrical qualities tell us what we want to know about the earthy tracks. Euclid's system builds up from theorem to theorem, proving each step before it takes the next one. In projective geometry we proceed no less logically, but our method is different. We are permitted to take a wider view and to move more freely about the wood, experiencing the manifold relationships between all it contains; we are no longer restricted to the single track.

In studying some of the aspects of collinear transformations, as we did in the last chapter, we saw in them a reciprocal relationship between *like* elements. In the type of correlative transformation we shall now be considering, the polar reciprocal transformation, the elements are changed into their *polar opposites*. This, too, is a linear transformation, the formless entities, point, line, plane, being changed into one another and not into actual forms; but it is a fundamentally different type, for now the intensive and extensive qualities are interchanged.

The Fundamental Polarity of Space

Let us return to the concept of Duality and Polarity, as summed up in the Axioms of Community in Chapter III. Point, Line and Plane may be created in interplay with one another; each may be composed of the other two; yet at the same time, each is an entity in its own right, a totality. We will state it again as follows (18):

Line as a whole

Line as a manifold of all the planes it contains.	Line as a manifold of all the points it contains.
Plane as a whole.	Point as a whole.
Plane as a manifold of all the lines it contains.	Point as a manifold of all the lines it contains.
Plane as a manifold of all the points it contains.	Point as a manifold of all the planes it contains (35).

As points are to planes, so are planes to points. The line is related equally to points and planes. Point, Line and Plane thus form a trinity, with point and plane representing the polar opposites, and line the intermediate, balancing factor.

We recall that this fundamental law of projective geometry is generally called "Duality", whether applied to two dimensions or to three. We shall make the distinction between the relationship of point and plane *in space*, which we shall call "*Polarity*", and the Duality of point and line *in the plane*. Moreover, we shall take into consideration the aspect of the principle of duality, to which reference has already been made, that of plane and line *in the point*.

Besides the geometry of the points and lines in a plane, there is a geometry of the planes and lines in a point; both are *two*-dimensional and they are the polar opposite of one another. Thus there is a polarity between point and plane with respect to the sphere, and also between the two two-dimensional geometries contained within the whole three-dimensional realm of the sphere.

Polarity: Between point and plane in space.
            Between the geometry of a plane and the geometry of a point.
Duality: Between point and line in the geometry of the plane.
Duality: Between plane and line in the geometry of a point.

We have become very familiar with the principle of duality as it holds for the geometry of the plane. The geometry of a point is new. It is far less easy to deal with pictorially, and it is perhaps for this reason that projective geometry has remained for so long inaccessible and is often preferably cultivated algebraically. (It is for this reason too that we have here so far dealt mainly with the propositions of projective geometry in the plane.)

It is interesting that the mathematicians were led to call the relations of point and line in the plane of a circle "*Pole and Polar*." Evidently, there must be a circle or circle-curve in the plane for processes which would otherwise come under the concept of *Duality* to deserve to be called *Polar*.

The presence of the circle-curve changes a process between *two* opposite elements into one in which there is an interplay between *three*. The circle-curve is not one of the simple, formless elements, such as point of line; it is a *form*, which in itself is an expression of duality. As such it plays the mediating part between the two contrasting elements in its plane; it *calls forth* the polar reciprocal transformations.

The part which is played by the *circle* in the plane is played by the *sphere* in space, while, as we shall see, it is the *cone*

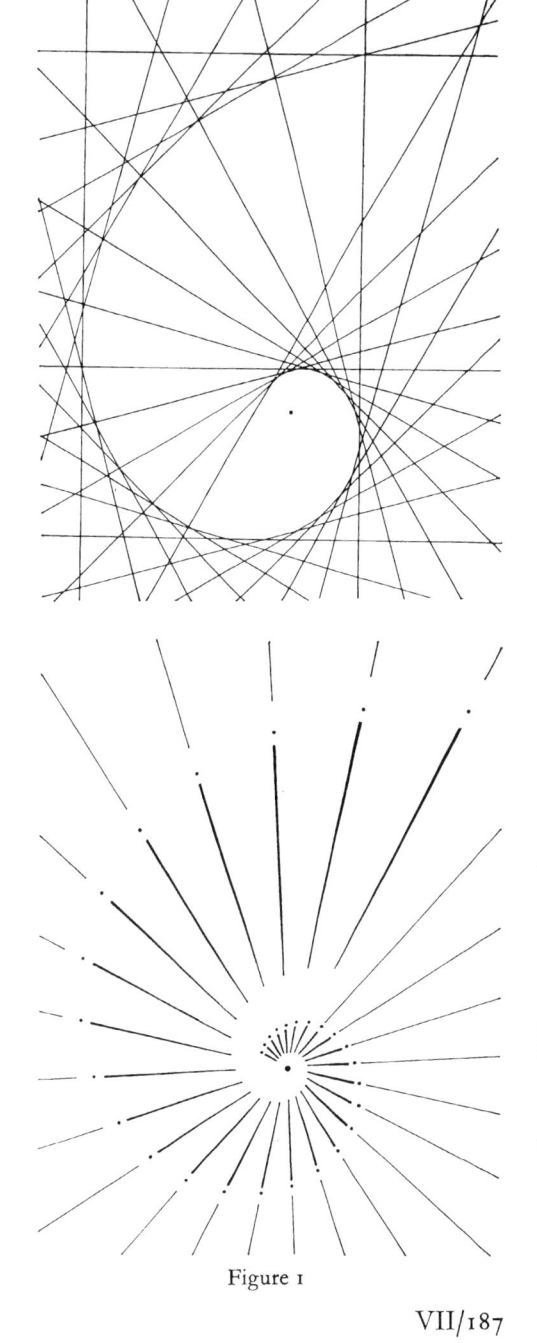

Figure 1

which plays the same part in the geometry of a point (p. 218).

Just as we have learnt to see point, line and plane as aspects of one another or as entities in their own right, so now we shall think in a similar way of curves and surfaces. (We will make a distinction between planes and curved surfaces, just as we did between straight line and curve.)

They may be more point-like or more plastic, but as curve or surface they participate in both aspects and play their part as such in the polar transformations. In the collinear or non-polar transformations we saw first single curves and then whole families being *created,* either pointwise or linewise or in both aspects at once. In the polar reciprocal transformations we see a process being brought about by the *presence* of a given curve in the plane or surface in space.

Pole and Polar with Respect to Circle-curves

Perhaps the most beautiful and the most important of all the properties which modern geometry reveals in the circle-like curves is that every such curve calls forth, in the plane in which it lies, a specific relationship between point and line. Given a conic in a plane, to every point of the plane the curve assigns a certain line, to every line a point.

Pole and polar line—or pole and polar as they are called—are not generally *in* one another, but there is one exception to this rule, namely, *the tangent line at any point of the curve is the polar line of that point; conversely the point-of-contact of any tangent line is the pole of the line.* We may therefore say that the curve in its linewise aspect (in other words, the sum-total of its tangents) consists of all those lines of the plane which bear their pole within them; so too the curve in its pointwise aspect consists of all those points of the plane which contain their own polar lines (Figures 1 and 2).

There are of course infinitely many points and tangent lines of the curve—points which bear their polar lines and lines which bear their poles within them. There are however infinitely many more points and lines of the plane, which are not on or of the curve. For these it will generally be so that if a point moves in towards the centre of the curve, its polar line moves outwards and vice versa.

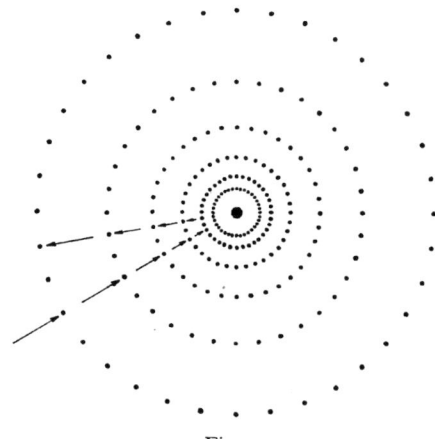

Figure 2

What do we mean by the "inside" and "outside" of the curve? It evidently divides the entire space of the plane into an interior and an exterior portion. We name them so from what naive feeling tells us in the case of the ellipse or circle. In Figure 3, containing ellipse, parabola and hyperbola, the inner space is unshaded, the outer is shaded. The curve is concave towards the inner side. The inner space of the hyperbola, we remember, though seemingly divided into two regions, is really continuous through the infinite.

How are the points and lines of the plane related to the inner and the outer spaces? If a point is in the inner space of the curve, the curve treats all the lines of this point in the same way, namely, every line of the point has two real points in common with the curve. Whereas if a point is in the outer space, the curve divides its lines into two different kinds, namely, those that have two real points in common with it and those which have none (the boundary between these two kinds being the two tangents from the point). The same applies the opposite way round. If a line is outside the curve, the curve treats all its points in the same way, namely, all of them have two real tangent lines in common with the curve. Whereas if a line strikes through into the inner space, the curve divides its points into two different kinds—those in the shaded part, from which two real tangents can be drawn to the curve, and those in the unshaded part, from which none can be drawn (the boundary between the two kinds being the two points which the line has in common with the curve).

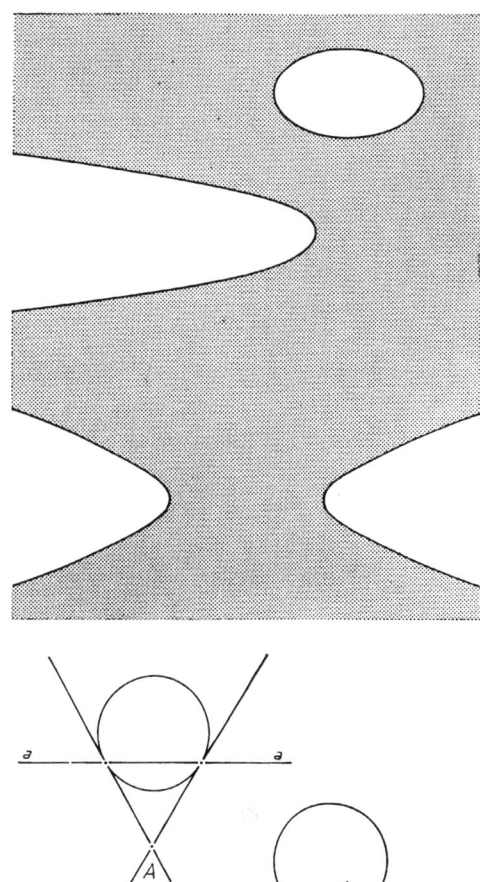

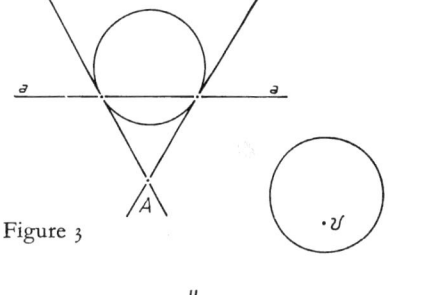

Figure 3

Thus we can say:

The outer space is a linewise space; this alone contains undivided lines.	The inner space is a pointwise space; this alone contains undivided points.
No line can be entirely contained in the inner space.	No point (considered as a point-of-lines) can be entirely contained in the outer space.

Stating it simply, we say: the outer (shaded) space is for the lines what the inner (white) is for the points. A more exact and detailed description would require us to reflect more critically on our conceptions of "inner" and "outer".

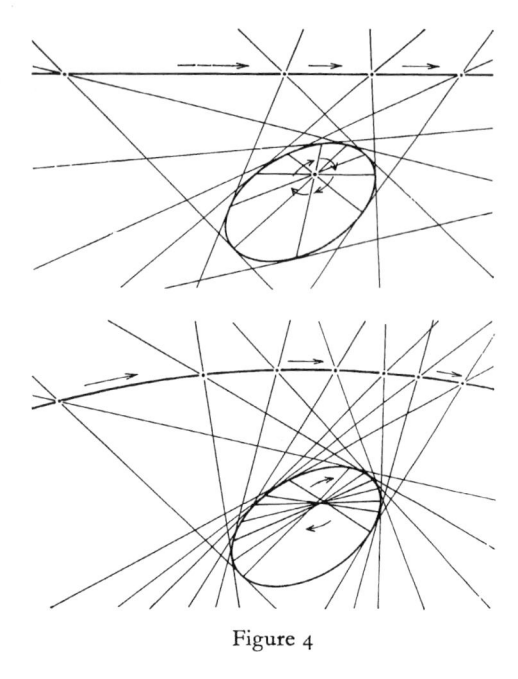

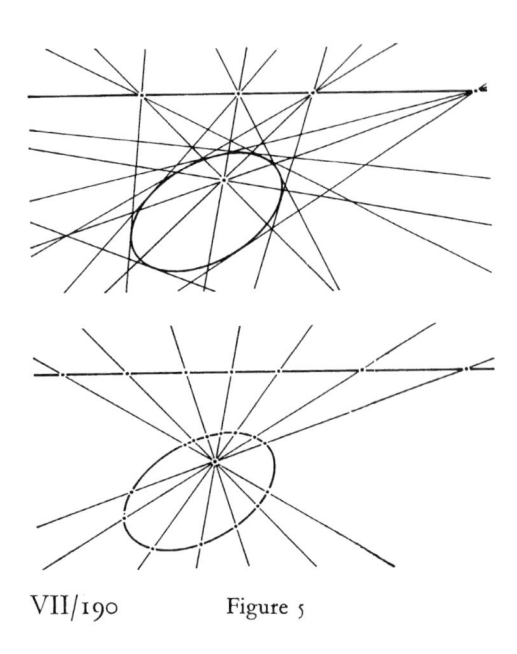

We should discover that we are always unconsciously giving preference to the pointwise aspect. The kind of space we instinctively think of, whether in the plane or in three dimensions, is a point-filled space—filled as it were with so many atoms. To be fair to the lines (and planes), we should have to develop the conception of a *line filled (or plane-filled) space*. We would then find that as the inside of the curve is the obvious interior thereof for pointwise space, so is the outside the natural "interior" for linewise space. Thus in a very exact sense the outer space is for the lines what the inner space is for the points, and to pass from the centric entity to the peripheral interchanges our ideas of "inner" and "outer". We speak of point-space and line-space (Punktraum and Strahlenraum) (36.)

For the present we will use the words "inner" and "outer" in the accustomed sense. *The pole and polar relation is then that an inner point has for its polar an outer line, an outer point an inner line.* Furthermore, we shall find that pole and polar line are so related to one another that the nearer the centre of the curve the pole is, the farther out into the infinite is the polar line. (The term "centre" must be qualified in the cases of the hyperbola and the parabola, pp. 197, 198). Conversely, if the pole draws very near the curve, so does the polar line, until they converge as tangent and its point of contact.

How does this relation of points and lines with respect to the curve come about?

We will begin with a line of the outer space and imagine a point to move along the line. From every position of the moving point, two tangents can be drawn to the curve. Each has a point of contact on the curve and these two points of contact have a common line, the *chord of contact*. As the point moves along the originally given line, this chord will also have to move. Its movement proves to be such that it always remains pivoted on a fixed point within the curve. This point within is the pole of the given line (Figure 4).

If now we begin with the point within the curve and seek its polar line, we need only interchange the ideas of point and line at every step. Imagining a line of the point within to move, we see that at every position it has two points in common with the curve. Each point is the point of contact of a tangent to the curve and these two tangents have a common

Figure 4

Figure 5

point. As the line moves round in the point within, the common point of the tangents will also have to move. We shall see that its movement must be such that it moves along a fixed line outside; it is the polar line of the pole within.

Let us now imagine the point in the outer space moving so as to describe a curve; the answering line in the inner space thereupon moves so as to describe a curve also. The less curved the outer form, the more curved the inner one will be, and when the outer one is infinitely flattened into a straight line, the inner one will be infinitely sharpened again into a point—pole and polar (Figure 4).

There is another way of arriving at the same point within, corresponding to the given line without (Figure 5). Again imagine a point moving along the given line, while at every moment two of its lines are tangent to the curve. We recall that for any two lines and any third line of a plane, there is an *unique fourth harmonic line* in the same point and plane. Here the two lines are the two tangents which the point has in common with the curve; the third is the line along which it moves. At every moment therefore we can look for the harmonic fourth. We find that this fourth line too pivots upon a point within the curve, the pole of the given line.

Interchanging point and line in the thought-process of this illustration, and starting with a line which turns about the given point within (Figure 5), we have a pair of points on the curve itself and the third the point within. Seeking at every moment the harmonic fourth point upon this moving line, it proves to travel along a straight line, the polar line of the given point within.

It will be easily recognised that the constructions given in these figures could be varied. If we set the pole farther inward, the farther out towards the infinite periphery the polar line is—and conversely. *When two tangents become parallel, the line common to their points of contact will become a diameter and this will be the polar line of an infinitely distant point. Thus for ellipse and circle the centre, with all the lines it contains, is the pole of the infinitely distant line of the plane, with all its points.* It is as pole of the infinitely distant line that in projective geometry the centre of a circle-curve is determined and defined (Figure 6).

If instead we take a line crossing the inner space, or a

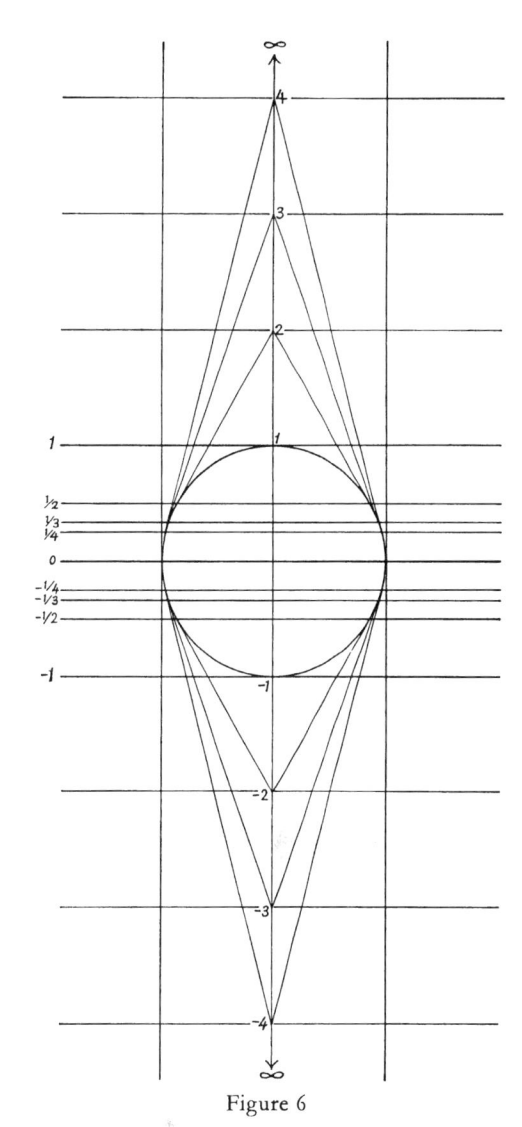

Figure 6

Figure 7

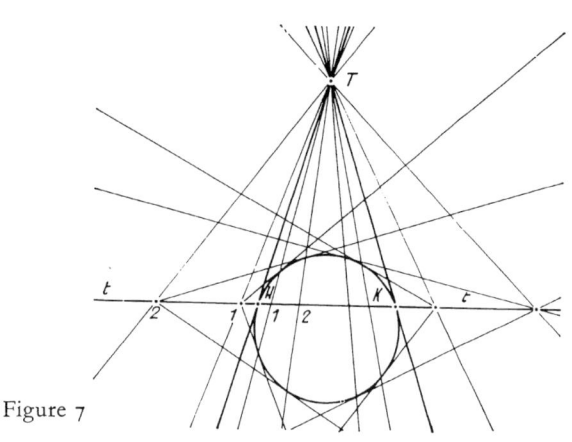

point in the outer space, we shall still be able to apply the same constructions, but we can no longer draw pairs of tangents from all points of the line, only from those of the outer segment. If we do so we shall find that the chords of contact all of them pivot on an outer point (T); so too will the fourth harmonic lines. Starting from an outer point as pole, the construction, carried out in the opposite way, will of course lead to the inner line, its polar. Not all the lines of this point have points in common with the curve (Figure 7).
Thus:

Every line in the plane of a circle-curve gives rise to a fixed point, its "pole", as follows:	*Every point in the plane of a circle-curve gives rise to a fixed line, its "polar", as follows:*
If a point moves along a given line, then (i) the chord-of-contact for the pair of tangents from the moving point to the curve, and (ii) the harmonic fourth of the given line with respect to this pair of tangents, will move in such a way as to pivot upon one and the same fixed point, uniquely determined by the line in relation to the circle-curve.	*If a line turns upon a given point, then (i) the common points of the tangents at the pair of points where the moving line meets the curve, and (ii) the harmonic fourth of the given point with respect to this pair of points, will move along one and the same fixed line—a line uniquely determined by the given point in relation to the circle-curve.*

From the theorem it emerges at once that the pole-and-polar relation is reciprocal, namely, that the definitions in the left- and right-hand columns are interchangeable. The reasoning applies, whether the polar line is outside the curve and the pole within, or vice versa.

It is clear that the chord-of-contact for an outer point is

itself the polar of that point, and for this situation the polar relationship is very easy to construct. This leads us to the third statement, which, although it would appear only to hold for the case when the pole is outside the curve, is nevertheless true for the opposite case, too, when we consider that when the line passes outside the curve it has imaginary points in common with it.

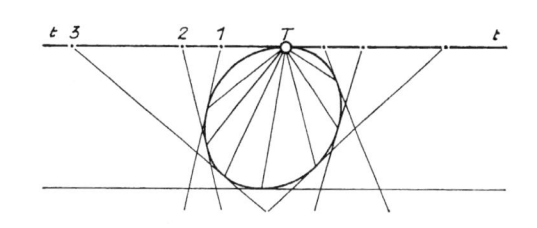

(iii) The pole of a line is the common point of the tangents at the two points in which the line meets the curve.

(iii) The polar of a point is the common line of the points-of-contact of the two tangents from the point to the curve.

This theorem makes it self-evident that the polar line of an outer point crosses the curve, and both from this aspect and from the harmonic property of pole and polar, it is clear that the polar line of a point within the curve must be entirely in the outer region. Harmonic pairs always separate each other; hence, if the pole is inside, every point of the polar line will be separated from the pole by the two points of the curve in which the ray from the pole along which the point lies strikes through the curve. In other words, all points of the polar line are in this case outside the curve.

For a point on the curve itself or for a tangent line thereof, all of these characteristics show that the tangent is the polar line of the point-of-contact and vice versa (Figure 8).

The last statement leads to the all-important truth:

(iv) If S is a point in the polar of T, then T is a point in the polar of S.

(iv) If s is a line through the pole of t, then t is a line through the pole of s.

Self-polar Triangles; Polar Conjugate Pairs

This reciprocity of the pole and polar relation leads to the points and lines of the plane grouping themselves very naturally into triangles having a special relation to the given conic (Figure 9). A triangle is by very nature "dual to itself", in that it contains an equal number of points and lines; here however, we see it self-polar with respect to the circle-curve.

Were we to look for the proof of all the relationships we have been describing, we should find it contained in what we already considered when we saw how any four tangents to a circle-curve and their four points of contact will form a

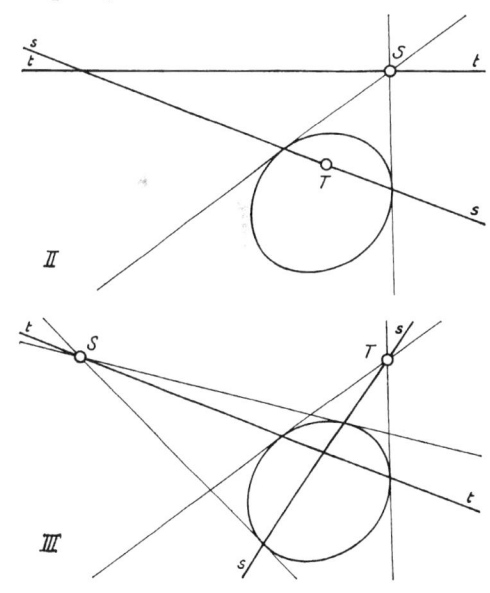

Figure 8

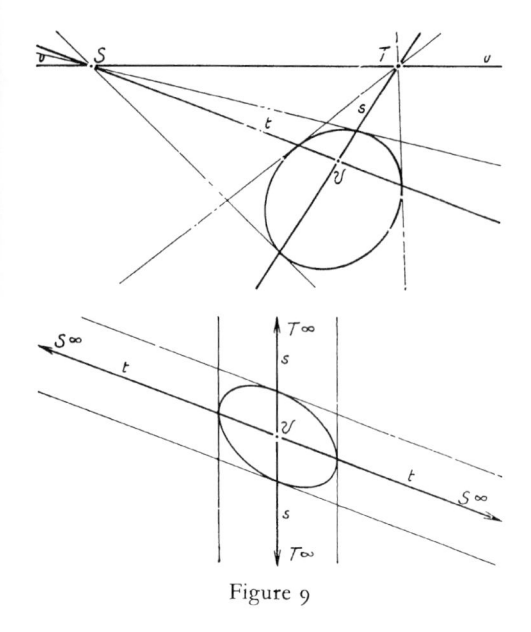

harmonic four-point and four-line, together with the diagonal triangle (p. 138). We find now that the diagonal triangle as well as being called the Harmonic Triangle is also called the Polar Triangle.

Through all these relationships it is clear that Pole and Polar relates closely to the Pascal-Brianchon constructions; moreover, the constructions call to mind immediately the Circling and Breathing Involutions. We see how the fundamental theorems of this geometry, while being independent and self-supporting, are nevertheless interwoven with one another.

There are infinitely many polar triangles for any given pole and polar, as the points and lines circle round in the pole and polar line. At every moment the polar triangle appears as a stage in a moving process.

A particular form of self-polar triangle is given (for the ellipse and circle) when the one point is, so to speak, as far inside as can be and the opposite line as far out as can be, namely, infinitely far away. Then the pole is the geometrical centre of the curve, the polar is the line at infinity of the plane in which it lies (Figure 9).

Every point (every line) begets another, its partner, by virtue of the construction; they are pairs which give rise to one another, and are described as "conjugate"; it is a one-to-one correspondence. We call a pair of points, each of which lies in the polar line of the other, in respect to a given curve, "polar conjugate points". Likewise, a pair of lines, each of which lies in the pole of the other, "polar conjugate lines" (Chapter VI, Figure 20).

When the pole is within the curve and the polar line without, the polar conjugate points and lines make a circling movement, the pairs chasing one another; in the other case however these elements always move in opposite directions, towards each other until they merge and then away again until they merge in the other direction. The laws we have studied concerning involutions appear again here.

There is, as we have seen, a transition from one case to the other, when the polar line becomes the tangent line itself and its point-of-contact the pole. For this case we may say:

Figure 9

For a line touching the conic, the point-of-contact is polar-conjugate both to itself and to all other points of the line.

For a point of the conic, the tangent line at this point is polar-conjugate both to itself and to all other lines of the point.

Here the conjugacy loses character, the one-to-one relation of polar conjugates breaks down. This is scarcely to be wondered at, considering how radically different are the circling and breathing movements, between which this case is placed.

It is interesting to realise that precisely the relation of a circle-curve to a line which lies right outside it and seemingly has no connection with it, is the one which leads us back most directly to the concept of a circle, as we know it from ancient geometry. This is in fact the case when the line has two imaginary points in common with the curve.

As the conjugate pairs of points circle round in the polar line, the conjugate pairs of lines pivot upon the pole, circling clockwise or anti-clockwise, like the spokes of a wheel, save that the angle between them does not as a rule stay rigid, but becomes now more obtuse in the one alternate segment, and therefore more acute in the other, and now again tending to balance the difference in pure right-angledness.

There is, however, one particular position of the pole in-

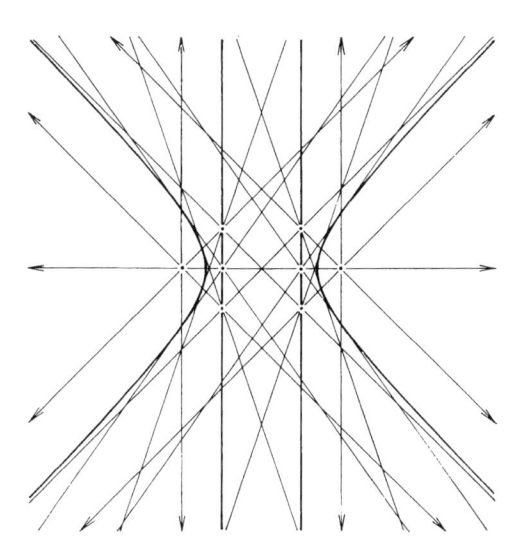

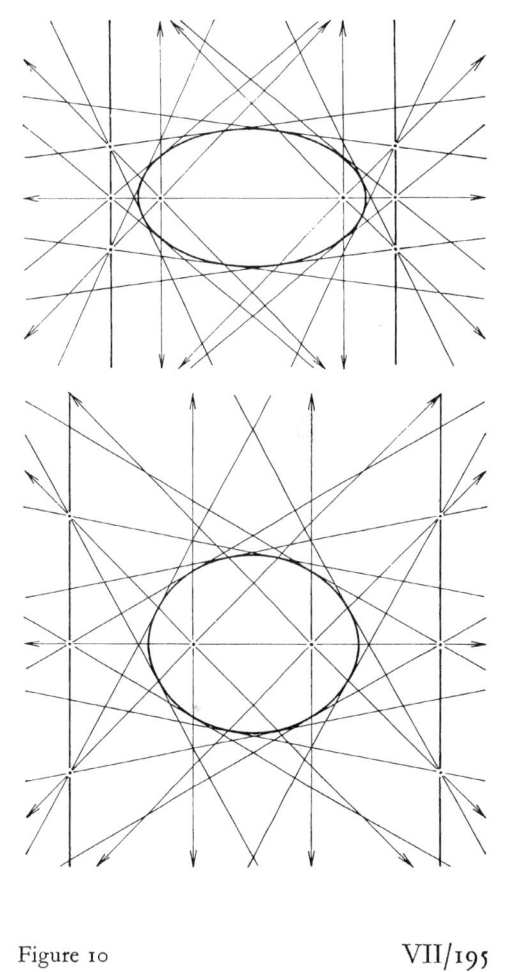

Figure 10

side the circle-curve, when this mobility of the angle between polar conjugate rays no longer holds, and the circular movement of polar conjugate lines remains right-angled all the way round. This in effect is what happens when the pole moves into the centre of the circle, into the one focus of the parabola, or into each of the two foci of the hyperbola or the ellipse. Focus and directrix (Figure 10) are but a particular instance of pole and polar; this is characterised by the fact that polar conjugate rays through the focus are always at right angles to each other. It is clear that when the two foci of an ellipse move towards one another it grows more rounded; at the last moment it loses ellipticity as the two foci merge, and it then becomes the perfect circle, where the polar conjugate rays are diameters.

Conjugate Diameters

The circle, as distinct from all ellipses, is characterised in the light of projective geometry by the fact that the polar conjugate pairs of points in the infinitely distant line are at right angles all the way round. The centre of the circle is the pole of the infinitely distant line of the plane; each diameter has its polar point at right-angles to itself on the infinitely distant line (*one* point, to be found in either of the two opposite directions); and the right-angled movement of polar conjugate diameters—as the wheel turns—corresponds to the right-angled movement of polar conjugate points in the infinitely distant line of the plane.

An ellipse has only one right-angled pair of conjugate diameters, namely, its main axes, and it has therefore only one pair of polar conjugate points at right angles in the infinite. The tangents at the points where the one conjugate diameter strikes through the curve are parallel to the other conjugate diameter, and vice versa. The common point of the axes is the centre of the ellipse, and this is the pole of the infinitely distant line, which is the common line of the poles of the two main axes. The form of movement of all the conjugate diameters in the centre and the conjugate points in the infinite determines the shape or "ellipticity" of the ellipse (Figure 11). A long and narrow ellipse will show the pair of conjugate

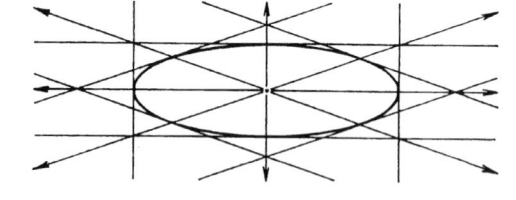

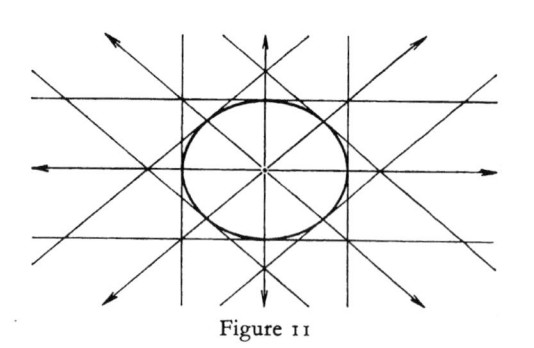

Figure 11

diameters at one stage forming a very acute and obtuse pair of angles. If the ellipse is more rounded the maximum acuteness and obtuseness of the angle will be so much less. Finally a perfect circle with its absolute right angled measure all round is in perfect balance; it has no special axes and therefore no special orientation within its plane, nor has it any special shape to differentiate it from other circles. In its seemingly rigid outer form the circle is the ideal summation, the balanced shape of an infinitely mobile process.

While the circle has such regularity of form, so, too, has the parabola, but this curve has no ordinary centre, and therefore no conjugate diameters. It is a curve *touching* the line at infinity, its point of contact with the line at infinity being the infinitely distant point of its axis, and therefore also of all lines parallel to the axis. The pole of the infinitely distant line which, in the case of the circle, is its mid-point and contains all the diameter lines, in the case of the parabola is coincident with the line at infinity. It contains infinitely many lines which are parallel to the axis. These lines are the diameters of the parabola. (See definition of centre, p. 192.)

Although we do not find conjugate diameters for the parabola, a beautiful and instructive instance of the polar conjugate relation, especially in its harmonic aspect, is given by the parabola if we consider lines parallel to the so-called axis. Given a parabola (Figure 12), think of any such "diameter" line, say t. One of the two points which this line has in common with the curve is infinitely far away; so, too, is the meeting point T of the tangents at these two points (H and K), for which the line t is chord-of-contact and therefore the polar of T. (One of these tangents is in fact the line at infinity.)

Hence if we draw pairs of tangents from any point of the line t outside the curve, their chords-of-contact will be parallel to the tangents at the common points H and K of the line t with the curve, and each chord is bisected where it meets the line. Moreover the pairs of points inside and outside the curve on the line t (1 and 1, 2 and 2, and so on) are equidistant from H (and from K) since in effect they are harmonic to H and K, and K is infinitely far away. We have here a well-known theorem:

The locus of mid-points of a family of parallel chords of a parabola

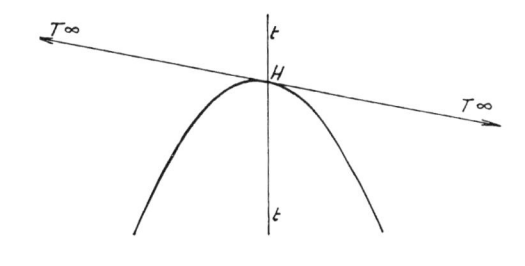

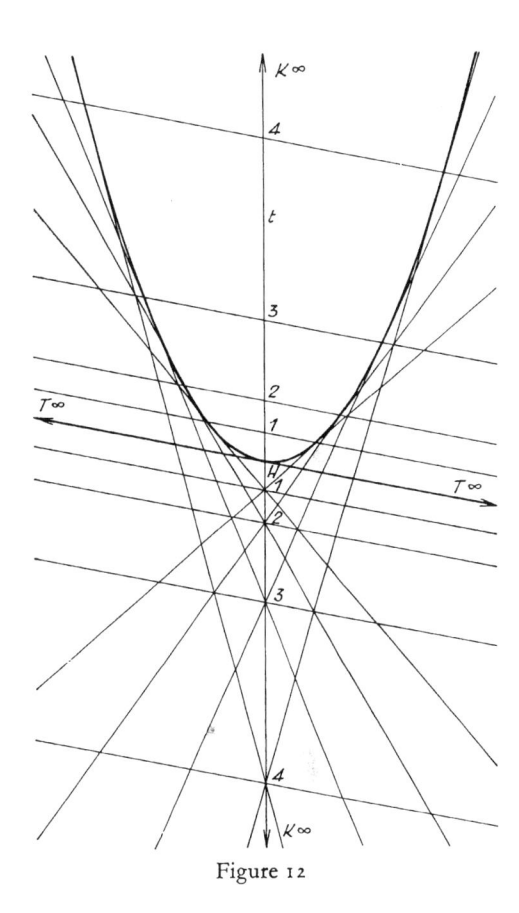

Figure 12

is a straight line parallel to the axis, meeting the curve at the point where the tangent is parallel to the family.

The tangents at the common points of any one of a family of parallel chords meet in a point of the same straight line parallel to the axis, the pairs of points on the line being equidistant from the points it has in common with the curve.

It is instructive to compare the construction in Figure 12 with the one in Figure 7. Turn the latter figure at right angles with t vertical, H below and K above; then move the tangent line TK into the infinite. The line t and the tangent at H retain their situations, while the tangent at K, together with T, moves to the infinite, until at last the ellipse becomes a parabola and all lines from T become parallel.

Note that the family of chords of the parabola will be the more oblique to the line t, the farther distant it is from the axis. The axis alone is at right angles to the corresponding family of chords which it bisects.

Although we do not actually speak of conjugate diameters in the case of the parabola, it is nevertheless interesting to compare it with the circle. The fixity of form appears in the circle in the angular regularity of the lines about its central point; in the parabola there is a similar angular regularity, but it takes the form of parallelism of lines in the points of the infinitely distant line of the plane. The parabola, like the circle, has no possibility of varying its shape.

Turning to the hyperbola, which is classed, like the circle and ellipse, as a "centric" conic, we find that we can repeat almost word for word what was said of conjugate diameters of the ellipse, although there is one important qualitative difference. The hyperbola has its "centre" (pole of the infinitely distant line) *outside* the curve. This curve *passes through* the infinitely distant line, and has therefore two points in common with it, so the involutary pairing of polar conjugate diameters will in this case not be of the circling, but of the breathing type (Figure 13).

There are two tangents touching the hyperbola in the infinite—they are called *asymptotes*—and they play the part of double- or guardian-lines. The centre of the curve is the common point of these two lines. Among the polar conjugate diameters which are in the breathing movement between the two asymptotes there will be one right-angled pair; these

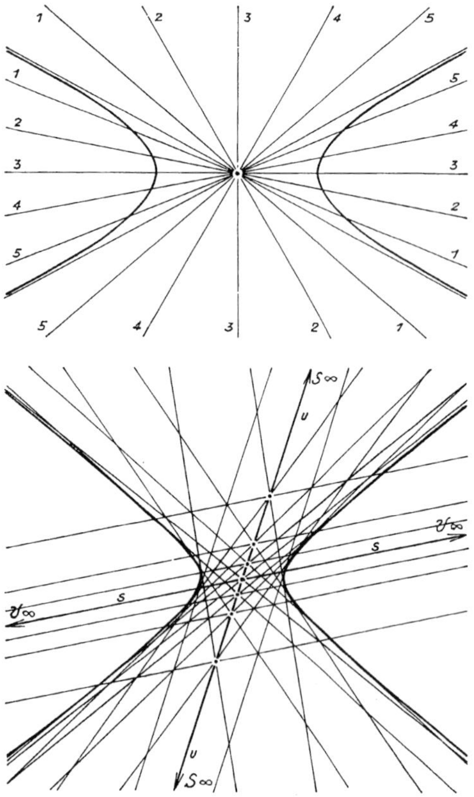

Figure 13

two represent the main axes of the hyperbola, of which the one that strikes through the curve bears the two foci. The pole of each diameter is the point-at-infinity of the conjugate diameter. Each pair of conjugate diameters of the hyperbola will be harmonic to the pair of asymptotes.

The form of the hyperbola is determined, like that of the ellipse, by the mutual relationship of conjugate diameters and conjugate point pairs on the infinitely distant line. A right-angled hyperbola arises when the right-angle rules the relationships between the asymptotes and their infinitely distant poles.

As with the circle in Figure 7, of a pair of conjugate lines in T, one has no points in common with the curve, while the other crosses it at two points which are harmonically situated as between pole and polar line (equidistant in the hyperbola, the centre being the pole of the infinitely distant line). The one that strikes through the curve is chord-of-contact for the two tangents drawn from its infinitely distant pole, which tangents are parallel to the conjugate diameter, as in the case of circle and ellipse. (Figure 13, top.)

The conjugate diameter, which has no points at all in common with the curve, may be compared with the line outside the curve in Figure 4 or 5. Pairs of tangents may be drawn from any of its points, and as the chords of contact will all of them pass through the pole, which is the point-at-infinity along the other diameter, they will be parallel to the latter and to one-another. Moreover, from the harmonic property of pole and polar, they will be bisected where they meet the conjugate diameter from the points of which the tangents were drawn. This is the relationship shewn in the oblique example in Figure 13.

Pairs of tangents may also be drawn from some of the points of that conjugate diameter which strikes through the curve; the pole and polar reciprocity shews here too that the chords-of-contact for these pairs of tangents will be parallel to the other conjugate diameter and therefore to one another. In spite of the difference in quality, each conjugate diameter fulfils a like function for its partner. The parallel chords of this second family of tangents will of course also be bisected by the diameter, for the points of which they are the chords-of-contact.

It will be clear to the reader that the pair of conjugate diameters u, s, form a self-polar triangle STU with the line-at-infinity (US = t) and the centre (us = T).

We have by no means exhausted the many beautiful aspects of pole and polar, but we have had a taste of the mobility of the thoughts and facts involved. The most fundamental concept we will take with us as we go on, namely, that *the circle-curve, in all its variety of form not only embodies the contrasts of point and line in itself, but calls forth this polarity throughout the whole plane in which it lies, transforming all forms therein into their polar opposites, and reflecting centre into periphery, periphery into centre.*

Polar Reciprocal Transformations; Correlations

It is the property of any circle-curve to transform not only points and lines in the plane into their opposites, but curves also. In the examples which follow, we shall be using the circle-curve itself as the transforming entity. Although for practical purposes it is easier to use the circle, *all* the variations of the circle may be used to bring about the correlative transformations.

We are now familiar with the method of finding either the polar line of a given point or the pole of a line, using the symmetrical situation about a chosen radius and passing from tangents to chord, or from chord to tangents according to need. The reciprocating circle is sometimes called the unit circle; we might also call it the all-relating circle, for it relates all the points and the lines of its plane in pairs in an ordered manner.

In the transition from the first to the second part of Figure 4 of this chapter, we see how, when a point no longer moves in a straight line, but traces a curve, its polar line will no longer pivot upon a point, but will also create a curve, this time linewise. This curve is then an envelope.

Let us now draw the arc of a circle linewise, choosing to take as its centre of curvature the centre of the reciprocating circle (Figure 14), and seek the pole of each line. From the point at which a tangent of the linewise arc crosses the radius of the reciprocating circle at right-angles, we draw the two tangents to this circle. The chord determined by the points of contact of the two tangents will also cross the radius at right-angles, thus determining the pole. Continuing round the linewise arc, we find that the poles of all its lines lie on a circle concentric with it and with the reciprocating circle. Clearly, had we begun with the innermost arc pointwise, by reversing the steps of the construction, we could have determined the outermost arc linewise.

This construction, based as it is on the right-angle and the symmetry about a particular radius, is one aspect—the symmetrical one—of the total, mobile and projective illustration of pole and polar as pictured in Figure 4. As a construction it must become second nature to us for use in what follows. It is, however, only the means to an end, namely, the finding of poles and polars, and it will by no means always be necessary to draw in all the tangents and chords, which are in fact simply construction lines. (The task is easy if one uses a large set-square with a right-angle marked on it. It will then be found that the most accurate method is simply to prick required points—points of contact of tangents and the points at which the chord passes through the circle—without necessarily drawing in all the lines.)

Had we begun with a larger outer curve, we should have been led to a smaller one within, and it is obvious that the innermost curve has only to shrink into the central point of the circle for the outermost one to expand and flatten into the *one* infinitely distant line of the plane.

Supposing, on the other hand, we think of a linewise curve within and a pointwise one outside the unit circle; in the extreme case the innermost curve must degenerate into a point of lines in the centre as soon as all the points of the outermost one have reached the infinitely distant line.

This phenomenon must be fully grasped. We have not only to do with a simple movement of expansion on the one hand and contraction on the other; the process is not fully described by saying that the one is centripetal and the other centrifugal. The important thing here is *the qualitative interaction of polarities which are interdependent*. Both poles have at one time an expansive, at another time a contractive tendency; they both move at one time outward, at another inward. In fact, it is the outward movement of the one which results in

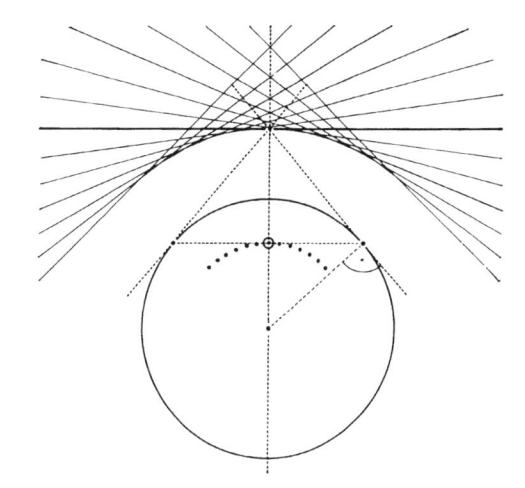

Figure 14

the inward movement of the other, and vice versa. Most significant is the characteristic *quality* of the movement and of the form it creates, for these are polar opposite qualities. We must experience to the full the qualitative difference between the outward or inward movement of a point and the outward or inward movement of a line.

Points move radially along lines or curves; points may shoot outward, like particles in an explosion, or they may contract or press together inward, as though towards a centre of gravity. (It is the pointwise type of picture which figures largely in the descriptions of the forces about which physics teaches.)

Lines and planes move very differently from the way points move. Unless they slide within themselves, they make sheering movements, moving at once along their whole expanded extent. In fact, they hover; one might say that they "plane", and in moving inward they mould and model a form, creating it plastically from without in a way which is quite foreign to points.

It is essential to know and fully to understand this fundamental principle of modern geometry. One should allow the inner harmony and reciprocal quality of the polar relationships to work upon one's soul without being tied to material and sense-perceptible pictures. The drawings require a high degree of mobile pictorial imagination. The underlying *idea* of the reciprocal transformations has a bearing on mysteries of creation to which science has not as yet penetrated; it leads towards the understanding of pointwise and peripheral metamorphosis in the realm of forms, but also of space itself. The correlative transformations are radical metamorphoses; what is centric becomes peripheral and the peripheral centric.

The French mathematician Michel Chasles (1793–1880), who at first took up the work of Poncelet and who is one of the greatest mathematicians of all time, glimpsed intuitively the significance of the Principle of Duality (or Polarity): "Ce principe, dis-je, pourrait jeter un grand jour sur les principes de la philosophie naturelle. Peut-on prévoir même où s'arrèteraient les conséquences d'un tel principe de dualité?"

These tranformations, then, rest upon the construction we

have just been considering (see especially Figures 6, 14). As a next step, we must note well the following: Not all curves in the plane of a reciprocating circle will have the symmetrical relation to it which we gave to our previous example, and not all mathematical curves are as simple in form as a circle-curve. We may draw, for example, a petal-shaped curve inside the circle, as in Figure 15. (It is drawn freehand within the framework given by circle, radius and right-angle. There are of course many ways of constructing curves, both projective and otherwise. With practice, however, it is possible to sketch a geometrical curve quite freely. We learn to recognise a mathematical curve as an entire form; it cannot, for example, consist of arcs of circles put together. It may be expressed by an algebraic formula, and even if it has abrupt characteristics, such as points (cusps), it will contain these organically in its otherwise flowing form. A true mathematical form may always be detected by its elegance, as indeed we have seen in the simpler forms of the circle-curves.)

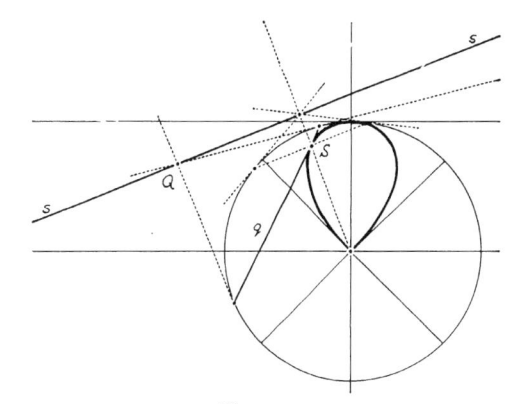

Figure 15

The petal-like curve we have drawn is in fact only part of a curve and we shall complete it later on. But now, in order to begin to find its reciprocal, we will seek, as before, the polar lines of a number of its points. These polar lines will envelope the reciprocal curve. But how shall we know where the curve will touch these lines, which are its tangents? In order to draw it exactly, we must know the points of contact. These will be the poles of the tangents at the points we choose on the petal-curve. (If s is the polar of S on the petal-curve, then Q is the pole of the tangent q of the petal curve in S.)

It is important to note that, having found the polar line of a point of the petal-curve, the point-of-contact of this line with the reciprocal curve will *not generally* lie on the radius of the reciprocating circle which crosses it at right-angles, as was the case in the situation of the concentric curves in our first example (Figure 14). This proves to be a difficult factor in the drawings and requires practice.

Now let us complete the inner curve; the result is a figure-of-eight (Figure 16), a natural continuation of the two directions in which the two sides of the petal-curve are pointing at the centre of the reciprocating circle. Two diameters of the

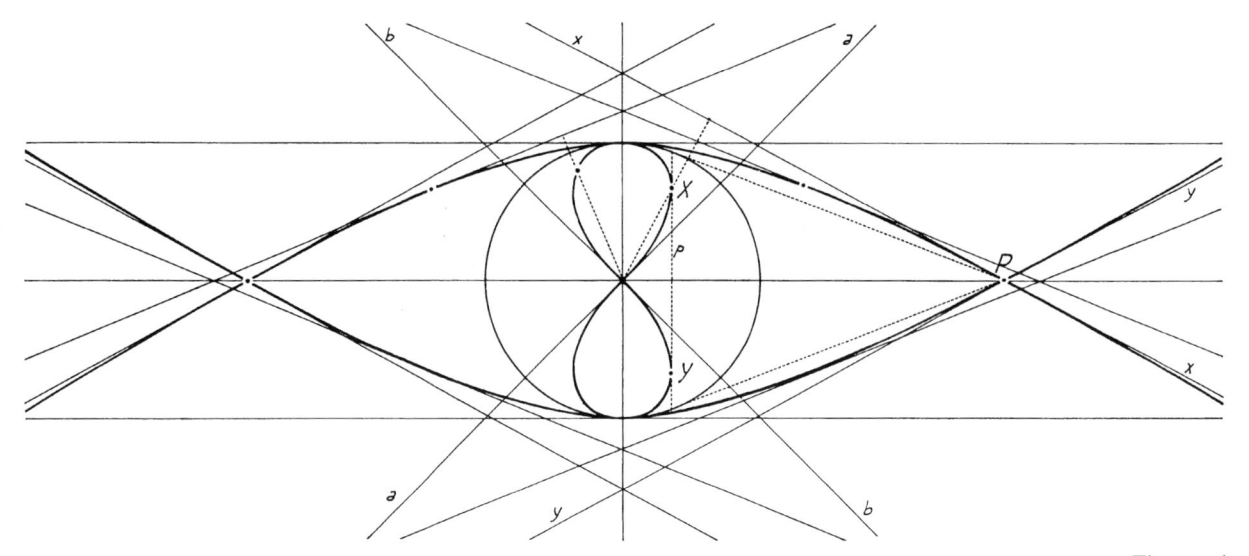

Figure 16

circle, at right-angles to one another, are tangents of the figure-of-eight curve, determining its directions at the crossing-point. We can now continue the transformation of the whole curve by determining further freely chosen poles and polar lines. (In the figure we have purposely not drawn in many tangents of the outer curve, in order not to overload the illustration with lines. With practice, we shall find that one or two lines and points of a mathematical curve give sufficient indication for it to be drawn in correctly, once we have recognised its situation and main characteristics.)

Comparing the two curves, we find that while the figure-of-eight touches the reciprocating circle twice from within, the new curve touches it twice from without. While the figure-of-eight passes twice (in two different directions) through the centre of the reciprocating circle, its reciprocal goes twice (in two different directions) through the infinite. We can be quite sure in which direction the curve goes to the infinite, when we note the direction of the tangents of the figure-of-eight at its crossing-point. The poles of these tangents (they are diameters of the circle) are the two infinitely distant points of the outer curve. Clearly, the two curves have contrasting characteristics.

Let us now consider the typical characteristics which may occur in a curve. The more complicated curves may show a number of characteristic form-elements other than the simple convex or concave forms of the circle-curve. There are four main special characteristics—they are called singularities— which may appear in curves. These are: a crossing point, a double tangent (or bi-tangent), a point-of-inflection and a cusp (Figure 17). (There may also appear what is called a Bamphoid cusp, shaped like a beak, and polar to itself.)

These singularities are polar opposite in character to one another in pairs. (Figures 17 & 18).

Figure 17

Crossing point. The curve passes through one point twice (at least) and has (at least) two distinct tangent lines in this point.

Point-of-inflection. The curve becomes infinitely flat and changes its direction at a point of inflection. A tangent which moves round the curve reverses the direction in which it is turning, when it reaches a point-of-inflection, while a point continues to move in the same direction.

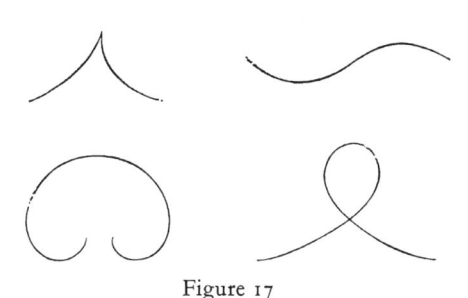

Double tangent. The curve touches one line twice (at least) and has (at least) two distinct points-of-contact in this line.

Cusp. The curve becomes infinitely sharp and changes its direction at a cusp. A point which moves along the curve reverses the direction in which it is moving, when it reaches a cusp, while a tangent line continues to turn in the same direction.

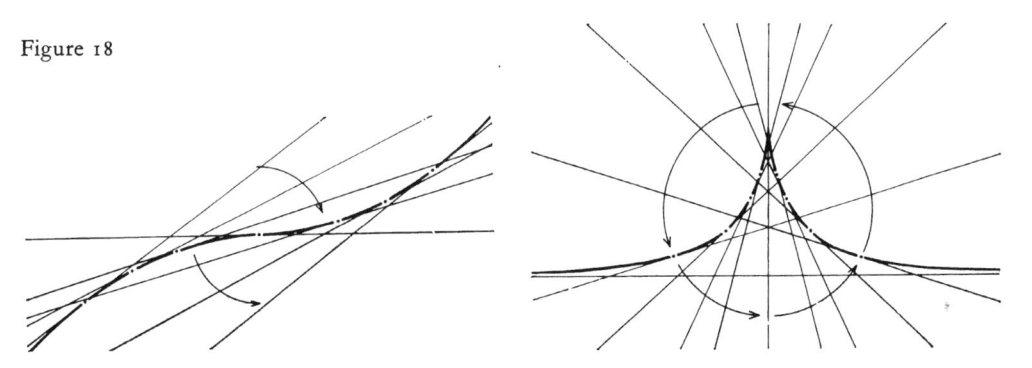

Figure 18

In the figure-of-eight curve and its reciprocal (Figure 16) are to be seen examples of the transformation of a crossing-point into a double tangent and vice versa. The figure-of-eight has two double tangents, symmetrically placed on either side; polar to these are the two crossing-points of the outer curve. In Figure 16, the tangent p is the polar line of the point

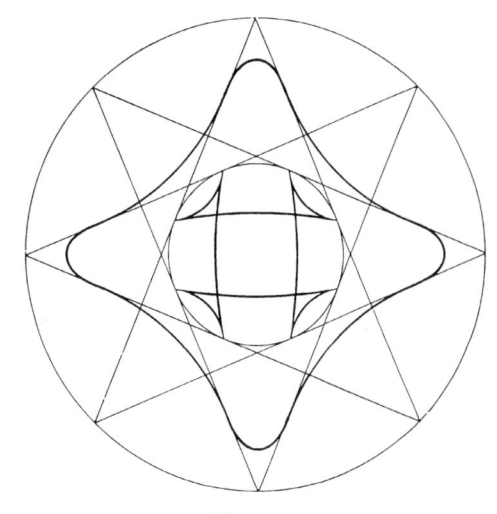

Figure 19

P, while the two tangents x and y of the outer curve in P are the polars respectively of the points-of-contact X and Y of the double tangent p.

We have however not yet taken into account all the special features of our example. The figure-of-eight not only has two double tangents and one crossing point; it inflects twice in this same crossing point, which is also the centre of the reciprocating circle. This is a rather special circumstance to which we shall return later. First let us take another example.

This time we will draw a curve which is entirely outside the reciprocating circle (both pointwise and linewise). For instance, in Figure 19 we proceeded as follows. We drew a circle as large as the paper allowed and an eight-pointed star within it. Then we sketched the symmetrically situated curve freely, which has rather the characteristics of a square, choosing to put its eight points-of-inflection one in each of the lines of the star. Then we placed the reciprocating circle with its centre at the centre of the large circle and of the inscribed octagram in such a way as to make the octagram lines its tangents. Thus, the eight lines are tangents both to the reciprocating circle and to the outer curve at its points-of-inflection. These are special situations, which we are free to choose, giving an interesting picture and also a good exercise in symmetry drawing.

Looking at the curve, we note its special characteristics and its relation to the reciprocating circle. It has four double tangents, eight points-of-inflection and there are eight places at which it passes at right-angles across a radius from the centre, in other words, eight symmetrical places. Now we try in thought to picture in advance what the reciprocal curve will look like. Rather than taking one tangent line or one point after another at random round the curve, we look at the special characteristics and contemplate the curve as a whole.

Taking first the eight places in which the curve crosses certain radii of the unit circle at right-angles, we find these symmetrical places easy to deal with, for the corresponding places in the reciprocal curve will also pass at right-angles across these radii and be symmetrical. Then we consider the four double tangents; these will give rise to four crossing-points in the new curve, each one being the pole of a double tangent. Moreover, the polars of the points-of-contact of the double tangents will turn out to be tangents of the new curve in these crossing-points; these tangents give the directions of the curve in the crossing-points.

The outer curve has eight points-of-inflection, the tangents at which are the eight lines of the octagram. As these eight lines are also tangents of the reciprocating circle, we must look for the points of the eight cusps in the points-of-contact of these tangents. (Tangent and point-of-contact are pole and polar on the reciprocating circle.) Furthermore, the polar lines of the eight points-of-inflection of the outer curve will be the eight cusp-tangents of the inner one, and these tangents will give the exact directions in which the cusps in each case are pointing.

It is a wonderful exercise, requiring considerable dexterity and mobility of thought and imagination, to transform curves in this way into their polar opposites. We sum up the whole curve, with all its characteristics and its situation with regard to the all-relating circle, which orders the whole plane in which it lies into pole and polar entities and qualities. Supposing we had set the outer curve in our last example further out in the plane; we should then have found that the cusp-points of the inner one would not have been in contact with the circle, but further inward towards its centre. A much more complicated but fascinating exercise would have been to have set the all-relating circle in an unsymmetrical situation with regard to the first curve. Varying the position of the centre of the circle and also its size will result in all manner of metamorphoses in the reciprocal curve.

Let us look at yet another example (Figure 20). A curve which is easy to construct is the so-called cardioid; its singularities are one cusp and one double tangent. (To construct this curve geometrically, one method would be to mark off equal steps on a circle and then draw circles with their centres on each point, each circle passing through one of the points. These circles will all be tangents to the cardioid and indicate its shape from inside.)

If a curve is such that one (or more) of its tangents passes through the centre of the all-relating circle, then its polar reciprocal will have one (or more) infinitely distant points.

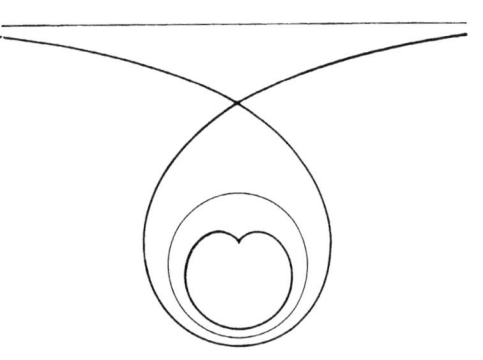

Figure 20

The cardioid in Figure 20 has a tangent, namely, the tangent at the cusp, which is a diameter of the all-relating circle, the pole of which is therefore a point at infinity at right-angles to this diameter. The polar line of the point of the cusp will be so situated that it points in exactly the direction in which this infinitely distant point is to be found, namely, at right-angles to the cusp-tangent. Thus we know the exact direction in which the reciprocal curve has a point at infinity. The curve will, in fact, touch the polar line of the cusp-point of the cardioid in the infinitely distant point of this line; this line—the polar of the cusp-point—is a tangent (asymptote) to the outer curve, its point of contact being its point at infinity.

The reciprocal curve to the cardioid in Figure 20 runs asymptotically into the infinite, touches its tangent there and returns from the other "end" of this tangent. In circumstances in which there is no singularity, as for example in the case of the asymptotes to a hyperbola, the curve will return on the other "side" of the line. In our present example, however, the curve returns again on the *same "side"* of its asymptote. What happens at that part of it which is polar to the cusp?

Here we are called upon to imagine a curve making an inflection in the infinite, which is difficult. We chose to set the cardioid into the reciprocating circle, so that the tangent at the cusp passed through the circle's centre; the reciprocal of this situation is an inflection in the infinitely distant line!

The double tangent to the cardioid turns into the crossing-point of the outer curve. In the lower part of the picture, both curves are rather circle-like, then, in answer to the double tangent in the cardioid, the reciprocal curve runs up to loop through its crossing-point, and then to continue outward on either side towards its point-of-inflection in the infinite.

The following illustration (Figure 21) will be a help in regard to this question of the point-of-inflection. Had we set an unsymmetrical cardioid into the unit circle in such a way that its cusp-tangent did *not* pass through the centre of the circle, the result would have been a point-of-inflection of the loop-curve (the polar point of the cusp-tangent) not at infinity. There is in Figure 21 a delicate inflection of the loop-curve at the point marked by the arrows. It would be an interesting exercise to distort the cardioid still more, so

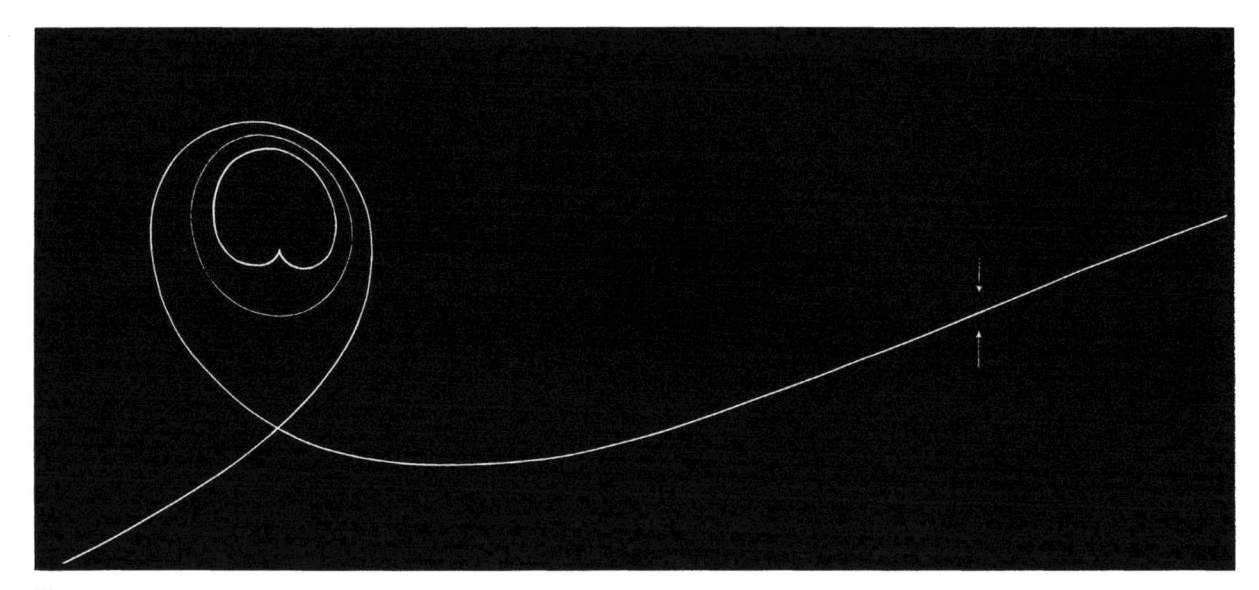

Figure 21

that its cusp-tangent is much nearer the circumference of the reciprocating curve (here an ellipse); the point of inflection in the loop curve would then be much further in and more marked. Figure 22 shows a curve which has a point-of-inflection in the infinite and one which has not.

We may now return to the example of the figure-of-eight in Figure 16, better equipped to understand what the reciprocal curve really does in the infinite. The figure-of-eight has *two points-of-inflection* at the centre of the unit circle. This means that the reciprocal must have *two cusps,* and that they must have their points on the line at infinity! The cusp points will be the poles of the figure-of-eight tangents at the centre, so we know exactly in what direction to look for them in the infinite. These directions are at right-angles to one another. We may picture the outer curve running from the crossing-point P, at which it is moving in the direction of the tangent at that point (say, the line x), gradually coming nearer to the line at infinity which it uses as its cusp-tangent. Its actual point of contact with the line at infinity is the point which the line b has in common with the line at infinity (the pole of the line a). When the point which is describing the curve reaches the cusp on the infinitely distant line, it will continue on the

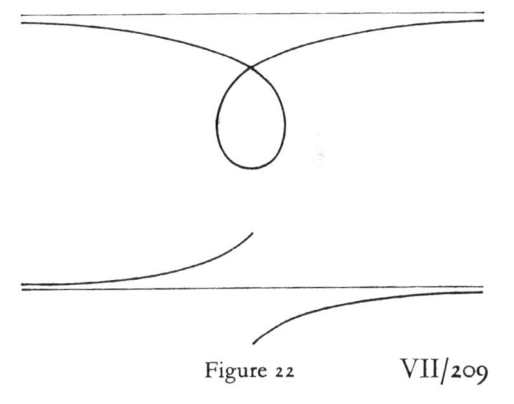

Figure 22

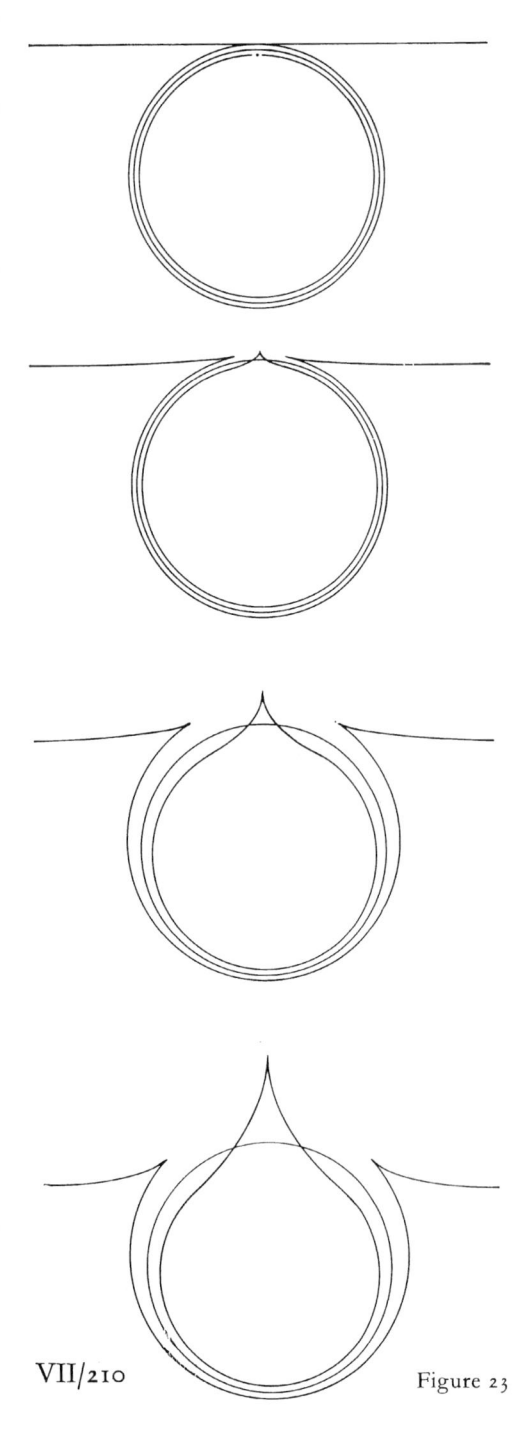

other side of the cusp-tangent (line at infinity), reappearing in the drawing on the other side of the line b. While the figure-of-eight flattens twice into an inflection in the centre of the reciprocating circle, the answering curve sharpens twice —invisibly—into a cusp on the line at infinity of the plane of the circle.

With all that we have now learned in mind, let us take another look at the loop-curve in Figure 20; it has a point-of-inflection in the infinite, as it runs out in the direction indicated by its asymptote and returns on the same side of this asymptote. Its sister curve, the cardioid, on the other hand, has its cusp-tangent in the centre of the reciprocating circle. Supposing we had set the centre of the reciprocal circle at the *point* of the cusp of the cardioid, we should then have to realise that the asymptote of the loop-curve (which is the tangent at its point-of-inflection) would be the infinitely distant line itself.

In dealing with these examples of the reciprocation of curves in the plane of a circle, we have met with all the basic facts necessary in order to carry out constructions of the most varied kind. The exercise brings unlimited possibility of following in detail the metamorphosis of any given curve into its polar opposite, according to whatever situation it may assume in relation to an all-relating conic. Let us not forget that these constructions may also be carried out, using the ellipse, hyperbola or parabola as transforming unit.

It should be understood that in the projective changes from one form of circle-curve to another we were concerned with *variations* of the same basic form, the circle, whereas in carrying out reciprocal transformations we have really entered the field of true *metamorphosis*. Metamorphosis involves a far more radical change of form than does variation, for it involves a more active interplay of fundamental polarities.

A number of examples of polar reciprocation of curves follow here, and it will be sufficient, on the basis of what we have so far practised, to give brief explanations as an aid to the reader in understanding them.

Figure 23 shows in four progressive stages changes in a curve, beginning with the simple form of a circle and one special point upon it, the reciprocal of which is the circle with one special line of it. The reciprocating circle is seen

Figure 23

between the two. The changes in the inner curve involve the formulation of a cusp, which then grows outward along the radius determined by its point. This progressive change in the inner curve was sketched freely, and then at each stage the reciprocal curve was found by construction and drawn in. The upward growth of the cusp is answered in the outer curve by the widening of the opening. As the cusp-tangent of the inner curve is at all stages the same diameter of the reciprocating circle, so the outer curve at all stages passes through the same point at infinity. The moment the inner circle in the first figure acquires a cusp, the curve will have a point of inflection on either side of this cusp; as a result the outer curve has two cusps. The single polar line of the single point in the first figure gives way to the two tangents of the cusps of the outer curve; these two cusp-tangents are in all stages lines which pass through the cusp-point of the inner curve; in fact they are the polar lines of the points-of-inflection of the inner curve. The two cusp-points of the outer curve correspond to the two flex-tangents of the inner one. While the

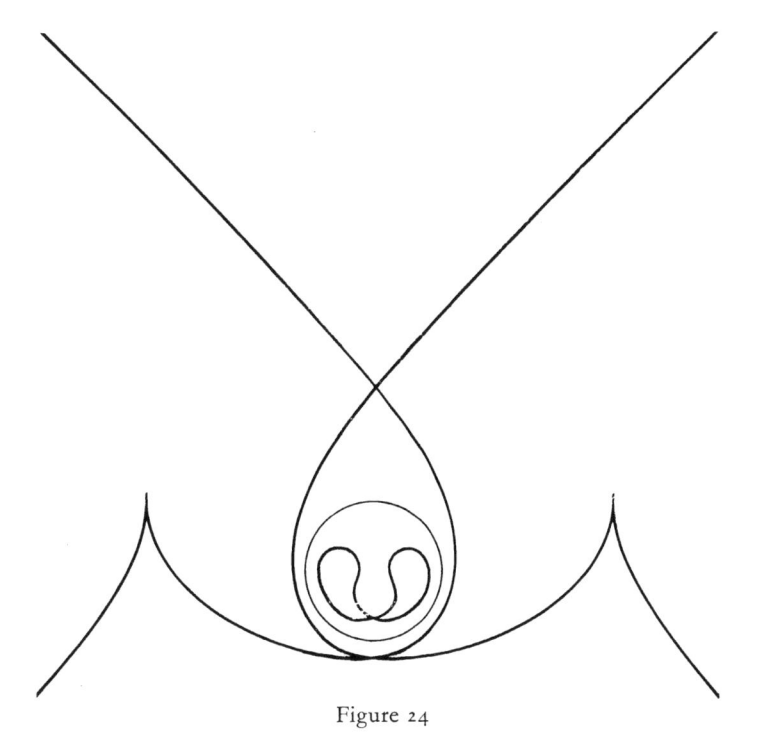

Figure 24

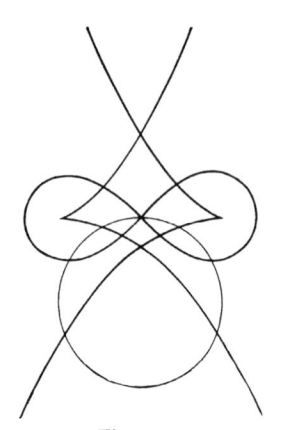

Figure 25

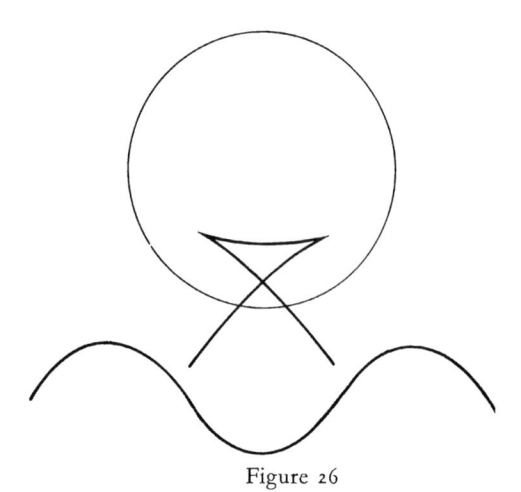

Figure 26

cusp point of the inner curve moves out along a radius of the reciprocating circle, its polar line, which is the asymptote of the outer curve, moves in towards the centre of the circle.

Figure 24. The looped form within the circle has two double-tangents, one above and one below; the outer curve has correspondingly two crossing-points. In the case of the looped curve, two of its tangents pass through the centre of the unit circle; thus the outer curve has two infinitely distant points. The inner curve has one crossing-point, which corresponds to the double tangent of the outer curve below it. The inner curve has two points of inflection and polar to these, on either side, are the cusps of the reciprocal curve.

Figure 25. The figure-of-eight has its crossing-point, which also contains two points-of-inflection, on the reciprocating circle; the reciprocal curve has two cusps, the tangents of which are united in the tangent of the circle at the crossing-point of the loop. Tangents of the figure-of-eight pass twice through the circle's centre; the reciprocal curve, considered pointwise, runs twice through the infinitely distant line. The figure-of-eight has two double tangents; the reciprocal curve has two crossing-points.

Figure 26. Two of the points of inflection of the wave-curve correspond in the reciprocal to the two cusps, while the double tangent of the wave changes to the crossing-point of the other curve.

Figure 27. It is left to the reader as an exercise to decipher the four illustrations of the full-page figure. In the illustration at bottom right it was the looped curve which was drawn first, while in the other three cases we first drew the curves which lie inside the reciprocating circles. In the illustration top left, the symmetrical situation of the flower-like curve gives six cusps which in pairs share three diameters of the circle as tangents. We find that the reciprocal curve runs out to the infinite in *three* directions. These three directions will be exactly given by the pairs of lines polar to the pairs of cusps opposite to one another, which are in fact the asymptotes of the curve (compare Figure 20, in which the asymptote, the line polar to the cusp-point, was also drawn in). The asymptotes are lines which belong to the curve just as intimately as do the actual points of the cusps, and one might

Figure 27

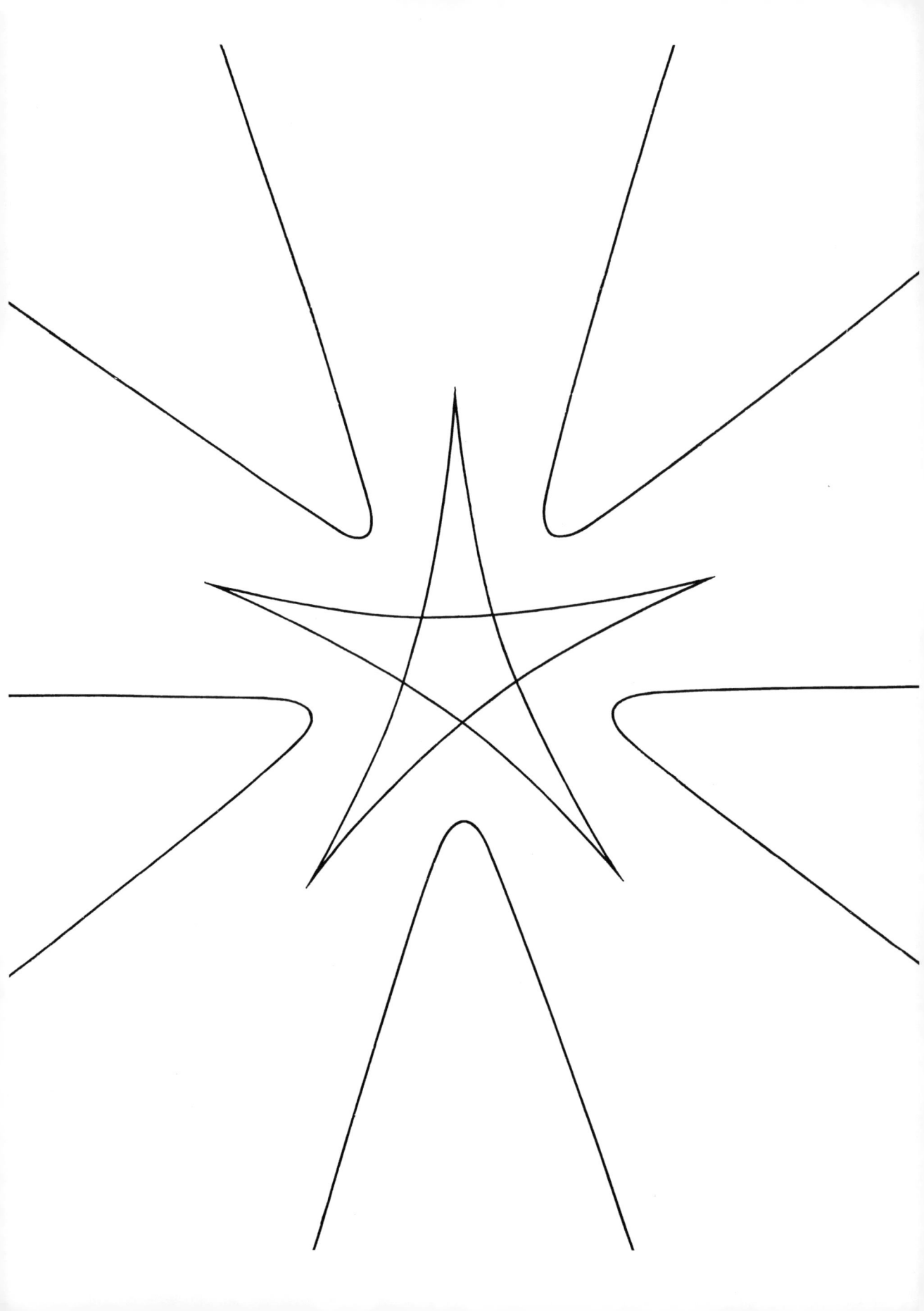

◀ Figure 28

well consider that the curve is not really complete unless these asymptotic lines are also included.

Figure 28 is a beautiful case of reciprocity, with its dynamic yet balanced expression of polarity. The whole field of polar reciprocal metamorphoses of curves contains archetypal geometrical thought-forms which transcend the simple symmetries of Euclidean geometry. As the Euclidean archetypes have given rise to all manner of patterns and designs in art, so the new geometrical thought-forms will inspire in the artist creations of a very different kind and quality. In the case of this illustration it would lead too far to describe the method of construction, except to say that the reciprocating circle is here an imaginary circle, and can therefore not be drawn in (25, 31). Although invisible, the imaginary circle works just as powerfully in the transformation as any ordinary circle.

Figures 29 and 30 show polar reciprocal curve families. In the family of curves in Figure 29, each member passes through four points; in Figure 30, all the curves of the family

Figure 29

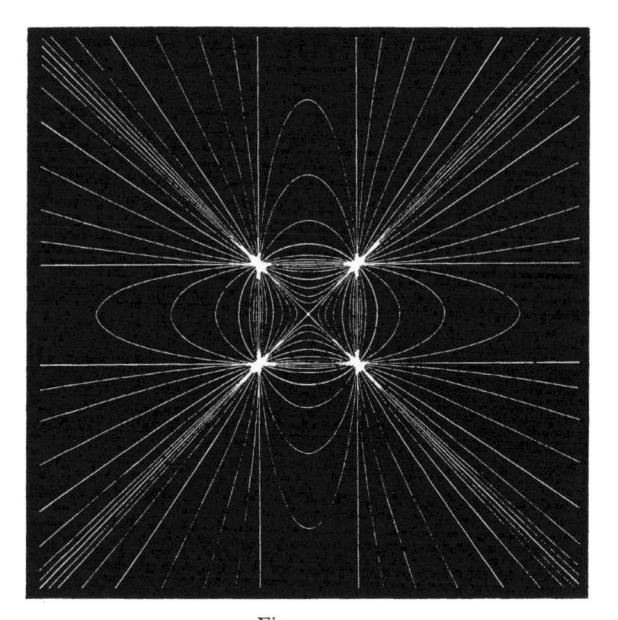

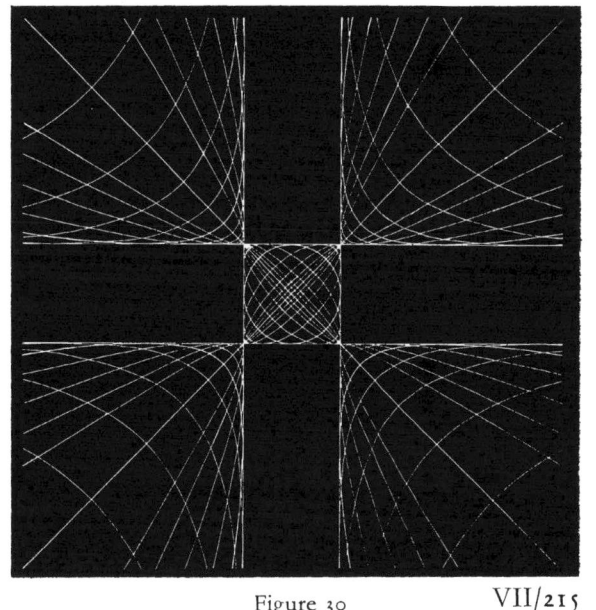

Figure 30

belong to the same four lines. Both are families of circle-curves.

Practice in making these polar transformations reveals how wide awake one must be in thought and ready for immediate reaction to quite slight and very subtle changes in the situations of the component parts of the construction. It is better not to have preconceived notions as to the result, for the slightest change in the shape of a curve or its situation in relation to the reciprocating curve may result in an amazing metamorphosis of the polar curve.

It has already been pointed out that there is a great difference between the projective transformations of the previous chapter, which involved variations in the forms of circle-curves, and the polar reciprocal transformations we have just been considering, where the change from one curve to the other is a radical metamorphosis. A proper understanding of metamorphosis requires a comprehension of the creative factors which bring it about, namely, the mutual interplay of fundamental polarities.

We must bear in mind that polar conjugate curves belong together, they are like an organism. They are "married" to one another and together they hold a balance, even in the most extreme of relationships. Contemplating the peculiar harmony of those with a centric symmetry—as between centre and infinite periphery of space—we may be reminded of the "seal" forms created by Rudolf Steiner to express the cosmic interplay, for instance, between spiritual spheres and the earth (55).

We may be reminded also of the type of symmetry forms suggested by him for use by young children in free form drawing, prior to their introduction to geometry.

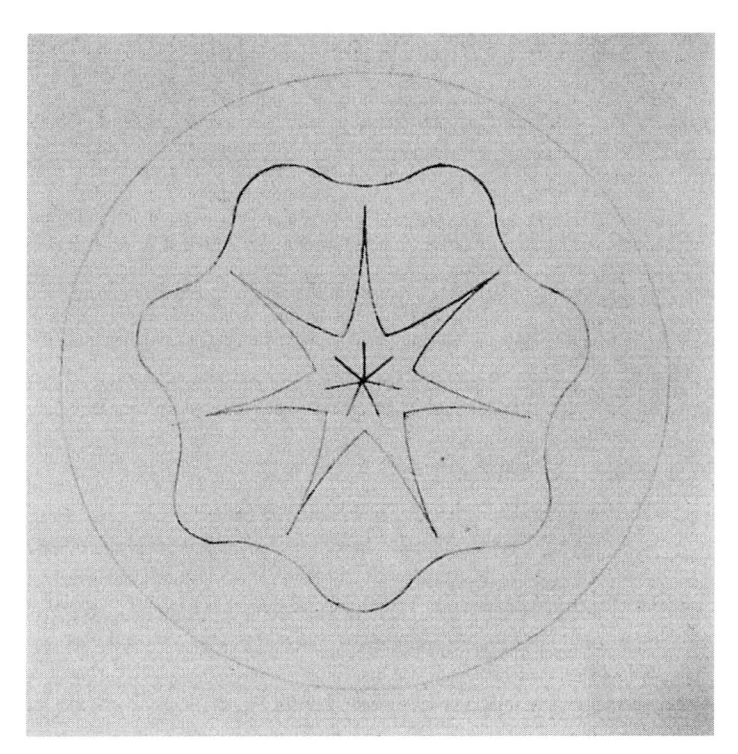

Saturn Seal, after Rudolf Steiner

VIII Polar Forms in Space

Figure 1

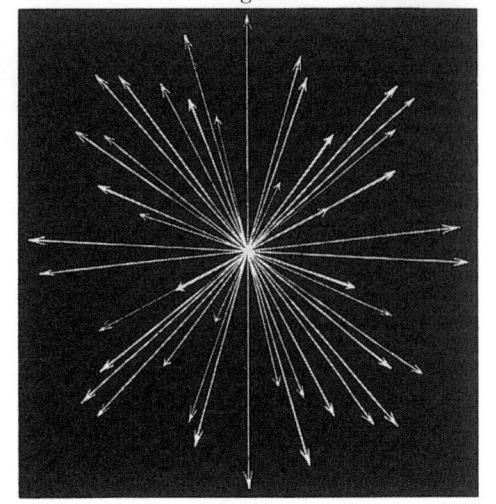

Figure 2

Figure 3

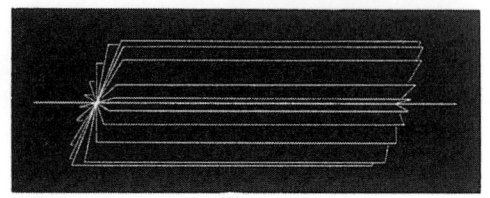

Figure 4

All that we have considered concerning point and line in the plane may be seen again in the relationships between lines and planes in a point. As the circle-curve brings about polar reciprocal transformations among the points and lines of the plane, so does the sphere (and its variations among the points, lines and planes of space.

Reference has already been made to the fact that besides the geometry of a plane, in which lines and points take part, there is the geometry of a point, in which the relative elements are lines and planes, all of which have the point in common (p. 187).

We have consistently practised the dual constructions in the plane, and our preoccupation with lines contained by points in the plane will be a help in dealing with the geometry of the point; that is to say, the point whose members are the lines and planes which pass through it, spreading out in all directions into the infinite distances of space.

The mathematicians have technical terms for these manifolds; the terminology stems from the German mathematician Christian von Staudt (25), in his classical work "Geometrie der Lage" (Nürnberg 1847). It is as follows:

Sheaf of lines or line-sheaf (*Strahlenbüschel*): the totality of all lines of space which belong to a plane (Figure 1).

Bundle of lines or line-bundle (*Strahlenbündel*): the totality of all lines of space which pass through a point (Figure 2).

Bundle of planes or plane-bundle (*Ebenenbündel*): the totality of all planes of space which pass through a point (Figure 3).

Sheaf of planes or planar sheaf (*Ebenenbüschel*): the totality of all planes of space which pass through a line (Figure 4).

They are old-fashioned terms from the nineteenth century which tend to fix the mind on the physical aspect of geometry, which is precisely what projective geometry intends to

overcome. We use a freer terminology, where possible (18):
A point of lines in a plane; or a star in the plane. (Linewise point in the plane).
A point of lines in space; or a star in space. (Linewise point in space).
A point of planes, or planewise point.
A line of planes, or planewise line.

Christian von Staudt (born in 1798 in Rothenburg on the Tauber) performed a tremendous feat of active thinking in the formulation of his "Geometry of Position". He showed with purely geometrical methods how the new geometry is not only founded entirely without the aid of measurement, but even develops the concept of measure out of itself. He himself stressed the value of powerful pictorial imagination, leaving aside the use of symbols, the manipulation of which requires a mechanical facility rather than spiritual energy.

Obviously the geometry of "sheaves" and "bundles" with their projective relations is difficult to picture and to illustrate, and it is not surprising that the textbooks contain few, if any, figures, and that it is often found easier to work algebraically. As we have said, however, the great development of geometry was in fact largely due to the introduction of the analytical method by Déscartes and Fermat in the seventeenth century.

Pure geometry though it is, von Staudt's book is difficult, lacking entirely in illustration and with an extremely short and compressed mode of expression. The nineteenth century thinkers did not make it easy. Adams (1), with great enthusiasm, tried to make it easier in that he set to work—to use his own words—"to disenchant the beautiful princess, hidden away in nineteenth century abstraction". Adams sought unceasingly to show forth the purity and grandeur of her truths in pictures and models and in lively descriptions.

If now we use finished, three-dimensional drawings and sketches as visual aids in what follows, the reader must use his inner activity of pictorial imagination, to unfreeze what can only be a momentary and partial illustration, bringing movement in at every stage, so as to grasp, not a finished form, but a whole *process*—a form in the making. Moreover, he must never forget that in reality, in this new geometry, *all lines and planes extend away into infinite space without limit*, interpenetrating one another and only thus creating forms.

We have learned how in the plane the points of a line may be perspectived into the lines of a point, and that when a circle is in the plane, there is a polar reciprocal relation between points and lines throughout the plane. So, too, the forms in a plane may be taken up by perspective into the lines and planes of an eye-point, and we shall now find that *when a cone is in the eye-point, there will be a polar reciprocal relation among the planes and lines of that point* (37).

All the laws we have been studying in plane geometry may be found again among the planes and lines of the point, and just as circle-curves arise in the geometry of the plane, so all manner of cone-forms arise in the geometry of the point. These cones, held, as it were, in an eye-point, should however not be thought of as one thinks normally of three-dimensional cones. Just as we have learned in thought to trace the circle-curve pointwise and linewise, always remaining in the one dimension of the line of the curve as it lies in the two dimensions of the plane, so now we learn in thought to sweep through the surface of a conical form linewise and planewise, that is to say, in lines and planes which always remain in the single point.

We shall see how this *intensive* field of the point is polar opposite to the *extensive* field of the plane, when we take into account the whole of projective space (Ur-space or Archetypal space) in its relationship to the spherical form. *This is a polarity of polarities.*

Polarity of the Extensive and Intensive Fields

In the plane:	*In the point:*
All laws and all forms which hold for lines in respect of points correspond to laws which hold for points in respect of lines—i.e. where the roles of line and point are interchanged.	All laws and all forms which hold for lines in respect of planes correspond to laws which hold for planes in respect of lines—i.e., where the roles of line and plane are interchanged.

It is the formulation of the principle of duality without reference to the sphere which allows of such reasoning. We must now recognise, however, that in reality it is the sphere which dominates this process. The threefold structure which arises when we think of the sphere—the mediating entity—between the poles, leads to a far more organic conception of the form-creating process and of space itself. The whole process involved in the principle of polarity considered, as we are now doing, as taking place throughout the *whole of space* and not merely in the plane or in the point, is due to the very existence and nature of the sphere (23).

Pole and Polar with Respect to the Sphere

As the circle (and its variations) in the plane relates point to line, so the sphere (and its variations) in space relates point to plane and pointwise line to planewise line. We bring to our aid all that we have learnt in regard to the circle and recognise what corresponds to this in three dimensions.

For every point in space there is a polar plane with respect to a particular sphere, and vice versa. If the point and the plane are not co-incident (tangent to the sphere), a whole cone of tangent lines and planes relates the one with the other, the pole with the polar and the polar with the pole (Figure 5), not merely two tangent lines, as in the case of the circle.

Follow in thought the reciprocal in- and out-breathing,

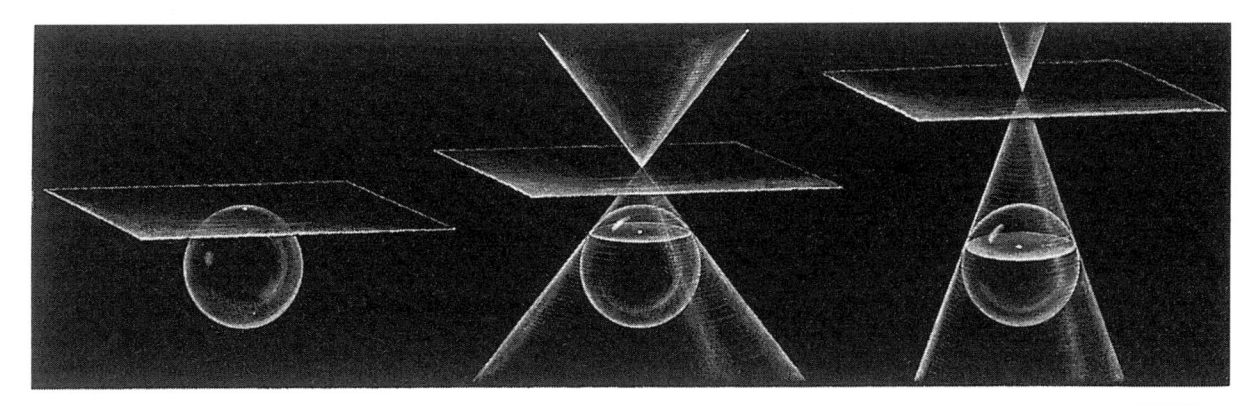

Figure 5

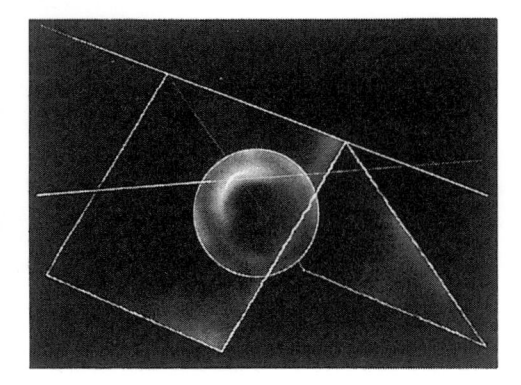

Figure 6

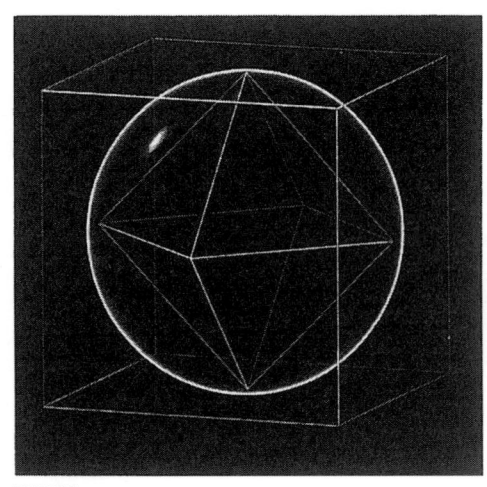

Figure 7

this time of point and polar plane. The points radiate outward or press together inward, while the polar planes hover inward or outward. The movement of the one calls forth the movement of the other. *The central point of the sphere has the infinitely distant plane of space as polar plane, and any plane through the sphere's centre has its polar point in the infinite at right-angles to it.*

The sphere not only brings forth a polar relationship between the points and planes of space, but also between all the lines of space. We may call this *the line-line polarity of space* (p. 233). Every line considered as a manifold of points has assigned to it in its relationship to the sphere a line considered as a manifold of planes. Figure 6 shows two such lines, one passing through the sphere, the other passing outside it. The two planes are tangent to the sphere; the line joining their points of contact is polar to their common line. Set this picture in movement in thought and let the common line of the two planes move out to the infinite. It will be clear that this will result in a movement of the tangent planes, such that their points of contact with the sphere will move round its surface carrying the line joining them towards the centre, until the moment will come when the two planes will become parallel. At this moment their common line has reached the plane at infinity of space and the line common to their points of contact will pass through the sphere's centre.

To every line passing through the sphere there is a polar line somewhere outside it, and to every line passing through the centre of the sphere, there is a specific line in the infinitely distant plane which is its polar. When the one line is determined by points, the line related to it by the law of pole and polar will be determined planewise.

We now meet again the regular forms with which we became familiar in Euclidean geometry, but now they appear in the light of the Principle of Polarity of projective geometry and show forth their polar relation to the sphere—their point-plane and their line-line polarity. For every point of a form inscribed into a sphere there will be the plane of a sister form circumscribing the sphere. Moreover, any line of the one form considered pointwise will be answered by a corresponding line of the other considered planewise. In Figures 7, 8 and 9 the regular, so-called Platonic forms are shown in their reciprocal relationships with respect to a sphere.

Drawings of polar opposite forms in three dimensions are not at all difficult if "parallel perspective" is used, that is to say, if no attempt is made to make a picture in true perspective. In parallel perspective, the lines in *any one dimension* are always of the same length and parallel to one another. It will be found that if one uses lengths in the proportion of 10:9:5 for the vertical, horizontal and forward-backward dimensions respectively, then turning the figure, making the appropriate small alterations to the right-angles as shown in the drawings, a sufficiently satisfactory picture will result. The drawing will then be absolutely accurate for the purpose for which it is here intended.

Cube (hexahedron) and octahedron are polar opposite forms (Figure 7). The one has as many points as the other has planes; both forms have the same number of lines. The points of the octahedron consist each of four lines and four planes; the cube planes are determined by four lines and four points. Comparing the planes of the octahedron with the points of the cube, we find a similar polarity, though here the number three pertains. Every point of the one form is polar to a plane of the other in respect to a sphere which surrounds the one form and is enveloped by the other. Every line of the one form corresponds (in the sense of Figure 6) with a line of the sister form. Whereas the cube consists of six planes, eight points and twelve lines, the octahedron has eight planes, six points and also twelve lines.

Figure 8 shows the tetrahedron to be self-polar. It has the same number of planes as it has points and contains six lines.

Figure 9 shows the similar polar correspondence between an icosahedron and a pentagon dodecahedron with respect to the sphere. The icosahedron within the sphere has twelve points, twenty planes and thirty lines, while the pentagon dodecahedron enveloping the sphere has twelve planes, twenty points and also thirty lines. Here the numbers ruling are five and three; whereas the dodecahedron's planes contain the number five and its points the number three, with the icosahedron it is the planes which are triangular and the points which are determined by five lines and five planes.

Naturally it makes no odds which form is within and which

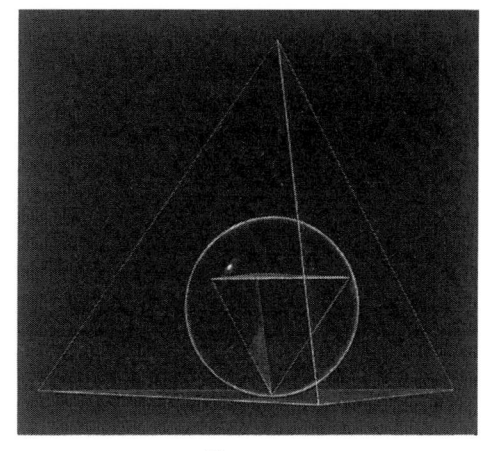

Figure 8

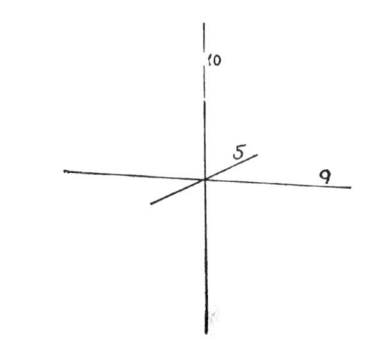

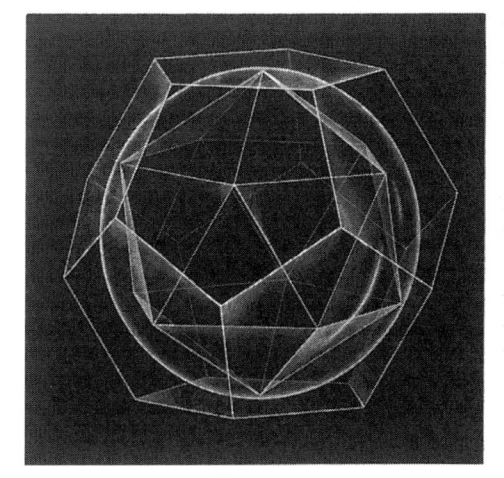

Figure 9

without, for the sphere brings about the reciprocal relationship either way; if the one form is inscribed into the sphere, the sister form will envelope it tangentially.

Figure 10 is an interesting illustration of the perfect balance between point and plane in these forms. It shows:

(1) The metamorphosis of an octahedron into a semi-regular form called a cubo-octahedron, which, as its name implies, has characteristics both of the cube and of the octahedron. In the cubo-octahedron, some planes are triangular and some are square; all its points are created by four lines.

(2) The metamorphosis of a cube into another semi-regular figure, the rhombic-dodecahedron, which also has characteristics of both cube and octahedron. In the rhombic-dodecahedron, some points are created by three lines and some by four; all the planes are created by four lines.

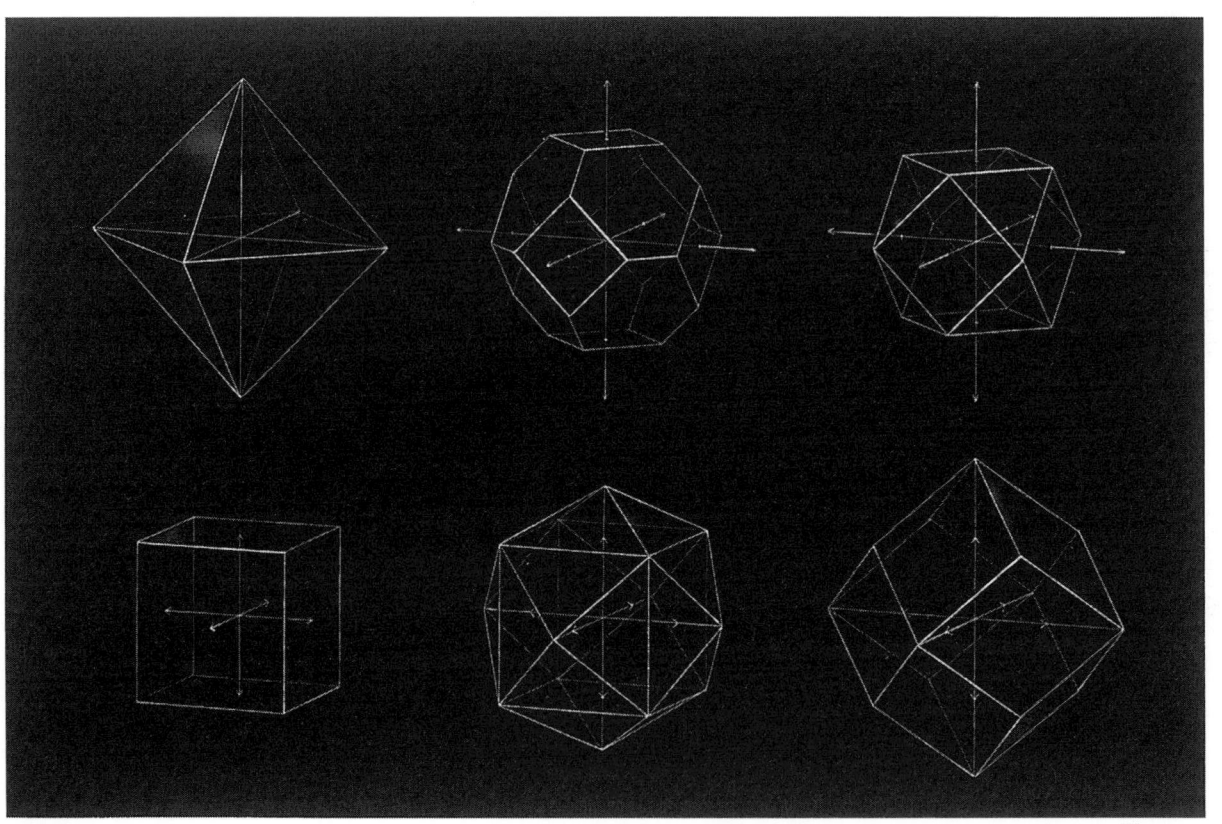

Figure 10

To create the first metamorphosis, imagine six planes, one at each of the octahedron points and perpendicular respectively to the octahedron axes. The interpenetration of these six planes would in fact create a large cube around the octahedron (as in Figure 7). Then imagine the six planes moving inward at all six points, keeping their perpendicular position in respect to the axis along which they are moving. At first a small square face will appear near to each octahedron point, which will grow larger as the planes move inward. In Figure 10 the resulting form is shown in two stages; first when the planes have moved in a third of the distance between the octahedron points and its centre and then when they have moved in halfway.

At the first stage (Figure 10, centre, above) the form has a strangely harmonious irregularity; at the second, the cubo-octahedron appears, when the large imaginary cube will have shrunk to such proportions that its edges will meet at right-angles with the edges of the octahedron and the interpenetrations of the two forms will be completely balanced. It is a helpful exercise to draw the interpenetration at this stage of cube and octahedron, when the corners of the one form protrude from the surfaces of the other respectively. The cubo-octahedron is the shape created by the planes of the two forms at this balanced stage of interpenetration (Figure 11 includes this situation).

The second metamorphosis, that of the cube into the rhombic-dodecahedron, is a good exercise in polar reciprocation, according to the principle of polarity. Instead of imagining planes moving inward along the axes of the octahedron, we imagine points pushing outward along the axes of the cube, at first creating four-sided pyramids on each cube face (Figure 10, centre, below), until, when the points have reached twice the distance of the cube planes from its centre, a balanced situation is once more reached. This time the octahedron edges have moved outward until they have come into contact with the edges of the cube. Instead of considering the *points* in which these edges meet in pairs, which are the points of the cubo-octahedron, we now consider the *planes* which they create, which are planes of the rhombic-dodecahedron. While the cubo-octahedron is formed by the 6 plus 8 planes of cube and octahedron, the rhombic-dodecahedron is determined by their 8 plus 6 points. Cubo-octahedron and rhombic-dodecahedron though not perfectly regular are polar reciprocal forms.

Figure 11 shows the balanced interpenetration of all the four forms we have just been considering. It is of course possible to carry out an operation similar to the one illustrated in Figure 10, using, for example, the pair of forms in Figure 9. There are many interesting ways in which the interrelationships of the various forms may be shown in drawings. (The method of drawing in parallel perspective is to be recommended.) It is a fascinating exercise to make a straw model in which all five regular forms are contained at once.

Now we will take a further step, for we are not merely interested in the complementary positions of points, planes and lines, when studying the polarities revealed by these forms, but especially in the way these polar processes are mobile in space. In Figure 5 we saw how the movement of a point is answered by the movement of its polar plane,

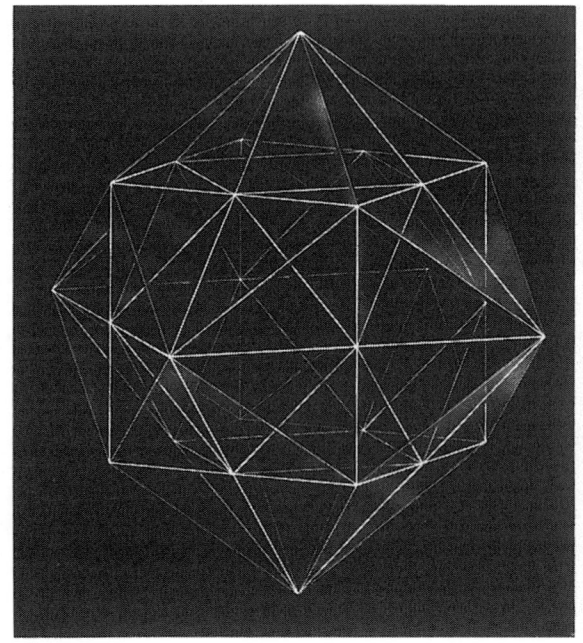

Figure 11

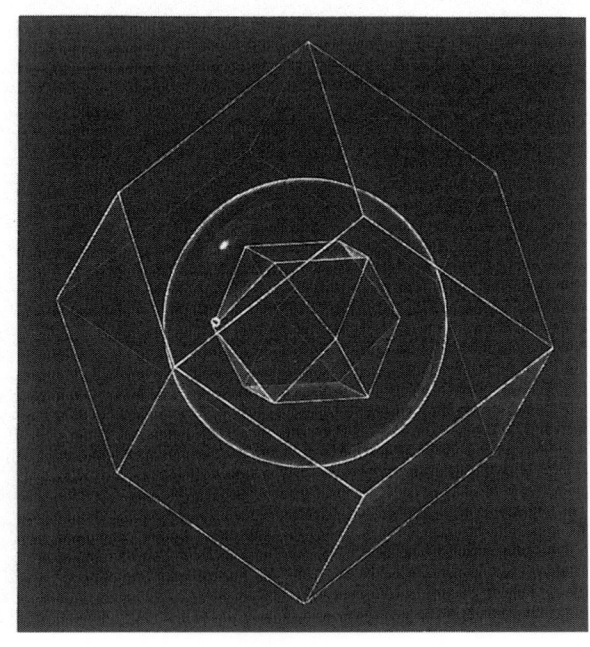

Figure 12

and vice versa; from this it will be clear that the movement of a whole structure, like a cube, created of planes, lines and points will be answered by the reciprocal movement of the whole form which is polar to it. Thus when a form inscribed within the sphere shrinks, the form which is polar to it, circumscribing the sphere, will expand (the opposite also being true). In Figure 12 the rhombic-dodecahedron surrounding the sphere has expanded, resulting in a contraction of the cubo-octahedron towards the centre. Figure 13 shows this process taking place in three stages with the cube and octahedron as examples.

When this process is followed to its extreme, the inner form degenerates into a single point of lines and planes (the twelve lines and six planes of the cube), while the outer form merges with the infinitely distant plane of space, thus degenerating into a single plane of lines and points (the twelve lines and six points of the octahedron).

It is necessary to follow this last thought to its fascinating conclusion. In the degeneration of the cube as we have been picturing it here, the planes and lines melt into one another in pairs; in the last resort the central point contains *three*

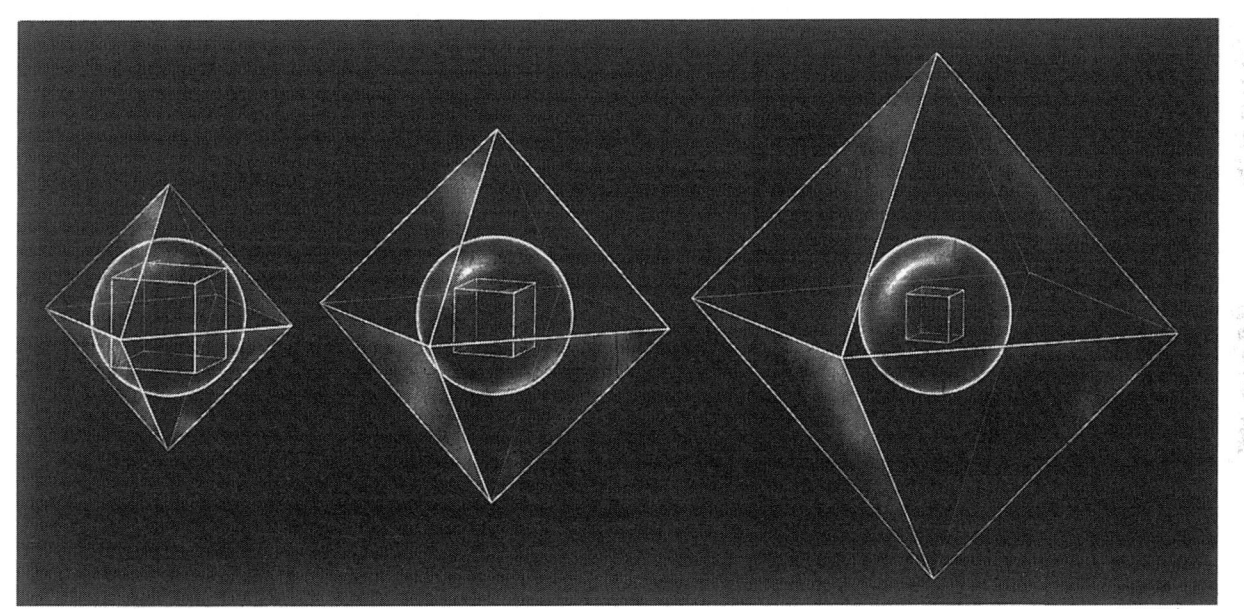

Figure 13

planes and *three* lines. On the other hand, as the octahedron expands, the planes and lines also melt into one another in pairs, *but they meet in the infinitely distant plane of space!* Whereas the cube points all come together in the central point, all the octahedral planes merge into the infinitely distant plane of space. Thus: *If an inner form, with all its points, lines and planes, contracts into the very centre of the sphere, degenerating into a point of lines and planes, then the outer form with its planes, lines and points will expand an all sides into the one infinitely distant plane of space. Figure 13 shows stages in this process.*

Figure 14

Now let us concentrate on Figure 14 (38) and bring it into movement in our imagination. (The sketches in Figure 15 are intended as a help in following the movements; picture them three-dimensionally and think of the whole plane as far as possible, and not merely of the part enclosed by ellipse or rectangle.) The apex of the cone in the outer plane is the pole of that plane passing through the sphere which is located by the ellipse; move the apex of the cone and this polar plane will also move. Let the point of the cone move in a line and its polar plane will turn round upon a line, an axis; the line of points traced by the apex of the cone (a pointwise line) will have as its polar opposite the line of planes passing through the sphere (a planewise line).

As soon as we give the point in the outer plane more freedom and let it move out of the line, its polar plane too will gain a greater freedom of movement and will *pivot on a point*. It is this point which is the pole of the outer plane itself. Taking any three positions of our moving point, so as to create a triangle of points in the outer plane, we shall see that the polar planes of those three points have a common point within the sphere. Answering to the triangle in the plane outside, we see a trihedron (a form created of three planes and the lines in which they interpenetrate) in the point within. We picture the whole form of the trihedron opening upward and downward to the infinite in both directions. *Triangle and trihedron are pole and polar with respect to the sphere.*

Take a further example. Instead of a triangle, let us imagine a circle being drawn in the plane outside the sphere. For simplicity, we will think of this plane as tangent to the sphere and will let the point in it circle round the point of contact with the sphere (the point at which a radius of the sphere strikes through the plane at right-angles). The form which is polar to the circle in the plane will be a cone with its apex in the pole of this plane. As the point describes a circle in this tangent plane, *its* polar plane, which passes through the sphere, will describe a circular cone with its apex in the pole of the tangent plane. The aperture of this cone will depend on the size of the circle.

The further out on the tangent plane we take the point which is describing the circle, the more steeply inclined will

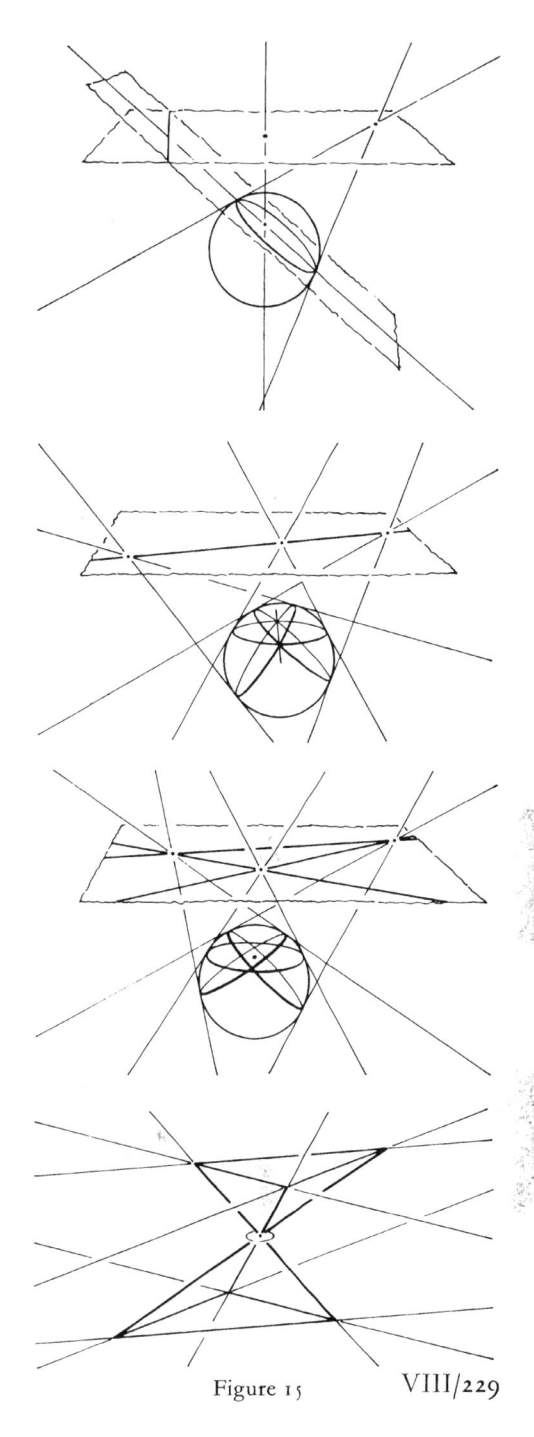

Figure 15

the polar plane of this point be, with the result that the aperture of the cone will be narrower. If the point is nearer the point of contact of the tangent plane with the sphere and therefore describing a smaller circle, its polar plane will be less steeply inclined and will create a more open cone.

Figure 16 shows a family of circles and their polar family of cones. The sphere is here not drawn in, but we should imagine it as described: the tangent plane is the horizontal plane containing the circles, while the cones all have in common the point of contact of this plane with the sphere.

There will be two extremes in this whole process. Either the circle grows vast and changes into the infinitely distant line of the horizontal plane, when the cone will close up and merge into its axis; or the circle diminishes into the point of contact of the plane with the sphere, when the cone will have opened out into a plane—the plane which is polar to the point into which the circle has melted. (To be exact, we should have to say that the cone covers the plane twice over when it degenerates into it.)

Figure 16

To complete this thought geometrically, we should remember that although we began by tracing the circle in the outer plane *pointwise,* it must also be considered *linewise,* so that, summing up, we may say:

Of the circles in the polar plane:
When infinitely large, their points are in the infinitely distant line of the plane, while their lines all become one with it. (This infinitely distant line is doubly covered by all the points and lines of the circle.) When infinitely small, their lines all lie in their midpoint, while all their points become one with it.

Of the cones in the point pole:
When infinitely closed, their planes are in the inner (vertical) line of the point, while their lines all become one with it. (This innermost line is doubly covered by all the planes and lines of the cone.) When infinitely open, their lines all lie in their "median" plane, while all their planes become one with it.

Just as the circles have a central point from which they expand out into the infinitude of the plane in which they lie, so the cones which are held in the point and open upward and downward, have a central plane from which they begin to contract towards their common axis, which is for them an infinitude within. We use the word *median plane,* pairing it with the mid-point of the circles, and thus differentiating between the two kinds of "middle".

It is a wonderful reciprocal interplay of movements between a pole and its polar plane and between line and line in their polar aspect with respect to the sphere. The ultimate line-line polarity is here expressed as between the (vertical) axis which passes through the pole and the cosmic "axis" which lies at right-angles to it in the infinite distances of space—the infinitely distant equator-line. Thinking of the infinitely large circle *pointwise* we find that its polar form is the inner axis considered planewise. Thinking of the infinitely large circle *linewise*, in the general case, when polar plane and pole are *not* tangential to the sphere, the infinitely large linewise circle will contain two parallel planes which determine the axis pointwise.

It is a most valuable—and for the purposes of Chapter IX— an indispensable exercise to practise all kinds of movements and variations of this construction in thought and imagination, so as to experience it through and through, with all its qualities and possiblities. The circles in the plane grow larger as circles do when a stone is dropped into still water; the cones close up around their inner axis as the leaves of some plants do at night. If, on the other hand, the circles contract into their centre, the cones will open out and flatten into their median plane. Apparent growth in the *extensive* realm of the circles denotes an apparent decrease in the *intensive* realm of the cones, and vice versa.

We may make endless display of forms in the polar plane, and all will be answered by the lines and planes of the pole, like a wingéd echo. If we think pole and polar with respect to a circle in the plane, the pole will answer with *a plane and a line in polar relation with respect to a specific cone*. Even the theorems of Pascal and Brianchon will come true again among the inscribed and circumscribed hexahedra of a cone!

We will be content with one more example. Think of a logarithmic spiral in the polar plane, its points and lines streaming inward and outward between the two infinitudes of the plane. The pole will answer with a great surface which furls and unfurls, spiralling between its two infinitudes: the cosmic horizon and its own inner "verticon" within (39). Figure 17 shows such a spiralling surface. One with

Figure 17

Figure 18

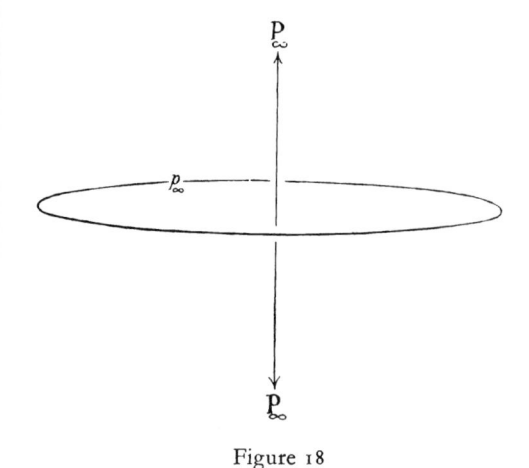

the polarity of central point and cosmic plane is the polarity of central axis and cosmic circle (Figure 18.)

All these wonderful laws are hidden away—frozen, as it were—in the five regular forms: cube and octahedron, icosahedron and pentagon dodecahedron, and the self-polar form of the tetrahedron. It is interesting to note that Rudolf Steiner, in the twentieth century, remarked that there are really seven such forms, for to the five should be added two more: "The sphere from within and the sphere from without", i.e. the pointwise sphere and the planewise sphere.

To this thought concerning the two types of sphere, we add the concept of *horizon* and *"verticon"*. We think not only of a vertical axis in physical space in relation to a distant horizon, but also of a great circle in the heavens in relation to an *inner* axis—an infinitude within—like the "spiritual staff" which Goethe saw in the innermost region of the growing plant (40).

Here we touch on another concept of the Imaginary. Every sphere turning upon an axis begets in relation to this axis a polar axis—a real but infinite circle at right-angles to this axis. The infinitely distant point of the axis is like the heavenly pole in respect of the cosmic equator-circle at right-angles to it. The sphere calls forth such a relationship in every conceivable direction of space. The mutual interplay of these polar points and lines in the infinitely distant plane, is called forth, as a polarity, by an invisible circle, which is called the Imaginary Circle, the archetypal circle (Urkreis) in the Absolute Plane of space. It is the entity which rules the three-dimensional right-angledness of our earthly space. Just as we saw how every circle-curve fills the plane in which it lies with a moving formation of polar triangles, so now we think of an invisible, imaginary circle in the plane at infinity, which does the same for three-dimensional space. Every time we picture the three great circles in the heavens made by the three dimensions of space, we are envisaging a polar triangle of the cosmic imaginary circle. This invisible circle, written in movement into the infinitely distant plane, is the cosmic counterpart of the three dimensions.

The Line-Line Polarity of Space; the Line Congruence

Our considerations would be incomplete, if we were to leave out entirely the most fundamental and free of all the projective space transformations, though we shall only take the first few steps into this marvellous field of raying creative lines—the field which in fact inspired the title: "Raying Creation of Worlds" (Strahlende Weltgestaltung) (41). Perhaps one day geometrical instruction in schools will include as fundamental this remarkable realm.

We have seen a whole geometrical discipline being built up on the first part of the last of the Axioms of Community: "Two lines of space, if they have a point they also have a plane in common. . ." But what of the second part? ". . . or else they have *neither* a point *nor* a plane in common." Does this negative part of the statement mean that here the axiom of community breaks down? Not at all!

It is indeed the more general case that two lines in space will be "skew" to one another; seemingly they pass one another by entirely, having no point and no plane in common. This applies in the most extreme situation for the two reciprocal lines we have called "horizon" and "verticon".

Any such pair of lines—however skew and however far apart—will be related projectively, *if we consider them both pointwise and planewise*. As a plane of one of the lines turns about it, it will engender a point which runs along the other line, through the infinite and back in either direction. There is a perspectivity between the planes of the one line and the points of the other.

Just as a perspectivity may take place between the points of a line and the lines of a point in a plane, so also between the two aspects of the line-manifold. Moreover, if several line-manifolds are related by perspective one with another, a projectivity will arise. The projective process taking place between line-manifolds results in the line-woven shape called a regulus.

Add to the two skew lines a third which is skew to both, then the planes of the first will meet this third line also. As a plane of the first line moves round, it will beget *two* points, one on either of the other two skew lines. This means that as the plane rotates, it will also always contain a *fourth line*, namely, the common line of these two points, which at

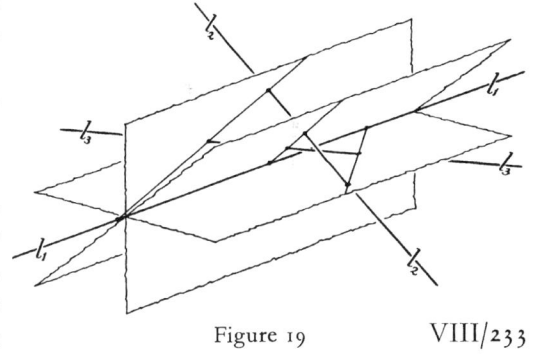

Figure 19

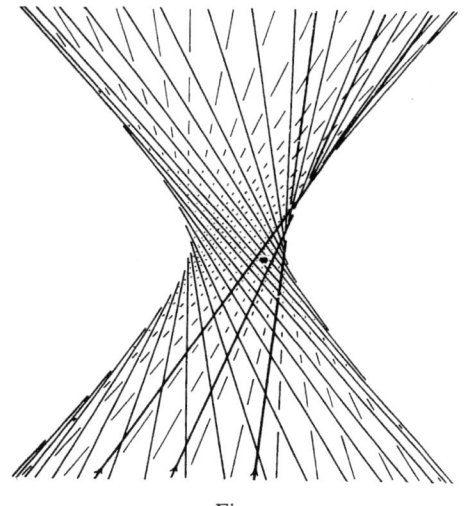

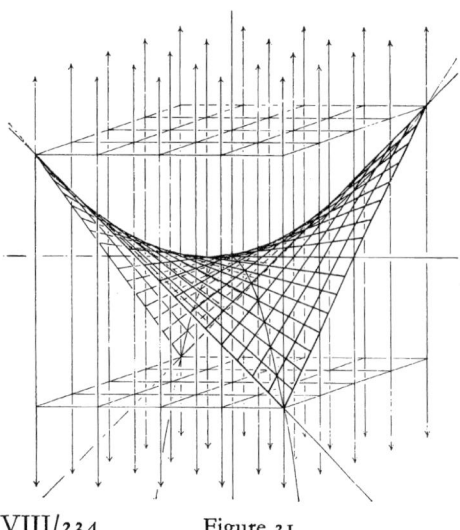

every moment will contain a point of each of the three lines (Figure 19).

Picture how, as the plane of the first line rotates, the fourth line will move, sliding, as it were, on the three given skew lines, as though on three rails through space. All the momentary positions it takes up in space as it passes through them will be skew to one another. This fourth line will swing through space like a cosmic merry-go-round, and all the time its points-of-contact with the three given lines will be circling through it!

The movement of this fourth line depicts an infinite manifold of lines, all skew to one another, moving right round and leading back at last to the starting-place. The projective relation between the three original lines and the fourth creates an infinite manifold of lines, which together plasticise a hyperboloid surface, opening upward and downward and returning into itself through the infinite (Figure 20).

Once the manifold is there, *any* three lines might function as the original three for a fourth line moving the other way round. If we can show that both "fourth lines" have a common point, we can also show that the self-same surface can be woven the other way round, forming a cross-weave, like web and woof. And so it proves to be. We see here another aspect of the close-woven property of space which we saw demonstrated for the plane in the ancient theorem of Pappos.

A variation of the hyperboloid form comes about when one of the three original skew lines is a line-at-infinity. The resulting form will then be a paraboloidal surface (Figure 21).

These line-woven surfaces, reguli as they are called, deserve a book all to themselves. They are far more fundamental and elementary than the sphere and its relations, and we would find again among them all the projective laws we have been studying in a yet more basic form. Although created from the same original elements and qualities of space as the crystal forms of cube, octahedron and tetrahedron, they differ essentially from them in character, for their surfaces extend into the infinite and have a far more intimate relationship with the plane at infinity.

The linewise hyperboloid is spanned between two "axes", one in the heart of it, as it were, while the other is the infinitely distant line at right-angles to the first. Setting a family of hyperboloids one within the other—which accords well with their nature—we see that the lines (generators) of the innermost one are only slightly inclined from the position of the inner axis (Figure 22). The closer to this axis the form approaches, the slimmer it becomes, until at the last moment it "slips through the straw and falls in"—that is to say, it degenerates into this axis.

On the other hand, as we move out from surface to surface, we find the generators becoming more and more inclined, until, if we were able to follow them right out to the infinite, we should see them all merge into the infinitely distant line at right-angles to the inner axis. The surface folds in upon itself all round and at this pole, too, it degenerates into a line.

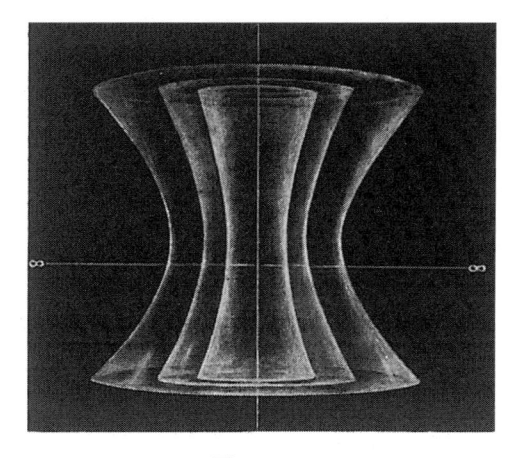

Figure 22

These forms—parts of them—are easily spanned with threads between boards, and if rotated, they reveal the beauty of their hidden laws in delicate spiralling movements, which show their kinship with the quartz crystal, with its major axis, and especially with the plant (Figure 23).

Were we to learn to know these surfaces more intimately, we could be led deeply into a most mysterious and hidden

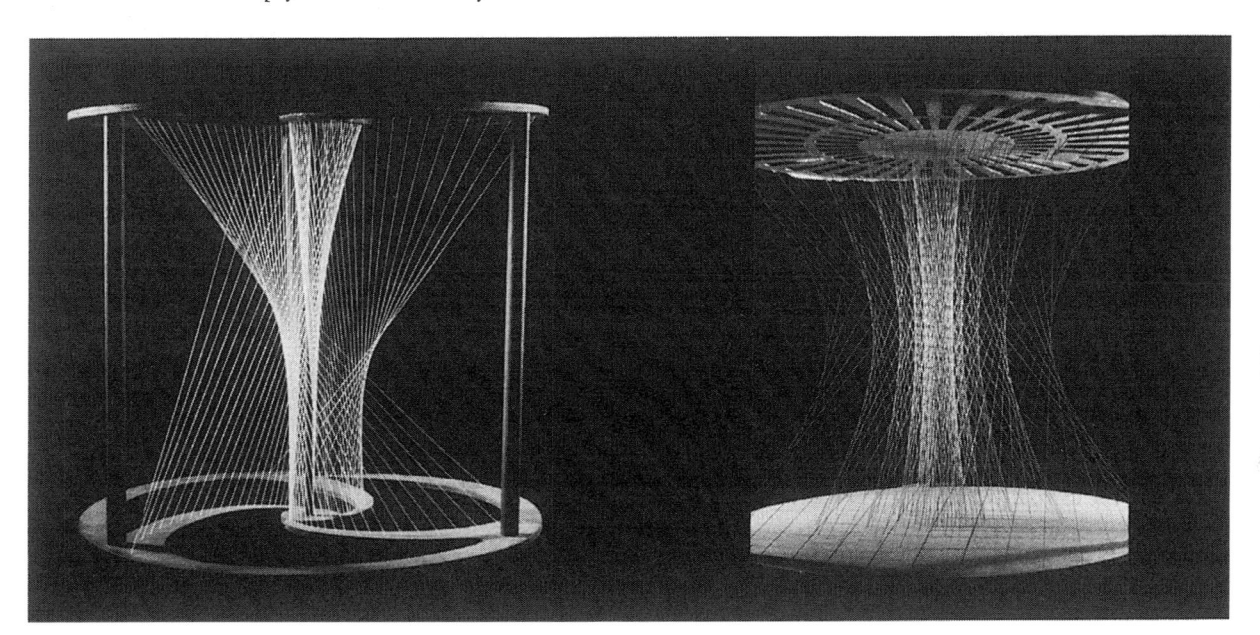

Figure 23

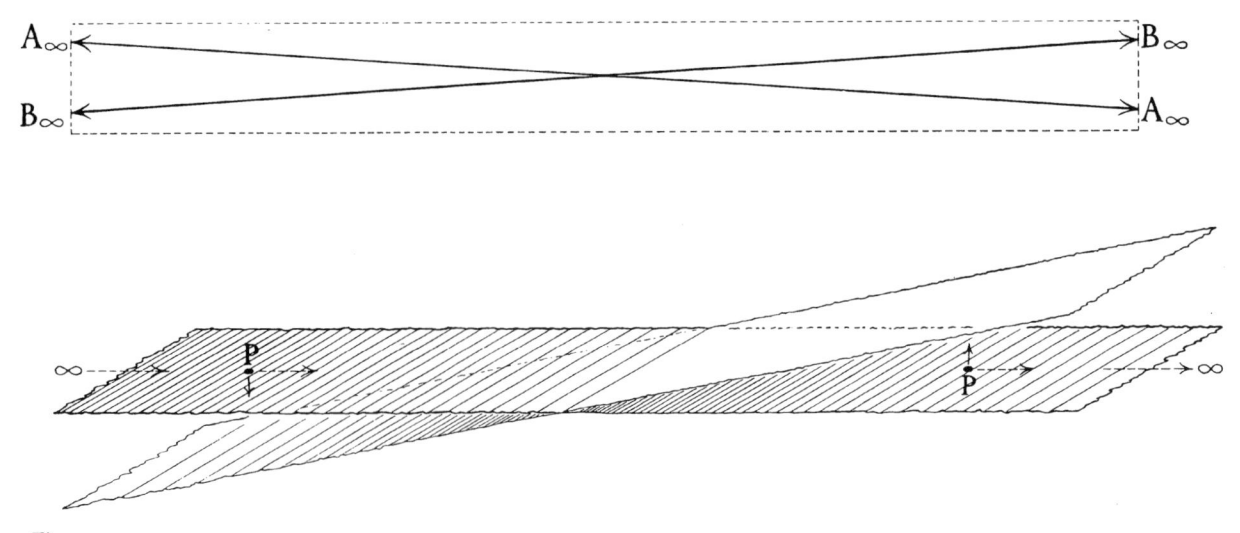

Figure 24

world of forms. We should meet, too, with the strange quality of the plane at infinity itself. Unlike a finite plane, the projective plane has not two sides, but only one; it is everywhere continuous. If one were to walk out to the infinite along a line in one direction, on returning again from the other side, one would be walking head downwards and would have to go out and back once more before one reached the place from which one started (Figure 24). The projective plane is a surface with only one side, continuously flat, and yet, paradoxically, it is twisted in itself! (42).

This lemniscatory quality of space was demonstrated towards the end of his life by the famous mathematician and astronomer August Ferdinand Moebius (1790-1868) to whom a great deal of mathematical discovery is due. His "Moebius leaf" is a familiar puzzle in popular books (Figure 25). The leaf is of course not continuous, but has edges, and the task was set at about the beginning of the century to create a closed model of the projective plane, with its strange attributes. Werner Boy (43) in Göttingen, setting out to prove it impossible, found that is was possible to create a surface which closes in upon itself without singularities (Figure 26). The model shows the lemniscatory quality which inevitably arises when the nature of space is investigated according to the laws of projective geometry.

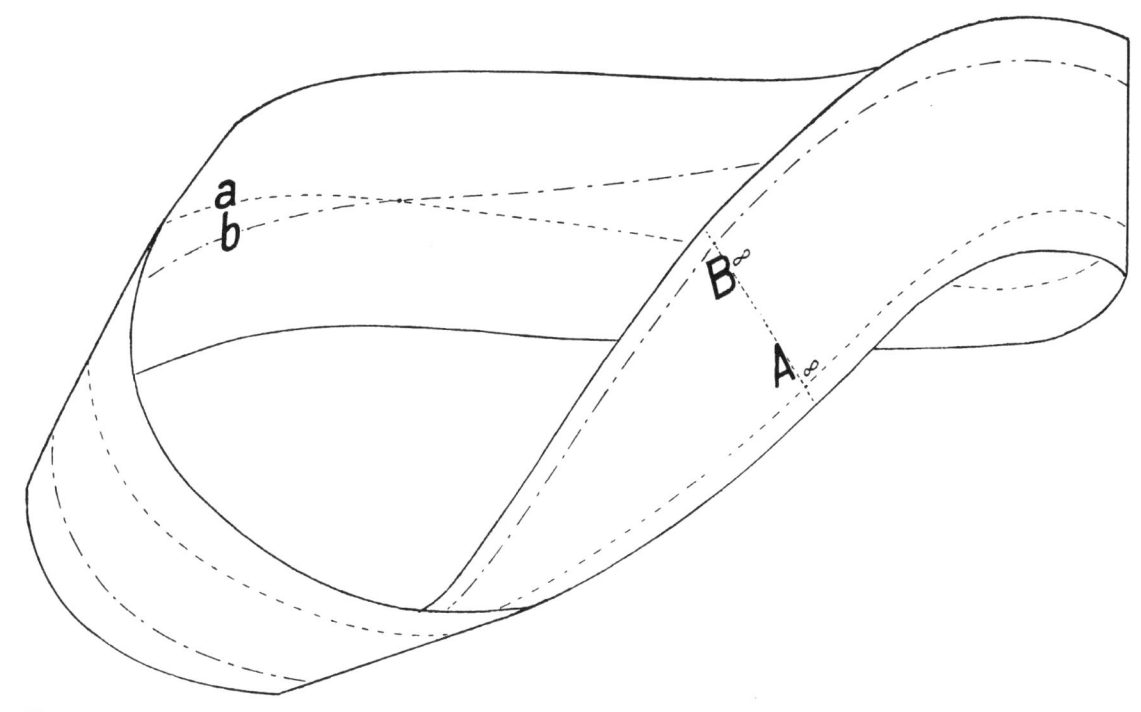

Figure 25

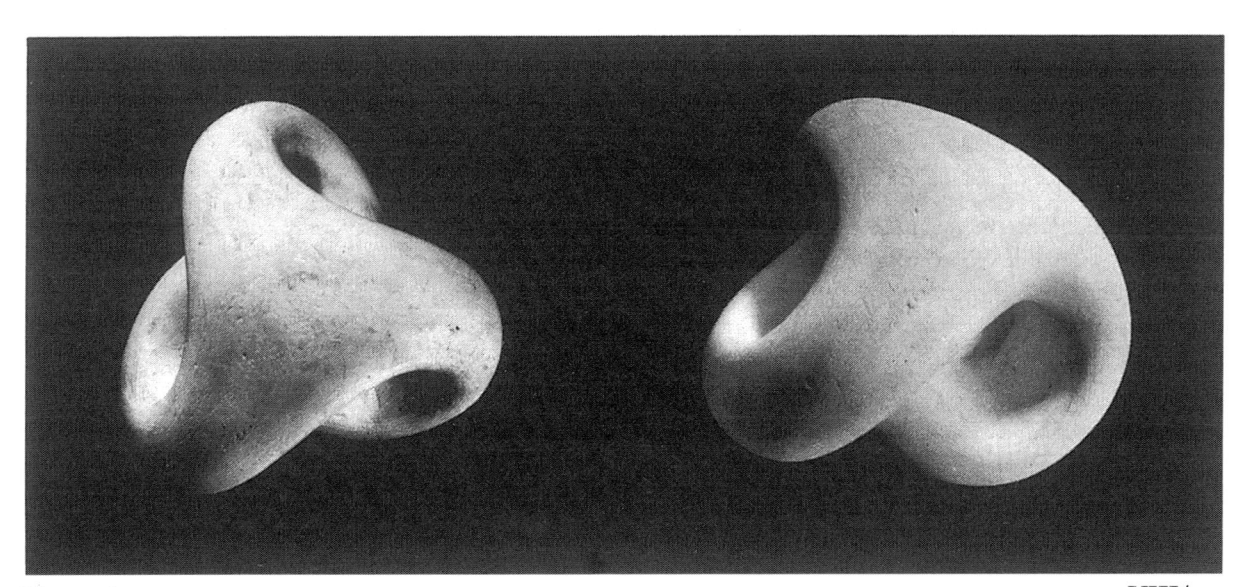

Figure 26

IX Geometry of the Twentieth Century

We have come a long way in our geometrical thinking since contemplating the geometry of the ancients in the early parts of this book. Although at every step our presentation could have been widened and deepened, for the realm is vast, we have covered sufficient ground to enable us to realise, looking back upon the path, that we have been led consistently to overcome existing barriers and limitations of geometrical thought concerning the creation of forms in space and even the very nature of space itself. We are no longer so bound in thought to the cross of the right-angle, nor held captive in the tomb of three-dimensional space.

From the consideration of geometrical concepts of ancient times, when knowledge of earth space was generally limited to the country a man lived in, we have progressed to the kind of thoughts entertained by geometers and astronomers at the beginning of modern time, when mankind was developing a consciousness of the whole earth. It is evident that the time is now at hand when men seek to gain a universal consciousness (44).

In the measure that man's experience of the earth and her spaces has grown, his memory of a divine world once experienced vividly has receded.

Yet science, led by mathematics, has long transcended the limits of the old spatial concepts, in the theories of modern physics. What is so interesting is that it is precisely in *physical* research that present day science has realised that there are areas beyond our naive experience unaided by instruments, where the simple three-dimensional concept of space no longer holds. Indeed, the spectacular developments and practical results of the physical sciences in recent years could not have taken place had this step not been made *in thought* by such men as Einstein, Rutherford, Niels Bohr and others.

As a result of the same versatility and detachment of modern thought, which enabled projective geometry first to liberate itself from the rigid laws of Euclidean space, the non-Euclidean geometries of the early nineteenth century were discovered, which have been partly applied, for instance, in the Theory of Relativity.

So closely related were the modern mathematical conceptions not only to the space in which we live, but to the forces we imagine to be active within it, that it broke in upon the world with great intensity, with the explosion of the first atom bomb in 1945, that those theories of Einstein's, which are contrary to all classical concepts, are no mere figment of the mathematician's imagination. The world had entered a new era.

The idea of the closed system of the material universe according to Cartesian principles, in which space and time are distinct, received a rude shock. At least in some respects the diverging pathways in mathematics from the time of Descartes and Desargues begin to come towards one another, in that the old idea of "Euclidean" space and the classical concept of the physical forces are seen to be only partially applicable. The relations of space and time and the behaviour of matter and of forces in space have to be rethought in modern time. The analytical method has begun to tear down its own barriers and limitations. The outward expression of this step is certainly as yet a materialistic one, but philosophically at least, a door has been partially opened.

Steiner repeatedly asserted that the true spirit of our time is reflected in modern scientific research, by no means always in its results, but in its spirit of search for knowledge and in the intensity, objectivity and freedom of thought and endeavour involved. He nevertheless called urgently for the development of yet more modern scientific and mathematical methods with which to meet the challenge of the materialism of the scientific era.

Rudolf Steiner's Indications concerning Space and Counterspace

Through Steiner's tremendous life work and in all his teaching there runs like a golden thread the theme of the interplay of polarities and the resulting threefolding of the whole, the trinity. Of his own work Steiner himself said that the entirely new theme, set for the first time into the world, and upon which he worked for forty years before speaking of it, is the teaching about threefold man (7).

Steiner's descriptions in thought and in artistic creation are woven through with the deeper concept of polarity. This does not mean merely *contrast* in the usual sense of the word, but a *qualitative polarity* in a far subtler sense. It is this more fundamental aspect of polarity which Steiner is continually describing in manifold ways.

Throughout the earlier years of his teaching he did this in a more general way; in later years, at the time of the founding of the practical branches of his movement—education, medicine, agriculture, and the like—he began speaking of it in scientific and educational lectures in more specific detail, in connection with the task of transforming and spiritualising the materialistic theories of natural science itself. In this connection he often indicated the mathematical task in relation to the widening of the concepts of space and of force (45).

The scientific lectures in which Steiner spoke urgently of the need to find a more spiritual conception of space and of the forces of nature are often very difficult to understand; they are full of indications, but he left the detailed working out of the material he gave to the mathematicians and scientists among his audiences.

For example, in the lecture course in the Hague (6), during the whole of which he tries to awaken an understanding for the need for a new mathematical approach to the phenomena of life, Steiner describes the contrast between the root and the flower of the plant. It is typical of many instances in which he brings new spatial concepts to bear on the understanding of the plant and of the forces which endow it with life, rooted as it is in the earth and yet growing towards the light. He describes the contrast between the root and the flower of the plant, saying that one can of course see in the root its obvious relation to the three dimensions of space, whereas it is not possible fully to justify the form of the flower according to the same spatial concept. To describe the form of the root in mathematical terms, he says, one would begin at the centre of a co-ordinate system, whereas this does not do in the case of the flower. There, instead of beginning with a central point, one would have to *begin in the infinite and come inward towards the flower.*

Georges de la Tour (1593–1652)

Circle of Innermost light.

Here, as so often, Steiner uses the words "centrifugal" in describing the way the forces work which are known to physics as operating in ordinary space, and "centripetal" for the way in which he pictures the life-forces which work into organic growth. *In order really to understand what Rudolf Steiner means here, it is not enough merely to think of movements taking place in opposite directions. We are asked to picture polar opposite types and qualities of movement in either direction.*

If descriptions like the above are brought together in thought with the principle of qualitative polarities such as we learn to know it from modern geometry, we shall come far nearer to an understanding of the modern spiritual scientific concepts Rudolf Steiner is trying to introduce.

Untiringly, Steiner attempts to make the scientist aware of the need for more imagination in coming to terms with the laws of the organic world. He is repeatedly evoking the picture of an interplay between polar opposite processes. He called the forces of life "etheric" or "ethereal"; he spoke also of "formative", or "universal" forces, in contrast to the centric forces of the gravitational field, linked more closely to material substance. These universal or "peripheral" forces he saw working into the gravitational field, *but not originating in it;* they are not in the first place to be found in the minute centres or particles of substance, but in a non-spatial, or—to use his own terminology—a *counter*-spatial field, the forces of which permeate matter.

This type of field or sphere, which Rudolf Steiner recognised to be proper to the ethereal or etheric forces he called a "counterspace" ("Gegenraum"). He wanted the scientist to think, not only of the type of force which works from centre to centre—a gravitational or other physical-type centre—but to conceive of forces bearing exactly contrasting attributes. Such forces are not linked to some centre of density in the way that matter or the explosive force (considered now in the old-fashioned sense) is related to gravity.

The ethereal forces should be thought of as functioning in planes or surfaces and originating peripherally. They develop not in a space which can be grasped in thought as formed from a centre outward, as for example from the centre of the earth, or from the mid-point of a three-dimensional cartesian system, but in a space which originates in the vast periphery of cosmic space. The foundation for such a space is, in fact, the plane at infinity of earth space.

In the above-mentioned cycle of lectures (6), the following statement may be found: "If you try to imagine a space created as from the starry heavens inward, it will not be a space depending on the three dimensions, but one of which I can only give you an indication in a picture. If, as I did yesterday, I describe three-dimensional space by setting up three lines at right-angles to one another, I would have to describe this other kind of space by drawing everywhere some kind of forms which would indicate planar forces ('Kräfte in Flächen, die sich nähern') which move in from all sides of the universe towards the earth, working from outside, plastically, upon the forms upon the surface of the earth." Rudolf Steiner relates such forces to the "etheric body", or body of formative forces, which in any living organism and in man, permeates the physical body, and without which it would be given over to the disintegrative processes of death.

Modern science, intent though it is on understanding and fully mastering the forces of life, has not been able to do so. When the biologist has gone as far as microscope or ultra-microscope will tell him, he looks to chemistry and physics for an explanation. Beholding the wonderful forms and patterns of living bodies, he tends to assume that the rational explanation for what he sees will be found in the realm with which the physicist and chemist are also dealing; indeed he moves from bio-chemistry to bio-physics.

The biological sciences have benefited enormously from the discoveries of physics, and chemistry, but they have also been drawn inexorably in a direction which leads *away* from a real understanding of the sources of life.

Tradition has always been overcome by courage. The physicists have had the courage of their convictions, with extraordinary results. An equally courageous step is required by the biologist: to realise and admit that *a living body is alive precisely when it is not functioning solely according to the known laws of physics and chemistry,* but is under the influence of laws which save it from the fate it would undergo if it were—laws which actually counteract the disintegrating processes of the lifeless realm.

We will include one more quotation relative to our theme,

Rembrandt (1606–1669)

Sun-space in Earth-space

this time from the book *Fundamentals of Therapy,* upon which Rudolf Steiner was engaged shortly before his death, together with Dr. Ita Wegman, to whom he entrusted the development of his medical movement: "It is to the earth that every earthly substance, or earthly process, owes its forces of the kind that radiate outward. It has these forces in common with the earth. It is, indeed, only as a constituent of the earth-body that any substance has the nature which chemistry discovers in it. And when it comes to life, it must cease to be a mere portion of the earth; it leaves its community with the earth and is gathered up into forces that stream inward to the earth from all sides—from beyond the earthly realm. Whenever we see a substance or process unfold in forms of life, we must conceive it to be withdrawing from the forces that work upon it as from the centre of the earth, and entering the domain of other forces, which have, not a centre, but a periphery" (45).

In the Hague lectures (6), the *space* in which such peripheral forces must be seen to be at work is described as one in which the three positive dimensions of ordinary space are cancelled out, one dimension after another becoming negative, until at last a *dimensionless central point is reached in a hollowed out space—not an empty space but a spiritually filled space, and not an ordinary point, but a spiritually laden one ("geistbeladen").*

The Mathematical Concept of Space and Counterspace

In 1933, some years after Rudolf Steiner's death in 1925, Adams published first in German and then in English the first statement of a new scientific concept based on projective geometry and answering Steiner's demand for a truer mathematical approach to the concept of space and counterspace and the related laws. The publication is entitled *Physical and Ethereal Spaces* and *Von dem aetherischen Raume* (1). It puts into mathematical form Rudolf Steiner's conception of Space and Counterspace (Raum and Gegenraum) and the gravitational and anti-gravitational forces.

In the following year, the Mathematical-Astronomical Section at the Goetheanum, under the leadership of Elisabeth Vreede, (1) published Adams' very substantial work, *Strah-*

lende Weltgestaltung, which brings together Rudolf Steiner's spiritual science with projective geometry in a very wide context. A second volume, giving in detail the development of the polar concepts, was in preparation; for a variety of reasons, this volume did not appear.

Between 1949 and 1960 there appeared three works in which Adams, together with the present author, again brought forward his mathematical description of physical and living spaces and their forces, this time in relation to the plant and with particular reference to *Goethe's Metamorphosis of Plants.*

Louis Locher-Ernst (1906–1962), Director of the Technical College of the Canton Zürich at Winterthur, published in 1937 a book entitled *Urphänomene der Geometrie,* followed in 1940 by another entitled: *Projektive Geometrie und die Grundlagen der Euklidischen und Polar-euklidischen Geometrie.* Locher-Ernst, at this time working quite independently from Adams, but inspired by his own contact with Steiner's teachings, was working at the same task and coming to the same conclusions. In 1957, when the two workers in this field were in contact with one another again, Locher-Ernst published a textbook on the geometry of space and counter-space, entitled *Raum und Gegenraum* (46).

The work of the two men is very different is style. Locher-Ernst's books are restricted to the mathematical statement, and have a more academic character; they make their very important contribution by reason of their strictly axiomatic form. Adams concerned himself especially with the application of the new conceptions in various spheres of life and research, and he took his start primarily from his observations of phenomena. For him, the important thing about the development of the mathematics was its use as a tool in scientific investigation. He attempted, for example, to rethink present-day theories in several aspects of physics, and he was interested also for this reason in the development of the theory of numbers to accord with the modern geometrical concepts. Some initial work of George Adams in this field appeared in the journal of the Institute of Mathematics and Physics, under the direction of the mathematician and physicist Dr. Georg Unger in Dornach (47), and in one or two articles in the journal *Elemente der Mathematik* (Basel). It would certainly be true to say both of Adams and of Locher-Ernst that they regarded their work as only a beginning.

Physical and Ethereal Spaces

In speaking of counterspace, Rudolf Steiner used the terms "negative space" and "counterspace", in his descriptions relating to the "etheric" or "ethereal", and often in relation to the Sun. He spoke also of "Sonnenhafter Raum", and George Adams preferred to use such a term, with its more spiritual connotation: ethereal space, sun space.

We call "physical space" the normal, three dimensional space of our naive experience. It is the so-called Euclidean space arising out of projective geometry through the inclusion of the idea of an infinitely distant plane as a unique archetypal plane with an imaginary circle—a unique archetypal circle-form—in it. By "physical space" we do not merely mean a given physical sense-perceptible space, but rather the type of space we experience in our inner imagination through our life in a physical body. This is the space upon which the Idea of the physical, mineral world and the physical human body is based, built as they are according to the three dimensions of space. From this ideal point of view we may call this space "physical".

As human egos, incarnated in a physical body, we experience physical space from within. Geometrically speaking, it is a "pointwise" space. "Geometry" in the classical, Euclidean sense means: to experience through ideas the ideal world-form from within. It accords also with the laws of this physical world that the myriad beings in it must live in bright array side by side.

Over against this we set an opposite kind of space, the archetypal entities of which are planes. We call ethereal or counterspace a space in which, with respect to physical space, the function of points and planes is in every respect interchanged. The formative laws of this space are determined by a single point—a single cosmic point—analogous to the way in which physical space is determined by a single cosmic plane. It accords with the laws of the ethereal world that in it all is in a state of interpenetration.

Adams calls "Urraum" or "Archetypal Space" the three-dimensional space of pure projective geometry determined by the fundamental and archetypal relationships between Point, Line and Plane (18). In this free projective space there is complete equilibrium between point and plane. It is the geometrical archetype common to both physical and ethereal space. If then, in this projective space, one singles out a *plane* from all the others, calling it the plane at infinity and experiencing it as such (including also the imaginary circle within it), one changes (methodically) the structure of projective space in such a way that it becomes physical space, built out of points. If, on the other hand, one singles out a *point*, determining it to be an all-relating "infinitude within" (including in it an imaginary cone), one alters the structure of projective space this time so as to make it into the geometrical archetype of an ethereal space.

The unique, innermost point which determines "negative space" and the plane at infinity of "positive space" have been called by Ernst Lehrs the "all-relating point" and the "all-encompassing plane" (allbeziehende Punkt; allumfassende Ebene) (48).

If in physical space we speak of an infinitely distant "cosmic" plane, in ethereal space we conceive of an infinitude within—a "cosmic" point. The plane at infinity of physical space contains lines and points, which we have called lines and points at infinity. They are the lines and points of the heavens (Weltenlinien, Weltenpunkte). The "cosmic" point, or point at infinity within an ethereal space contains lines and planes, which for this type of space are lines and planes at infinity. This is no ordinary point, such as the central point of a cartesian system in physical space.

Our study of projective geometry has trained us to think of points as bearers of lines and planes, but unless we eliminate all traces of the old geometrical way of thinking, we shall not be able to experience what this ether-geometry means, and it will become abstract and senseless.

A point such as the infinitude of an ether-space, with the lines and planes it contains, takes on more the quality of a star than anything which has to do with the centre of three-dimensional space. The space which exists by virtue of such a star-point is formed by its surrounding planes. Such a star-point is not the centre of an infinitely large sphere; on the contrary, the nearer the enveloping planes hover towards the star-point, the "larger" the ether-sphere will become, which is formed by them. (The family of circles in Figure 2, Chapter VII, gives some impression of this in two dimensions.) We have conceived of a space created planewise and spherically and *not* radially, in which the star-point or "star-centre" is the infinitude within, towards which the inward-moving, space-moulding planes tend.

(It is in the nature of language that words have often to be used in different senses. The word "sphere" means a round ball. We speak of the infinite distance of space as spherical, because to begin with we experience it as such. Inasmuch, however, as we begin mathematically to recognise this infinitely distant periphery to be not in the nature of an immense ball, but a plane, we impart to the "plane" something of the quality of "sphere"—of something peripheral, and it is justified in our ether-geometry to use the word also in this sense. Much that we call "spherical" from a more cosmic aspect is indeed related to the ethereal nature of the plane.)

In physical space, we know the centre of the three-dimensional axes to be the *point of departure* for pointwise measurement outward towards the infinitely distant plane, the Absolute of space. In ethereal space this infinitely distant plane is the *plane of origin* of the ethereal planes, and their "star-infinitude" within is the Absolute for this type of space. The planes, as they depart from the one infinitely distant plane, tend in towards their goal; the "star" is the centre of a "hollow", spherical type of space. By "hollow" we do not mean "empty"; on the contrary, this is an *ethereally filled* space.

The geometrical concept of such polar opposite spaces, based on all we have practised in the preceding chapters, can become crystal clear; ethereal space must become entirely light-filled. The pictures we form of it in our thought and imagination must have shaken off the weight and density of earth forms and have become transparent, imbued with light. Passing over in thought from physical to ethereal spaces, we have to overcome our inborn pointwise, centric and radial way of thinking and reach to the opposite extreme.

In order really to experience the geometry of ethereal space, we must practise thinking of the plane as an entire whole. As long as we still cling secretly to the old way of thinking: "the plane consists after all of a lot of points", we shall not be able to experience the ether-geometry. One should not only try to experience the plane as an undivided whole; one should try to imagine oneself within it and as an ego-being to become completely identified with it. As an ego-being one should no longer be centred only in this or that point—no longer fixed to a centre but widened out into a periphery—into an outspread plane.

Once having learned this, however, we must look for the various modes of interaction of two such extremes. This is of the utmost importance. We shall perhaps be reminded of how the field of darkness and the field of light are in manifold interplay, creating colour. There are two dangers for the man of today; he may die into the darkness and the cold, or he may die into the brilliant light and the scorching heat. Only by taking the middle way can he live, for all life owes its existence to the interplay of great cosmic polarities.

Summing up once more we may say:

Physical Space is pointwise in character and it is centric. It has as its infinitude the infinitely distant plane, the Absolute of physical measure, with its imaginary circle.	*Ethereal Space is planewise in character and it is "spherical." It has as its infinitude a "star-point" within, the Absolute of polar "measure", with its imaginary cone.*

Once it had been realised by the earlier mathematicians that underlying the rigid space of the earth with its Euclidean principles of measurement, there was the more mobile, metamorphic space of projective geometry with all its point-planar symmetry, it was felt to be unsatisfactory that there was no polar equivalent to the plane at infinity itself, and the specialised metric arising from it. We mentioned that other non-Euclidean spaces are derivable from modern geometry. Professor A. N. Whitehead called one of them Anti-space, and there are occasional references in the earlier literature to the kind of space we are now describing, among all the other possibilities, only to be left aside as seemingly of no particular interest or use (49).

Yet this is precisely the kind of space which we need, to give ideal interpretation and clarity to the form-phenomena of life and growth. It also supplies the mathematicians with their missing counterpart to the plane at infinity of Euclidean space. To quote Locher-Ernst in *Raum und Gegenraum*, George Adams was the first to perceive clearly how to formulate mathematically the concept of a counterspace, including also the nature of its measure.

It is important to realise that for physical, Euclidean space the Absolute is permanently given, and that any point in this space may be regarded as a centre from which to establish a cartesian system, such as the three-dimensional axes with their planes at right-angles to one another. *The Absolute point of an ethereal space, however, may be situated anywhere in physical space,* for ethereal spaces permeate physical space, wherever life is to be found; and these ethereal spaces all share the same planar origin—the infinitely distant plane of *physical space*. It is a magnificent reciprocity. Within physical space, which is fixed under the firmament of the plane at infinity, the mobile ethereal spaces come and go in space and time, wherever life flourishes or decays. *The ethereal or sun-spaces have the infinitude of physical space as their origin and source, and a particular "star-point", embedded somewhere within this space as their Absolute—the infinitude into which they spend their living forces.*

In physical space:
One Absolute cosmic plane with which many points (earth-centres) are related. Point-centred forces radiate outward towards a peripheral infinitude.

In ethereal space:
Many Absolute "star-points" with which one cosmic plane (ground of worlds) is related. Peripheral forces are received inward by an infinitude within.

This is a wonderful concept; in fact, it is true to life. In physical space, which is spanned as it were beneath the cosmic canopy of the star-strewn heavens, the ether-spaces come and go in constant flow. They are present wherever life flourishes; they disappear when life ebbs away. The infinitude of physical space is the origin or foundation of ether-spaces. Some particular "star-point" which becomes embedded in physical space is the Absolute of an ether-space and the infinitude into which it pours its living, ethereal forces.

We will include here some sketches which should help to make these first steps of the geometrical concept clearer. If we want to make some geometrical construction according to Euclidean laws, we take the line or plane at infinity as given, and unless we are making a perspective construction (as for example, in the harmonic nets), we may not draw it in. When, however, we begin to construct the counterpart in ethereal space, we must denote the situation of the infinitude within; this we will do with a star. We will take the first few steps in two and then in three dimensions, positive and negative, and we shall soon recognise some familiar figures.

Two positive dimensions: *Two negative dimensions:*

Figure 1

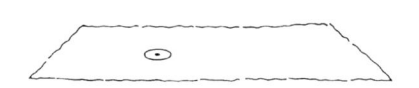

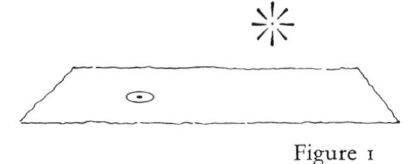

A point in a plane.   A plane in a point (Figure 1).

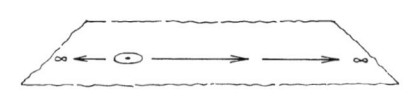

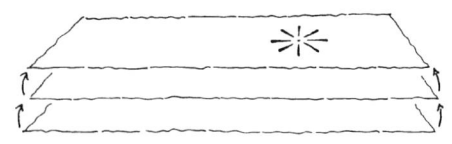

Figure 2

The point moves towards the Absolute plane.   The plane moves towards the Absolute point (Figure 2).

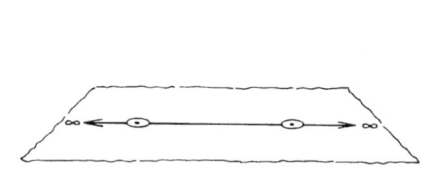

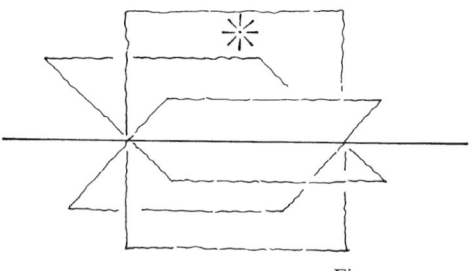

Figure 3

The common line of two points has a third point in the Absolute infinitude without.   The common line of two planes has a third plane in the Absolute infinitude within (Figure 3).

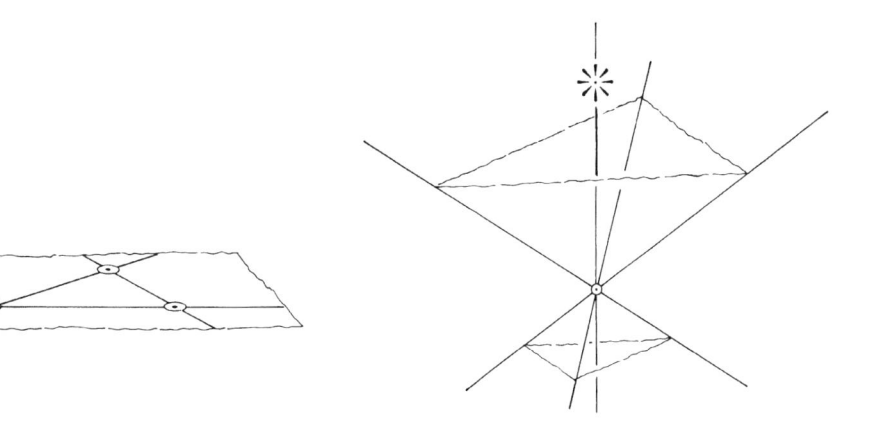

Figure 4

A triangle in the plane; each of its lines has a point in the line at infinity of the plane.

A trihedron in the point; each of its lines has a plane in the line at infinity of the point (Figure 4).

Considering Figure 4: In the left-hand sketch, each line of the triangle has a point in a line of the Absolute plane of positive space. Dual to this (right-hand sketch): each line of the trihedron has a plane in a line of the Absolute point of the negative space. These three planes (which are not included in the sketch) interpenetrate one another in the line at infinity within.

We know from our previous studies (p. 229) that corresponding to the triangle's centre in the positive, extensive field of the plane, the trihedron, in the negative, intensive field of the point, has a "central" or *median* plane. The positive space movement of the triangle's points outward towards the infinite, is answered in the negative space picture by an angular sweep of the planes towards the star. Alternately, the contraction of the triangle into its central point accords with the angular sweep of the three planes of the point away from their Absolute and towards their median plane. As the one shape dies into the point of contraction, the other dies into a plane of expansion.

The circle in the geometry of the plane and the cone

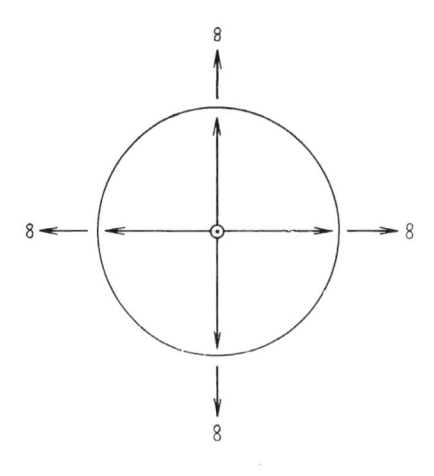

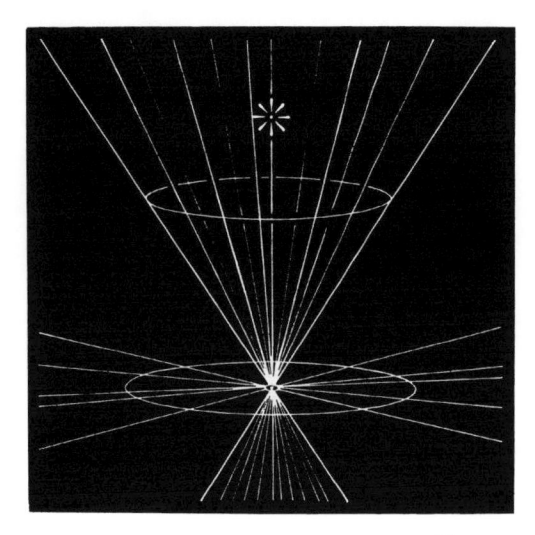

Figure 5

in the geometry of the point, are typical forms expressing an archetypal polar phenomenon. The circle is poised between its centre and its Absolute, and the two-dimensional cone is poised between *its* "centre" (the median plane) and *its* Absolute (the verticon). We might call these two-dimensional spaces: the "space of the ring" and the "cone space". As in positive space, a vector is the directed distance between points (pictured typically by an arrow), so in negative space a vector is the angular sweep between planes, for which the arrow is *not* an adequate expression (Figure 5).

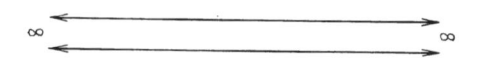

Figure 6

In positive space:
Two lines are parallel if their common point is in the Absolute plane.

In negative space:
Two lines are "parallel" if their common plane is in the Absolute point (Figure 6).

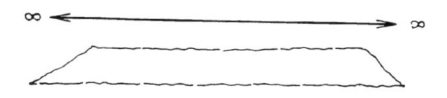

Figure 7

A line and a plane are parallel if their common point is in the Absolute plane.

A line and a point are "parallel" if their common plane is in the Absolute star-point (Figure 7).

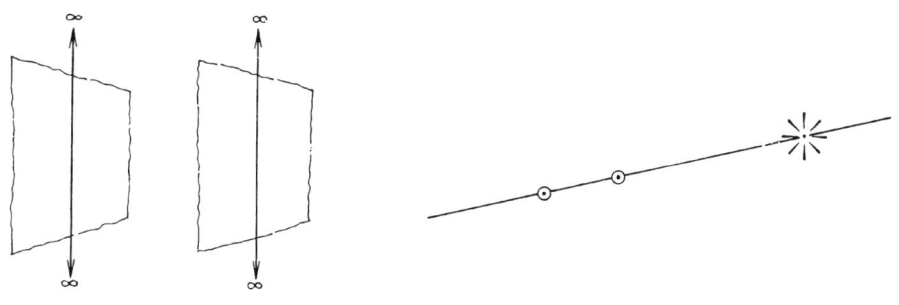

Figure 8

Two planes are parallel if collinear with the Absolute plane.

Two points are "parallel" if collinear with the Absolute star-point (Figure 8).

The question will arise: is it sensible to use the word "parallel" here, still more, when it comes to speaking of "parallel points"? Locher, for instance, sets "centred" points over against parallel planes. Going back, however, to the origin of the word "parallel" in the Greek, we find that it actually means "together" or "side-by-side". In this sense, Adams' use of the word is justified; it emphasises the symmetry of the concepts, while avoiding the use of the word "centre", which is typical for earth-centred space.

Three Positive and Three Negative "Dimensions"

In the illustration (Figure 9) we have cut a cube into equal octants; it has its centre at what we shall in the corresponding picture call the Absolute point or star. Each of the eight cubical octants shares with the others the centre of

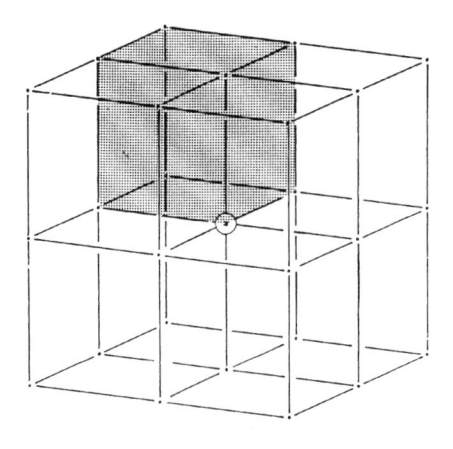

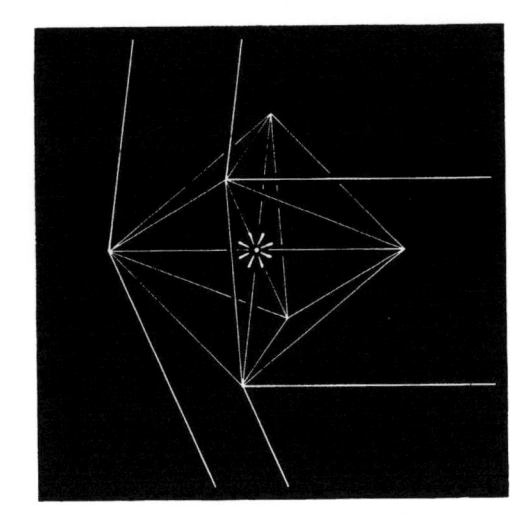

Figure 9

the original cube as one of its corner-points and they all have the Absolute or celestial plane in common.

In the corresponding negative counterpart (Figure 9) we have an octahedron, the star of which we have put at the point which for the cube was its central point. The octahedron's median plane (that element which corresponds to the mid-point in physical space) is infinitely distant—in fact it is to be found where, in the physical space of the cube we would look for the cosmic plane. We have only been able to draw in part of one of the eight octahedral octants, for, unlike their cubical counterparts, which keep themselves nicely to themselves, although they share the centre of the form, these octants open out into space and interpenetrate each other on all sides. Each octahedral octant shares the celestial plane (the octahedral median plane) with the seven others, using it as one of its planes. All the octahedral octants have their Absolute, the star-point, in common, while the cubical octants have a similar relationship to their cosmic plane. In the positive space picture, the large cube shares one of its corner-points with a corner of one of the octants; in the negative space form, the original octahedron shares one of its planes with a plane of one of the octants.

It is characteristic of the cubes that they fill a space from a central point outward with regular step-measure, leaving empty the space which divides them from their celestial Absolute. The octahedron picture is very different and would

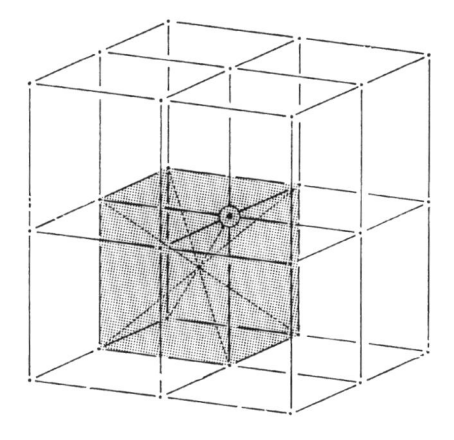

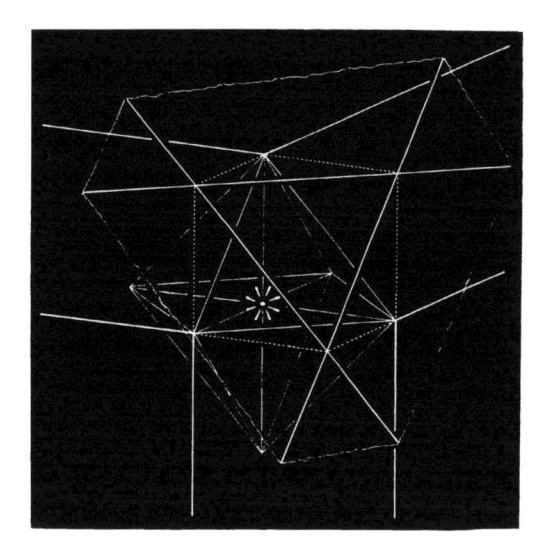

Figure 10

be still more so, were we to draw in all the octants which interpenetrate one another on all hands. Their planes weave throughout space, from the celestial plane inward. One of the planes of each octant moulds from without and hollows out the innermost octahedral space, which is set like a jewel around the Absolute star of this ethereal type of space.

The following illustration (Figure 10) shows the correspondence of the mid-point of one of the cube octants with the median plane of the octahedral octant.

The four diagonal lines joining opposite corners of the cube octant determine its centre. Correspondingly, the four lines common to opposite planes of the octahedron octant determine its median plane. One of these four lines is the line at infinity of the median plane.

If we bring these constructions into movement in the imagination we shall gain an entry into the different qualities and resulting measures of the two polar opposite space-creating processes, the pointwise and the planewise. The cubic-octahedral polarity is of course only one formal expression of this positive-negative polarity. In the sense of Rudolf Steiner's expression: "sphere from within, sphere from without", we must experience the characteristic qualities—centric, radial and convex on the one hand and peripheral, plastic planar and concave on the other.

These are elementary examples. It is of course by no means

necessary to place the Absolute of an ethereal space in the centre of a physical one, as is the case in the two examples shown. The centre of the earth may well be thought of as the centre of a great archetypal ethereal sphere, inasmuch as the Earth herself shares in the cosmic forces of life. In general, however, just as seeding centres or germinating spaces are to be found throughout earth-space in the living kingdoms, so too, in our geometrical constructions, we may choose to set the star-centre at any place we please. A moment's reflection will tell us what a vast field of study it is upon which we have entered, with all the possibilities of metamorphic interplay of formative spaces.

Gravitational and Anti-gravitational Forces

The study of the laws of this negative Euclidean type of Space, with the planar movements involved in its forms and transformations, gave Adams access to a theory of the anti-gravitational forces, of which Rudolf Steiner always spoke in this connection. If the physical forces are measured according to the thought-forms of point-centres, areas and volumes, the ethereal forces must be described according to their own nature. This is a realm of research which we can only touch upon here (50). In describing the opposite types of force in the two types of space, Rudolf Steiner used the expressions "Schwere" and "Leichte". We have our English words "weight" and "lightness", or "gravity" and "levity". The latter is a word with connotations which we do not wish to include here, but we nevertheless use the word. It must be understood that "lightness", in the way we use it here, does not merely mean the absence of weight; it qualifies a *force*. The shot from a gun may have the quality of lightness in contrast to a heavy cannon-ball; but both are in the same physical space and obey the physical laws—*both centrifugally and centripetally.*

The ethereal forces also operate in these two directions, but with an entirely different effect; they are linked to phenomena for which the word "plane" gives a truer picture, in contrast to "point". The ethereal planes hover inward *and* outward. For lack of a better word, Rudolf Steiner often used the word "suction" to describe the activity of the ethereal type of force. The ethereal planar forces have a moulding, formative power, and at the same time they draw or suck the substances which have come under the sphere of their influence away from earth gravity. The levitational force is polar in all respects to the force of gravity.

It is a problem that we have as yet no words with which adequately to describe forces for which science has so far no concepts. We need not apologise however for coining new words or phrases in order to express new scientific concepts, for this is indeed a prerogative particularly of modern science and technics today. We do however plead that the terminology we use be taken in the strictly mathematical and scientific sense in which it is used. Mathematical concepts have a regulative influence on scientific thought, and once the balanced duality of spatial theory known to the pure mathematician has penetrated to the scientists, questions as yet unanswered—or not yet even asked—will no doubt find their answer. However fine and rarified a type of force the scientist today discovers, his mode of thought, although it may have allowed him to progress from the concept of "particle" to that of "wave", still dictates the idea of force in terms of radiation from a centre outwards.

The scientist needs the mental equipment to enable him to think of a type of force borne upon a vast or even infinitely extended layer or plane, or originating from some outer layer or skin and penetrating into an interior space or cavity. He should, to say the least, have another peg upon which to hang his question, let alone his answer. He will find that in the universe there are potent forces in polar antithesis to the centric ones—gravitational, electro-magnetic and the like. In observing phenomena, a man naturally notices what he is wont to think, and things escape his notice even if he sees them, if the idea that is in them is foreign to his mind.

The influence of modern geometry may be seen in the works of a number of contemporary artists, eg—Naum Gabo, Antoine Pevsner, Richard Lippold. Above all, Barbara Hepworth, especially in her later works, brings to expression the "non-spatial" spaces.

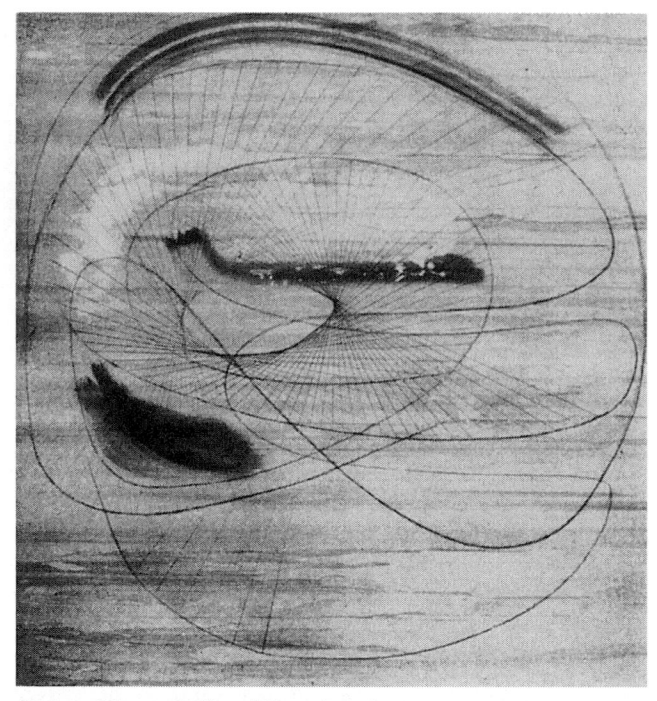

Barbara Hepworth, Curved Form (1956)

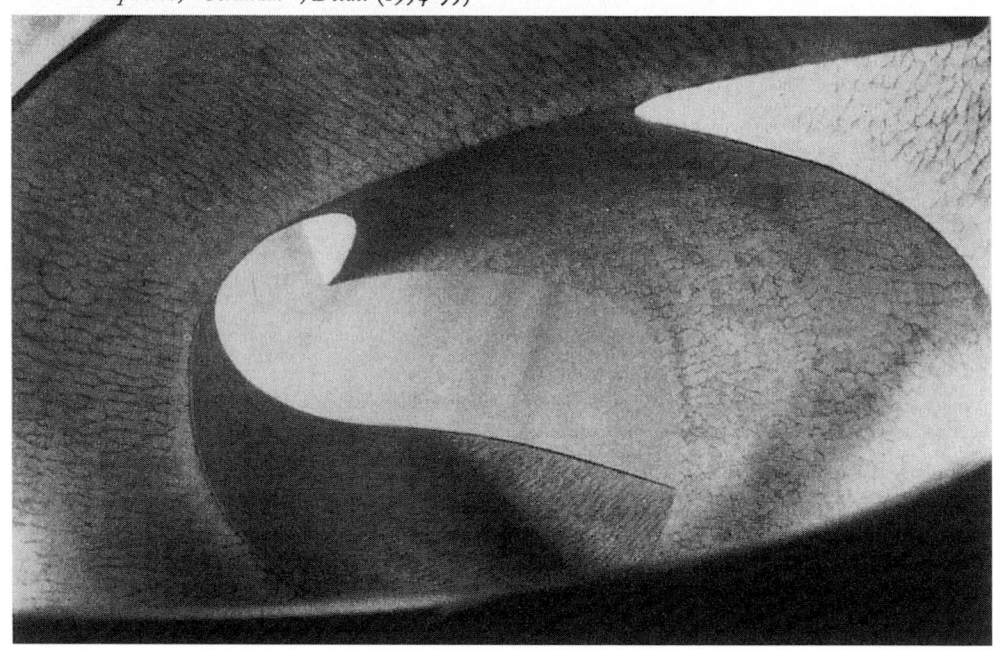

Barbara Hepworth, "Corinthus", Detail (1954–55)

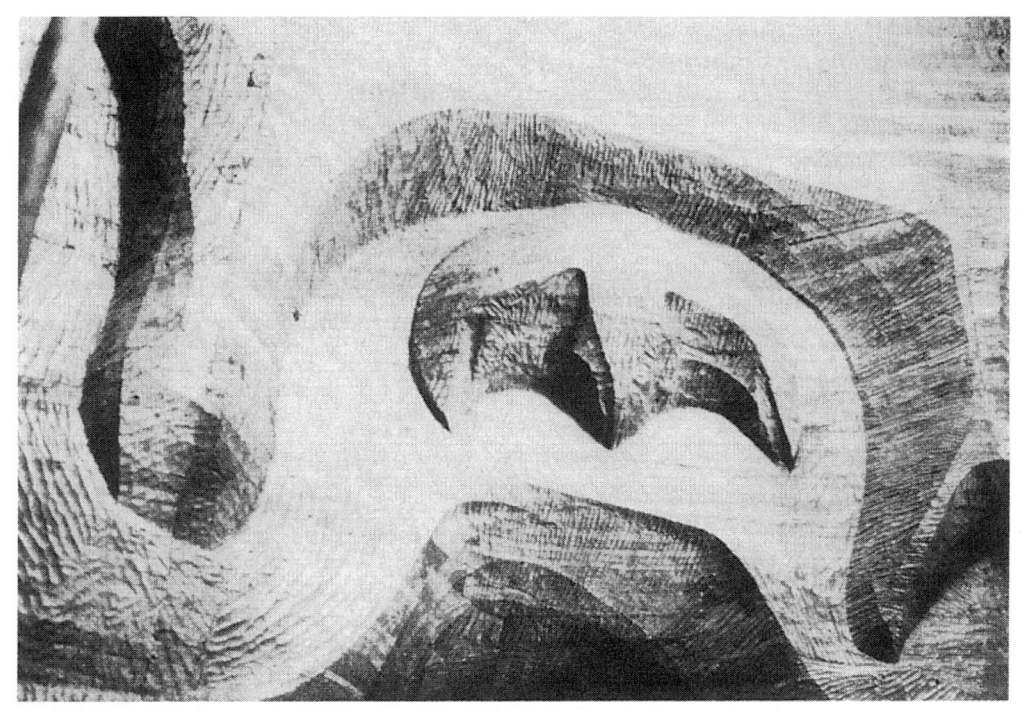

Rudolf Steiner, Architrave (Detail) in the small cupula of the first Goetheanum (1915)

Rudolf Steiner described the doubly-curved leminscatory surface as being "the archetypal phenomenon of life"

Sun and Earth

> *Suchst du das Grösste, das Höchste,*
> *Die Pflanze kann es dich lehren.*
> *Was sie willenslos ist,*
> *Sei du es wollend—das ist's.*
>
> Schiller

The tremendous advances in the technical sciences of quite recent years have presented the researcher in many fields with phenomena of which the earlier scientists had no idea. Here, as well as in other respects, mankind is crossing a threshold of discovery, with all the accompanying upheaval of tradition. We have already remarked that courage is required, and that it is indeed to hand; the scientist has penetrated far into the universe of stars and deep into the mysterious interior of matter. So far, however, he has taken his earth-space with him, and made his observations on the basis of his analytical formulae, and of thoughts applied to material phenomena alone.

In the material consciousness man sees only the reflection of what is really an inner light in all phenomena of sense-perception. Allied to material phenomena alone, thought is held spellbound by its own shadow. Thought, conscious only in the immediate wake of sense-perception, is like the wave that breaks when dashed against the shore, shining as foam in the moment of its destruction. But thought is actually ethereal in origin—it is *of the light;* and it must awaken in its own primal element, for man is about to awaken to the ethereal light that pervades the life of surrounding Nature, and his own life. It is to the courage of Rudolf Steiner that modern man owes the pathway *in thought* which transcends material consciousness and it is in the very progress of science that he will take this step.

Rudolf Steiner made extreme demands on the mobility and openness of mind of his hearers. He first introduced the idea of a counterspace probably in January 1921, and applied this concept to the nature of the sun (51). From then on, he often described the sun as polar in the universe to the earth, both spatially and in regard to its forces, saying that it would be nearer the truth to picture the sun not as a great ball of matter, however gaseous and refined, but as a place in the universe which is like the focus of a hollow type space with the forces characteristic of its nature—in fact, a counter-space, as we have attempted to describe it here. If the earth is a material focus, the sun is an ethereal one.

"What we call the physical constitution of the sun cannot be understood with ideas borrowed from life on the earth. We gain no access with these ideas. The only way is to find adequate concepts to match the results of experimental observation, which, up to a point, are certainly quite telling. . . . You can picture that the so-called interior of the sun is of such a nature that the phenomena it manifests are not pushed out from the centre, but that the processes take effect from the corona inward to the chromosphere, atmosphere, photosphere; instead of taking place from the inside outward, the direction is from the outside inward. The processes. . . take place inward, becoming lost, so to speak, as they tend towards the centre, just as what goes out from the earth loses itself in outward-tending spheres. . . . Only by entering into it qualitatively in this way, and by being ready to develop in the widest possible way a kind of qualitative mathematics shall we make any progress."

In many ways and in the deepest of senses, Rudolf Steiner tried to awaken understanding for the spiritual interaction of the forces of the sun with those of the earth. The sun is the giver of life on the earth, yet as external sun, it is also the great dealer of death. We must seek to understand the life-giving power of the sun, not externally, but in more spiritually scientific modes of thought *here on the earth* adequate to the purpose. In its spiritual aspect, the sun gathers into itself the cosmic forces of fixed stars, constellations, planets . . . to endow the earth and the beings of the earth with all the varied qualities of life. For where life of any kind finds its place upon the earth, there, in that place, we may look to find a sun-space—a focus like a living hearth—receiving and pouring forth again into manifold developing forms, the spiritual forces of the sun.

In the spring of 1947, George Adams was walking in Regents Park in London, as the hedges were bursting into leaf. Seeing the budding green of the young shoots, the thought suddenly struck him that *this is* an ethereal space,

Figure 11a: *Rhododendron*

Figure 11b: *Canna indica*

according exactly in its gesture of form with the geometrical concept we have called the "cone-space"—the two-dimensional negative space cone.

The morphology of the plant, less complicated than the animal form with its ensouled development, shows clearly and simply—once one has lived long enough with the concepts which help one to understand the gesture of its unfolding forms—that *it grows out into earth space from an ethereal space, the space of the shoot.* Once rooted in the earth, and while growing upward and getting physically stronger and more robust as it does so, the plant actually develops from above downward. Unfolding from the growing tip downward towards the earth from which it springs, *it creates with its own organs the ethereal space in which it grows.*

We will describe this phenomenon as George Adams saw it, largely in his own words (52).

It is a wonderful paradox of Nature that the upward-shooting plant brings forth materials and forms which both in use and in appearance are proverbial for their radial, penetrating power, yet there is little of this quality in the way they first come into being. The upright stem does not *thrust* its way into space like an arrow or spearhead. The upward-growing power of the shoot is indeed one of the mightiest phenomena we know, and the eventual outcome of it is a thing of strength in the realm of earthly pressures and tensions—formed into pillar and pile, spoke and ramrod for human use from ages past. Yet it was not with this earthly-radial quality that the growing shoot made its way up and outward.

Describing it exactly as we see it, the typical phenomenon at the growing-point is the very opposite of a spearhead. The growing tissues are delicate and watery; what we behold at the tip of the growing shoot is concave and not convex; it is a hollow space we nearly always see. The actual growing-point of the stem is deeply hidden amid the young enfolding leaves. Their gesture is as if to guard, there in the innermost of the "empty" space between them, a hidden treasure with protecting hands (Figure 11). The youngest leaves reach upward, sometimes singly, sometimes in pairs and very close together, sometimes in whorls forming a hollow cone, first deep and steep, thence gradually opening and flattening.

At first each single leaf is concave on its inner side, making the hollow space in some cases more conical, in others more spherical and cup-like, as in Figure 12.

Leaf after leaf, whorl after whorl with further growth expands and comes away, opening more or less towards the horizontal; meanwhile within them other, younger buds have grown to take their place. So long as the shoot is growing, the gently guarded hollow space is there.

This concave gesture of upward growth is an essential feature of the impression we receive from the green plants that bedeck the Earth around us. The plants live by the light coming to the Earth from the Sun, from cosmic spaces. Pictorially, it is as though each single shoot were reaching out to receive and hold its portion of the light. Leaf and leaf-bearing branch, as they grow older, tend out towards a planar and even horizontally flattened form (Figure 12).

With the unfolding of leaf and branch is associated another quality which we perceive and feel in the phenomenon of plant-life above the soil, though science hitherto has lacked the corresponding concept. The leaves, as we said, tend to unfold towards a plane. They are, in a sense, *planar organs*. It is not only the crude quantitative fact that they develop a far greater surface-area than thickness; in their whole quality, function, and morphological gesture they reveal that the character of "plane" belongs to them, just as the character of "point" belongs to every earthly object by virtue of its mass and weight—namely, its centre of gravity. In countless instances, the fully opened leaves of plants—often the branches, too, which bear them—make manifest the plane, or rather, countless planes, one above the other. We see it when the sunlight falls through the young leaves in the beechwoods in May and June. A myriad planes seem to hover in the sunlit air. The impression we thus receive from the outspreading leaves is one of buoyancy and lightness. They seem to be upborne.

At the tip of a vegetative shoot the nodes and internodes are crowded together; often the leaves of many nodes combine to enfold the hollow space above the growing-point. Internodes lengthen out quickly as the plant shoots upwards, but at the tip a younger sequence of unfolding leaves maintains the form. They open up and away, maintaining the

Figure 12: *Field Cress (Lepidium campestre)*

Figure 13a: Wild Rose (*Rose canina*)

inner space in continuity, like the streaming flow of water through a vortex.

When in its further metamorphosis the plant comes to flower, the gesture of a hollow and enfolded space is all the more enhanced. The flower-bud enwraps a space more tightly closed, and when we see it open to a flower it is as though the space were now poised in silence. What the growing tip of the vegetative shoot suggested in an ever-changing form—in an enfolded space, ever unfolded and renewed again from within—this is now brought to rest in the flower-chalice, maintained as long as the blossom lasts. And in this "chalice", something hitherto unmanifest about the plant is now revealed. One is inclined to say: if the hollow space tended by the young unfolding leaves, going before the apex of the stem as it grew upward, was not just emptiness but had a deeper meaning, its presence indicating a real sphere of forces scientifically still to be defined, then in the flower something more of this ideal space has been made visible. What hitherto induced the upward and unfolding growth, yet in its quality remaining latent, has now revealed its essence in another way. The material, sense-perceptible part of the plant has united with it more deeply than hitherto (Figure 13).

Figure 13b: Mallow (*Malva Sylvestris*)

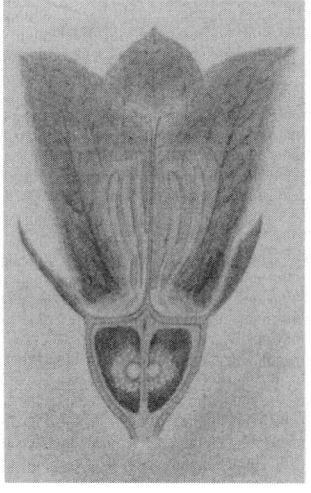

Figure 13c: Harebell (*Campanula rotundiflora*)

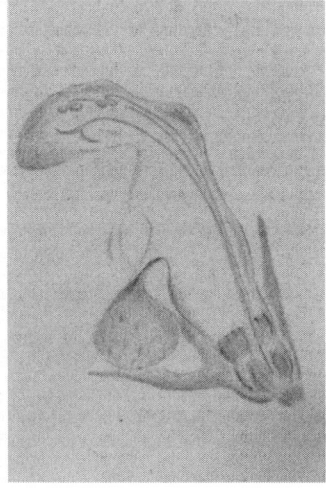

Figure 13d: Dead Nettle (*Lamium album*)

The flower opens in glory, showing forth the plant's individual being and essential beauty. It is as though, as the plant grew, there were a focus of life and growth which was not at first claimed by the material, watery-earthly body. Receiving from the forces of this focus, the plant tends up towards it, enveloping it with its green leaves, which unfold and come away from it in turn. The flower then envelopes it more closely, pausing, and seeming to come into a nearer relation with whatsoever has been hidden here—making it manifest in colour, form and fragrance.

In the flower, too, the petals often open out to a plane or even curl back beyond it. Yet it is the form of the chalice which is typical of the flower. It takes on innumerable variations and metamorphoses; sometimes the flower opens to a shallow hollow, as in the wild rose, or it deepens to a bell or a tube or even appears in the metamorphosed forms of hood and horn. In all these forms, the archetype is the enveloped inner space, the chalice of the flower.

This hollowed space of the blossom holds and hides more within it than did the space of the green shoot. In calyx and corolla, the metamorphosed leaves circle around the innermost heart of the ether-space, as though to emphasize the peripheral and unified essence of the sun-space, which now they envelop most intimately.

The transition from leaf-bearing shoot to fully open flower is clearly like the crossing of a threshold. The plant now comes into a new and more qualitative relationship to its sun realm, the cosmic sphere containing the sunlike star—the infinitude within—which bears its archetype. The flower reveals in the glory of beauty, something of the quality of the beneficent virtue, which at last will be incorporated into the fruit and stored in the seed as promise for the following year.

In the hollowing space of the blossom, the plant almost completely withdraws from external space and meets most intimately the focal realm of hidden forces, concentrating the potency of its being in the innermost realm of the seed. When the flowering process and with it the "fertilisation" is complete, the ovary grows right into the living focus, or draws the virtue of it down into its substance. When at last the plant comes to fruit and seed, it has united its earthly substance with the ideal sphere by which it was endowed

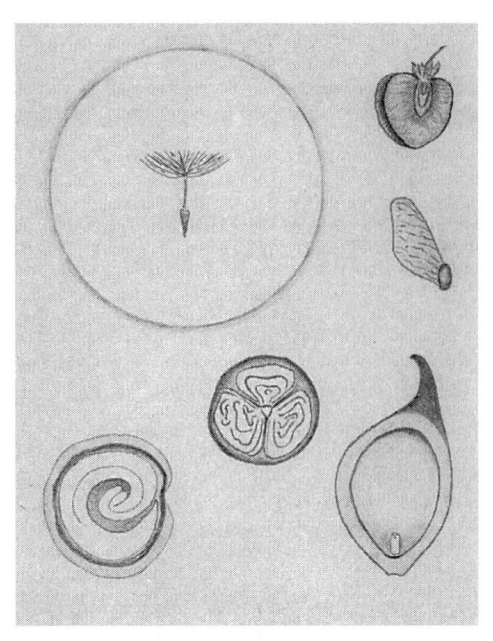

Figure 14a: *Taraxacum officinale* (top left). *Spergula arvensis* (left below). *Convolvulus arvensis* (centre). *Rumex acetosa* (top right). *Acer campestre* (centre right). *Ranunculus repens* (right below).

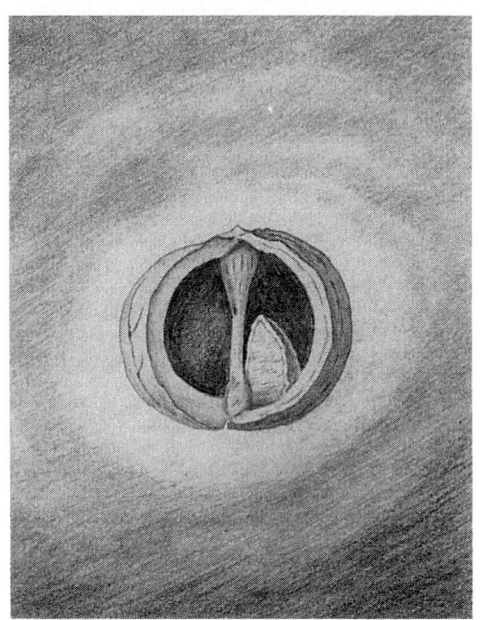

Figure 14b: Brazil Nut (*Bertholletia excelsa*)

throughout its vegetative life. Now the sap and growing tissues fill the whole volume of the fruit; the "Apple" is formed, spheroidal fruit-forms in innumerable variations of their archetype, the sphere. Heavily laden, the shoots are weighed down with their fruity burden, and the seeds are formed, bearers of the focus of cosmic life for future generations. In seed-vessels, the plant brings forth all manner of shapes, winged or feathered for flight through the air, or hard forms, sometimes reminiscent of crystalline formation, strong to contain the germ or the spiral of living substance within (Figure 14).

Without such "star" or "sun" spaces within the spaces of earth, there would be no life upon the earth. In the higher plant we see such spaces functioning for the greater part in the light and air, showing forth unending metamorphoses of outer form. The upward-spiralling, foliage-bearing part of the plant is the true plant, its mercurial, healing qualities, as symbolised by the ancient "Staff of Mercury", being the harmonious outcome of the interplay of light and darkness, sun and earth. Here the plant is two-dimensional in its ethereal space. Below, in the earth, together with its salty nature, the plant roots take on earthy, three-dimensional shapes; above, opening to the spheres of cosmic light and sulphur, the plant *negates all earth dimensions,* telling of the realms from which it comes.

These dimensionless spaces of life are more of the quality of Time than of Space. One might call them "Time-Spaces". They come and go amid the stream of cosmic rhythms which play around and through the Earth.

Rudolf Steiner often used the figure-of-eight or "lemniscate" to illustrate his descriptions concerning the interplay of polarities in all manner of ways. This figure may of course be traced by the locus of a point. It may, however, also be constructed in the interplay of two families of circles in growth measure, in which case the curve arises as the uniting link between polarities and the two loops and their foci are polar opposite in quality. This is not a projective construction, but it is a true picture of the interplay of the cosmic polarities with their breathing reciprocity (Figure 15). In making such a construction, other curves arise also, which,

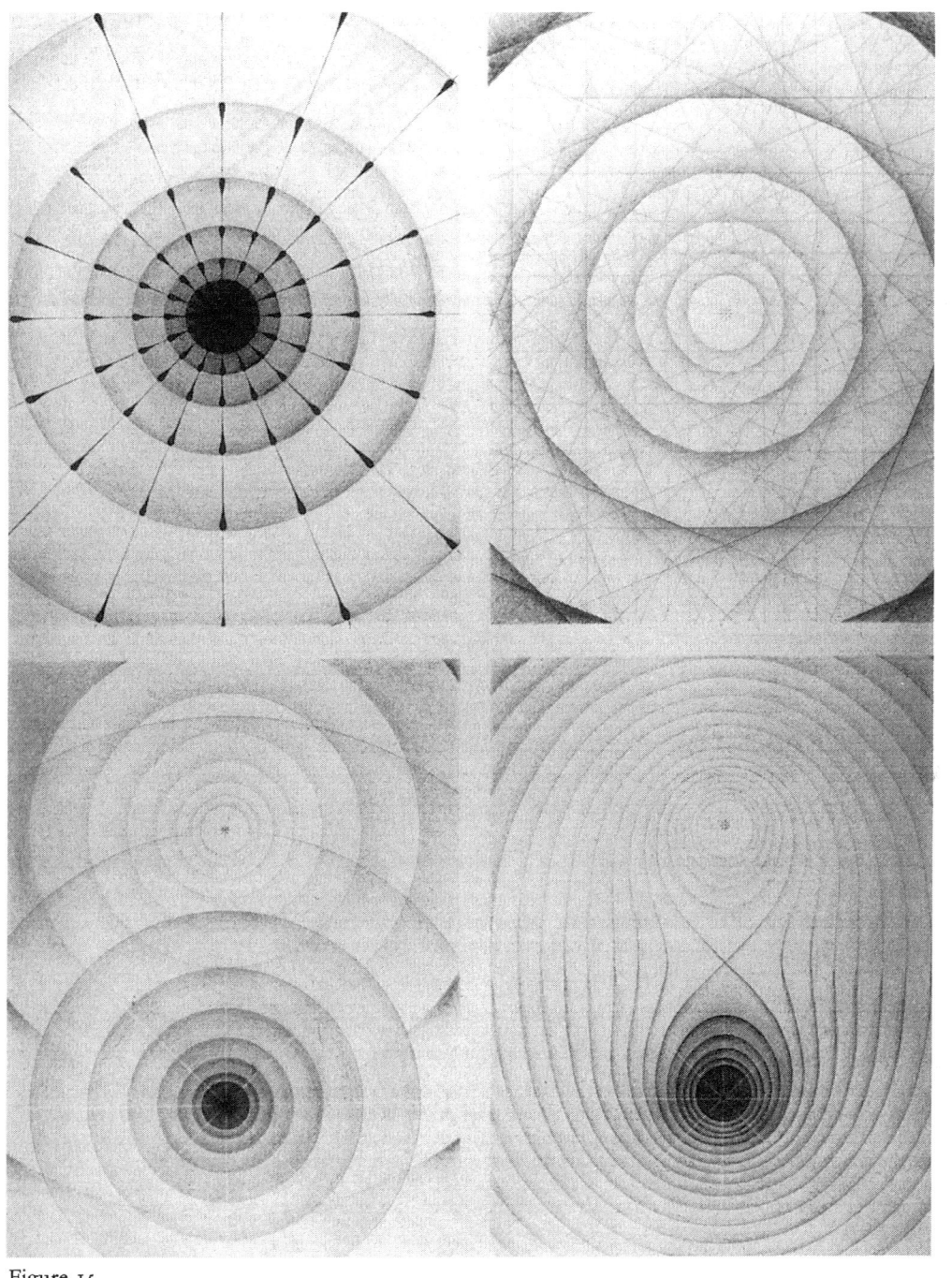

Figure 15

together with the lemniscate itself, form a whole family. These are the so-called curves of Cassini (53).

This two-dimensional picture of interpenetrating circles may be developed to represent a full spatial-counterspatial process, if one imagines the interpenetration of "physical" and "ethereal" spheres. This is not merely a conceptual picture of a form but of a creative process arising between two opposite poles. In the process of its unfolding, it reveals gestures of form which call forth in the imagination thoughts and concepts which are invaluable in the attempt to approach with some understanding what takes place between the physical and the ethereal.

Bring the two centres or foci together into one point, and we still have the lemniscatory process *dynamically* present, although spatially the picture is simply the picture of concentric spheres—"sphere from within, sphere from without". The physical and ethereal spaces will in this instance have a common centre and also a common plane; they are co-centric and co-peripheral, as we saw it with cube and octahedron.

This picture of a lemniscatory space is spiritually of the utmost importance, and it is a significant exercise, to make the necessary turning inside-out involved in picturing the transition from one space into the other. It is the kind of change-over which all beings have to make who come into earthly space from the living spheres of the spiritual, and then one day pass out again beyond our sight, from the world of visible phenomena to the invisible world of spirit.

The ethereal- or sun-spaces in living forms on earth are not of lasting duration like the rocks of the earth, they come into being and pass away again. Moreover, although they usually manifest as a space which is to some extent rounded off and enclosed within physical limits, they are not in reality closed but open to the cosmos. If one speaks of boundaries or limits to these forms, one might say that these are not to be sought in the sheaths by which such spaces are created—leaves, petals, living skins or membranes, but in the plane at infinity of our space; and they hold their infinitude within them. The sun-spaces in living forms are only seemingly enclosed; enveloped by living organs, they are open to the spiritual forces of the cosmos.

The thoughts and illustrations which follow are simply indications of how the archetypal concepts arising from the basic ideas of a new morphology may be applied in other spheres. Needless to say, as put forward here, these are no more than indications, but they concern realms of serious research.

In the living processes of an ensouled form, the "star" indwells the vital, watery substances, influencing the far more complicated foldings and enfoldings of animal tissue. This morphological field, especially in the early stages of embryological development, or in the development of individual sense organs, such as eye or ear, may well be approached in the light of these concepts. Figure 16 is a drawing by Haeckel from his *Anthropogenie*, it pictures a human foetus of five months in the enveloping membranes.

Even the waters and the meteorological phenomena of the earth herself reveal in their spiralling vortices the outward expression of the presence of ethereal forces, where in the streaming media the enfolding surfaces create hollows which act as focal regions for their reception. Figure 17 shows the delicate forms made by the boundary layers in the water of a vortex rising vertically in a tank of water (viewed from above). Figure 18 is a photograph, taken from the Satellite Kosmos 114, of two spiralling storms above the Indian Ocean. It is from the rhythmic flow of watery movement and not alone from particles of earth, that all finished form arises; and in the forms which life has left behind, we see so often petrified pictures of that watery movement and the ethereal force-spaces long departed. Figure 19 shows beautifully the spatial-counterspatial forms of a shell (54).

In the higher spiritual processes of man, too, such spheres must function—in head and heart—in which the warmth and light of the spiritual Sun may be reborn, if he so wills it. Into that consciousness which sees in all clarity the material forms of earth, there must dawn the light of the Thoughts which brought about creation. Through his own individual endeavour man today must awaken to the spiritual nature of thought—to those thoughts of which he can truly say: "It thinks in me".

In the ancient mystery centres of the East, men learned to descend into the grave in order to live with the Cross. In

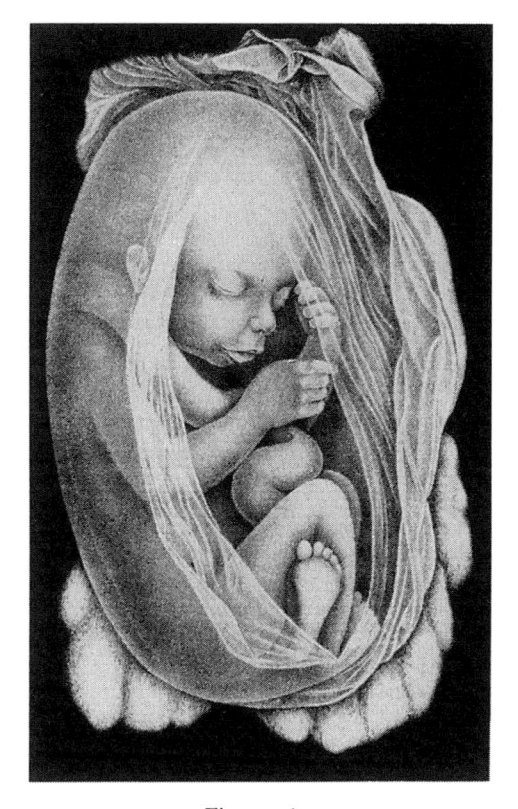

Figure 16

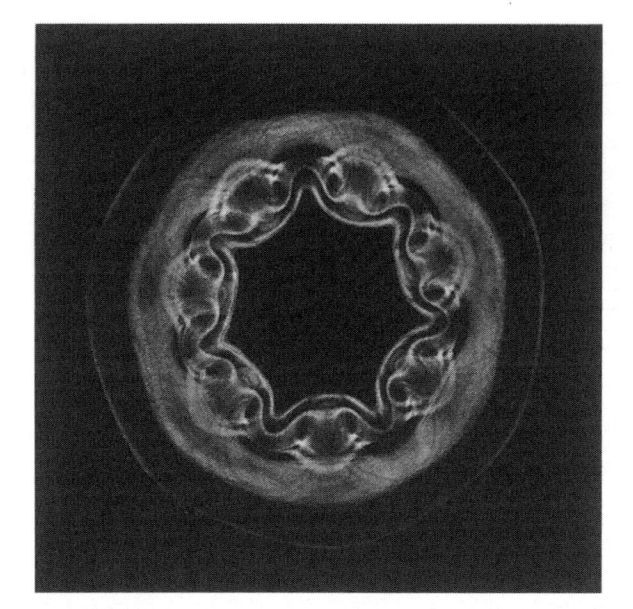

Figure 17

Figure 19

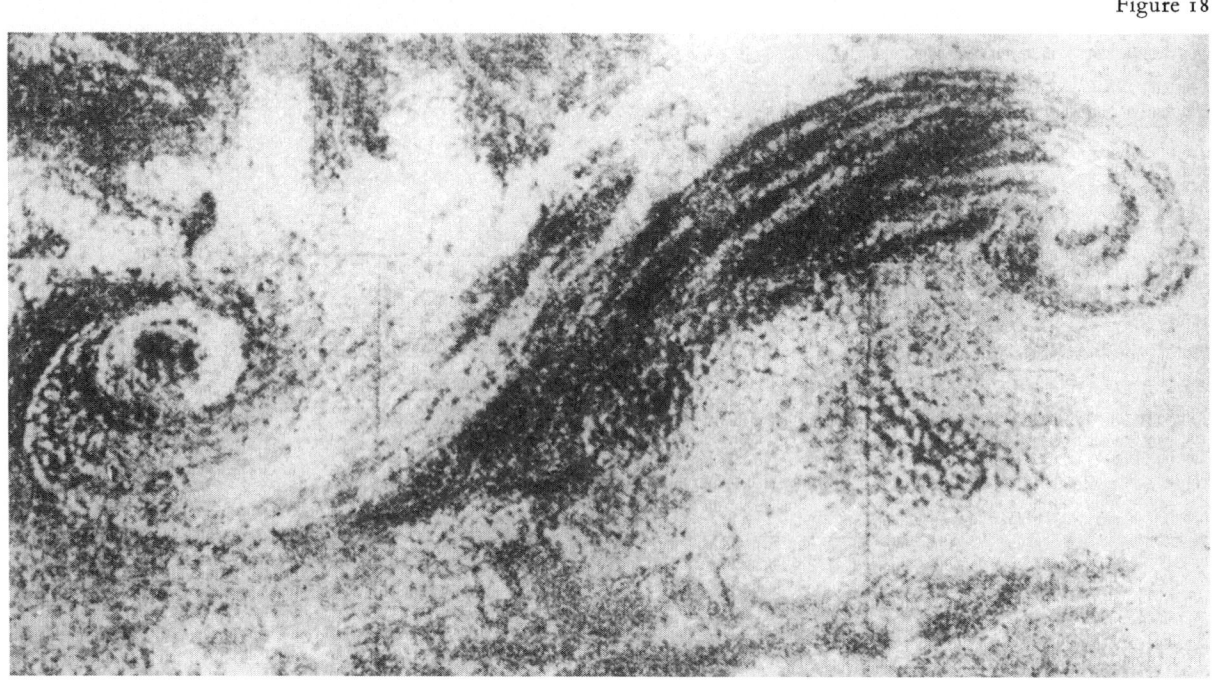

Figure 18

the Hibernian Mysteries of the West, which foreshadowed the Mysteries of the Future, the Druid priest experienced the spiritual Sun in the shadowy spaces of the stones. So the Circle appears upon the symbol of the Cross. It is the *circle from without*. As the plant would teach it today, so the ancient priest experienced in the *seemingly* enclosed and finite space the working of the spiritual forces of the Sun. Think of the hidden, inner world of the developing seed-vessel!

Rudolf Steiner created a Temple Space on Earth, which was built according to the living flow of organic shapes, rather than in the geometrical style of building inspired by rigid forms (1, 55). This building was like a nut-shell. "Just as the shell of the nut has received its form from the same forces which formed the kernel within, and just as one can only experience the nature of the nut-shell according to the nature of the fruit within, so *this shell,* too, had to be an enveloping form to contain all the art and knowledge which pulsates within."

The first Goetheanum was not to have walls which closed it off from the outside; its walls were to give the feeling that they were transparent and open to the widths of space. Standing in this Space, a man felt himself in accord with the whole cosmos. Streaming movement abounded everywhere, all held in balance. The forms grew out of the earth and there was living force of growth in them; the inner Space was flooded with colours and the Heavens looked in from above.

This Space on earth was devoured by flames, so Rudolf Steiner laid the seeds of the Mysteries of the Future in the hearts of human beings, that in the new social forms men are seeking today Sun Spaces may arise. In circles of men the forming forces of the Sun may be received and recognized, if man himself wills it to be so.

Concerning the striving of the mathematicians of the early nineteenth century to overcome the rigid and subjective Kantian conception of space, George Adams writes in *Strahlende Weltgestaltung* that it is clear from the writings of Gauss that he foresaw that in this regard humanity had reached a threshold at which mere intellectual thinking can lead no further. The moment has come when the true reality of Being may only be attained in the free activity of spiritual

Imagination. Space not only exists of necessity, but contains the impulse of freedom. It is the creation of a freedom loving Being. Gauss once uttered the thought that finite man should not presume "to want to consider something infinite as given to him to be encompassed by his ordinary way of understanding". Again, he writes significantly: "I come more and more to the conviction that the essence of our geometry cannot be proved—at least, neither by nor for *human* understanding. Perhaps in another life we shall come to other views about the nature of space, views which are as yet inaccessible to us."

It can be said today that the possibility exists for man even in this life to come to an understanding of the true nature of things not bound by the necessity of external proof—a knowledge to be attained by the individual through the activity of spiritual Imagination.

The times change; the free spirit of man must penetrate new worlds. May coming generations exercise their thought and imagination in the field of modern geometry to throw off the shackles of a concept of space which is rooted in pre-Christian times, in order to be able to perceive and to create future civilizations, future worlds.

In the above-mentioned scientific lectures to teachers, Rudolf Steiner insisted that in our time new beginnings must and are being made towards a truer scientific understanding of the Sun. "This possibility is even present in our time in a most intensive way, where the effort is simply being made to bring analytical geometry and its results into relation with an inner experience of projective geometry. This is certainly only a beginning, but a very, very good beginning."

Let us nurture this good beginning, like a young plant.

Notes and Bibliography

*Reference has been made to all English translations which exist to date in print. In most instances, manuscript translations of Rudolf Steiner's lectures not in print are available for reference in the Library at Rudolf Steiner House, 35 Park Rd., London, N.W.1.

1 George Adams Kaufmann was born as a British subject in Poland on February 8th, 1894 and he died in Birmingham, England, on March 30th, 1963. In 1940 he changed his name to George Adams.
As a student and then research graduate of Christ's College, Cambridge, during the First World War, he became convinced that the dominating, and as he saw it, one-sided manner of thinking which led to the theories of science at that time required a counterbalance. He rejected the deeply rooted monism of scientific thinking, which can only lead to the atomistic theories and recognised that the analytical method alone does not give access adequately to the great variety of phenomena in the living world. He was confirmed by Rudolf Steiner (2)—whom he went to meet at the Goetheanum in 1919—in the direction his scientific thoughts were taking. He very soon abandoned his own scientific career and devoted the rest of his life to the movement inaugurated by Rudolf Steiner. He made remarkable on-the-spot interpretations into English of over 100 of Rudolf Steiner's lectures and did a great deal of translation.
Adams took part wherever possible in the scientific work at the Goetheanum, in the Section for Natural Science led by Guenther Wachsmuth (1893–1963), in the Medical Section led by Ita Wegman (1876–1943) and he was particularly active in the Section for Mathematics and Astronomy, led by Elisabeth Vreede (1879–1943). It was in this time of collaboration with Dr. Vreede that his book *Strahlende Weltgestaltung* was published by the Goetheanum. (Second edition Dornach 1965). Other works during this time are: *Space and the Light of the Creation*, London 1933, *Von dem ätherischen Raume*, Stuttgart 1964, and *Physical and Ethereal Spaces*, London 1965.
In 1947, Adams founded, together with Michael Wilson (researcher in the field of Light and Colour, with special reference to Goethe), the Goethean Science Foundation, Clent and Forest Row, England, an institute for the study of spiritual principles in science. During this period George Adams published, together with Olive Whicher, the following books:
The Living Plant, Stourbridge 1949.
The Plant between Sun and Earth, Stourbridge 1952.

Die Pflanze in Raum und Gegenraum, Stuttgart 1960.

Pflanze, Sonne, Erde (Folder of coloured illustrations with German and English text) Stuttgart 1963.

After the war, Adams worked in connection with the Mathematical-Physical Institute founded in Dornach by George Unger (47) and he also took part in the work arranged at the Goetheanum by Louis Locher-Ernst (46), then leader of the Section for Mathematics and Astronomy.

In 1960, together with Theodor Schwenk (54), with the medical doctor Alexander Leroi (1906–1968), with Georg Unger and other friends, Adams founded the Institut für Strömungswissenschaften in the Verein für Bewegungsforschung (Herrischried, Germany). During the last three years of his life, Adams devoted himself to the problem of the purification and the revitalisation of water by methods derived from a knowledge of spatial-counterspatial forms and forces.

2 Rudolf Steiner, the founder of the anthroposophically orientated spiritual science, was born on February 27th, 1861 in Kraljevec (Austria). He studied in Vienna, became editor of Goethe's Natural Scientific Writings in Kürschner's "Deutscher National-Literatur" in 1883 and from 1890 to 1897 he worked in Weimar at the Sophien Edition of Goethe's works. Then he moved to Berlin and worked there as writer and editor, developing his anthroposophical world conception through his writings and widespread lecture activities in middle and northern Europe, at first in connection with the Theosophical and later the Anthroposophical Societies.

He built the Goetheanum in Dornach (Switzerland), a Free University for Spiritual Science (Laying of the Foundation 1913). It was destroyed by fire in 1922 and on the basis of a new model by Rudolf Steiner was built again in concrete (1925–1928). Rudolf Steiner died in Dornach on March 30th, 1925. Among his main works are:

Philosophie der Freiheit (1894), Dornach 1962 (Eng. tr: *Philosophy of Freedom,* London 1964).

Goethes Weltanschauung (1897), Dornach 1960 (Eng. tr: *Goethe's Conception of the World,* London 1928).

Das Christentum als Mystische Tatsache und die Mysterien des Altertums, (1902), Dornach 1959 (Eng. tr: *Christianity as Mystical Fact,* New York 1961).

Die Geheimwissenschaft im Umriss (1910) Dornach, 1966 (Eng. tr: *Occult Science,* London 1963).

For the reference on page 15 see: *Die Geistige Führung des Menschen und der Menschheit* (1911), Dornach 1963 (Eng. tr: *Spiritual Guidance of Man and Humanity,* New York 1950).

Die Kunst des Erziehens aus dem Erfassen der Menschenwesenheit (7 lectures 1924), Dornach 1963 (Eng. tr: *The Kingdom of Childhood,* London 1964).

3. Dirk J. Struik: *A Concise History of Mathematics*, New York and London 1965.
4. Rudolf Steiner: *Mein Lebensgang* (1923–25), Dornach 1962 (Eng. tr: *The Course of my Life*, New York 1970). Chapter 3.
5. J. L. S. Hatton: *Principles of Projective Geometry*, Cambridge 1913.
6. Rudolf Steiner: *Die Bedeutung der Anthroposophie im Geistesleben der Gegenwart* (6 lectures 1922), Dornach 1957. See also: *Philosophie und Anthroposophie* (essays 1904–1918), Dornach, 1965, first essay.
7. Rudolf Steiner: *Von Seelenrätseln*, 1917 Dornach 1960. *Theosophie*, 1904, Dornach 1961 (Eng. tr: *Theosophy*, London 1970). *Die Kernpunkte der sozialen Frage*, 1919 Dornach 1961 (Eng. tr: *The Threefold Social Order*, New York 1966).
8. Rudolf Steiner: *Erziehungskunst, Methodisch-Didaktisches* (14 lectures 1919), Dornach 1966 (Eng. tr: *Practical Course for Teachers*, London and New York, 1937). *Seminarbesprechungen und Lehrplanvorträge* (1919), Dornach 1969 (Eng. tr: *Discussions with Teachers*, London 1967). *Gegenwärtiges Geistesleben und Erziehung* (14 lectures 1923), Stuttgart 1957 (Eng. tr: *Education and Modern Spiritual Life*, London 1954).
9. Rudolf Steiner suggested to carry out, in the sixth school year, simple drawings in projection and shadow-throwing. The child should have an idea of the way shadows of bodily objects appear on plane and curved surfaces. In the seventh year simple drawings in interpenetrations should be dealt with. The important thing is that shapes should be studied which arise as the result of changing relationships. Then in the eighth school year exercises in perspective drawing should be gradually taken further into the realm of art.
Hermann von Baravalle: *Geometrie als Sprache der Formen*, Stuttgart 1963 (Eng. tr: *Geometrical Drawing and the Waldorf School Plan*, New York 1967). *Perspektive*, Stuttgart 1952 (Eng. tr: *Perspective Drawing*, New York, 1968).
Alexander Strakosch: *Geometrie durch übende Anschauung*, Stuttgart 1962.
10. Zeno of Elea (450 B.C.), pupil of Parmenides, tried to overthrow his master's teaching by creating stories containing paradoxes concerning the relationship between time and space in the world of phenomena. The best known are the race between Achilles and the Tortoise and the story of the Flying Arrow. They show that a finite length can be subdivided into an infinite number of smaller, finite lengths, and call into question the thought of Pythagoras, who described space as a sum of single points. Aristotle showed where the fallacy in Zeno's thought lay.
11. There are three stages of geometry: metrical, affine and projective. The most universal of the three is projective geometry. The other two are derived from projective geometry when certain determining factors are included, which reduce the free, metamorphic forms of projective geometry to fixed forms with some kind of measure.

In affine geometry the fixity of form is partially achieved, in metrical geometry it is complete. There result the two kinds of "non-Euclidean" geometries, which depart from the familiar kind of space which it is natural for us as human beings to think. The relationships of these different geometries may be shown as follows:

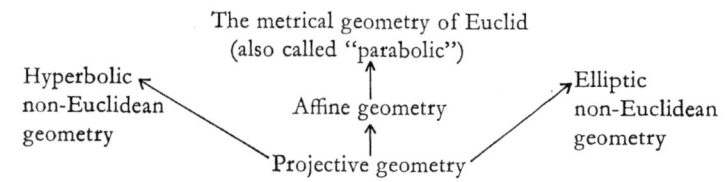

```
                    The metrical geometry of Euclid
                        (also called "parabolic")
                                ↑
Hyperbolic ←                                          → Elliptic
non-Euclidean       Affine geometry                   non-Euclidean
geometry                        ↑                     geometry
                       Projective geometry
```

The lectures of Felix Klein in particular deal with the historical and biographical connections and with the relationship of projective geometry to other fields of modern mathematics: *Nicht-Euklidische Geometrie,* Berlin 1926; *Entwickelung der Mathematik im 19. Jahrhundert,* Berlin 1926; *Elementar-Mathematik vom höheren Standpunkte aus Geometrie,* Berlin 1925 (Translation: *Elementary Geometry from an Advanced Standpoint,* New York, 1932).

Claire Fischer Adler: *Modern Geometry; an integrated first course,* New York 1967.

12 The new geometry came to birth in interplay between east and west. Arthur Cayley, born near London, spent the first seven years of his life in Russia. Gauss, in north Germany, was a friend of Johann Bolyai, who lived in Siebenbürgen. At the same time, but independently, Lobachevski worked in Russia. The Swiss, Jacob Steiner, was foremost in developing the new geometry into actual *synthetic* or *projective* geometry. Poncelet was called to Russia from France and on his return gave his original and seminal work to the mathematicians of the west. Poncelet, Brianchon and Michel Chasles worked at the perspective and polar-reciprocal transformations. Important French geometricians were Monge, Legendre, Carnot. The German Genius was active in the further development of the new concepts; August Ferdinand Moebius, Christian von Staudt, and Felix Klein, together with Arthur Cayley were responsible for the teaching concerning the raying, light-filled archetypal formative principle in space and for the clear concept of the heavenly plane at infinity, the so-called "Absolute" of our metrical space, from which all earthly forms derive their measure. Not to be forgotten are Hermann Grassmann and Julius Plücker and the English mathematician J. J. Sylvester. A. N. Whitehead: *The Axioms of Projective Geometry,* Cambridge 1906. Bertrand Russell and A. N. Whitehead: *Principia Mathematica,* 2nd edition. Bertrand Russell: *Principles of Mathematics,* New York 1937.

13 T. L. Heath: *The Thirteen Books of Euclid's Elements,* Cambridge University Press, New York 1945.

Euclid's fifth postulate is as follows: If a straight line falling on two straight lines makes the interior angles on the same side less than two right angles, the two straight lines, if produced indefinitely, meet on that side on which are the angles less than the two right angles.

14 Strictly speaking, if the radius of a circle becomes infinitely long, the circle changes into the infinitely distant line of its plane, *twice overlaid*. This may be recognised in the projective picture of a family of concentric circles, as in Chapter VI, Figure 41.

15 If the radius of a sphere becomes infinitely long, the result will be: (a) Two planes, namely, one in the finite together with the infinitely distant plane, if a point of the sphere remains in the finite, while its centre moves to infinity. (b) A double-plane = the infinitely distant plane twice overlaid, if the centre of the sphere remains in the finite. This description regarding the sphere must therefore be taken with care; it is above all important to understand that the community of all infinitely distant points has the characteristics of a plane.

16 Rudolf Steiner: *Initiaten Bewusstsein* (11 lectures 1924), Dornach 1960 (Eng. tr: *True and False Paths in Spiritual Investigation*, London 1969). Third lecture.

17 Rudolf Steiner: *Das Michael Mysterium* (*Anthroposophische Leitsätze*) (1924–25), Dornach 1962 (Eng. tr: *The Michael Mystery*, London 1956).

18 *Geometrical terminology.* In this note, we indicate for the benefit of the mathematically trained reader the relation between forms of expression to be found in this work and terminology in general use. It is not our intention simply to discard the old terminology, but we consider it desirable to find new ways of expression in order to overcome the remnants of a one-sidedly pointwise experience of space. It is essential to take seriously the polar formation of archetypal space, and to use words, the quality of which will awaken a feeling for the polar types of spatial experience.

Line. The word "line" is always used in the sense of a straight line and not a curve, for the line is membered of planes as well as of points.

Geometry of a point. For the entity which is polar to a plane with its points and lines, namely, a point with its lines and planes, it is common to use the word "bundle". (Both are two-dimensional entities.) Polar to the geometry of the plane is "the geometry of the bundle"; we speak of the *geometry of the point*. The idea must be awakened that there exist not only extensive, but also *intensive* spaces. The fact that the lines and planes of a point (which are to be thought of in their entirety) are members of the point, is for projective space just as true as its polar opposite, that lines and points are members of the plane. We do not need to use the clumsy word "bundle" but can quite truly speak of the *geometry in a point*.

Circle-curve. "Circle-curve" means any plane curve which can be transformed projectively into a circle. As the name "conic section" applies only to curves of the second degree, i.e., curves which result from plane sections of a second degree cone, we sometimes use the word "circle-curve" as well as "conic section". All real conic sections have the same projective properties as the circle and are related in this sense to the circle.

Guardians. There have been from time to time different names for the elements of a transformation which return into themselves or in the last resort remain at rest. As well as calling them "double points", the words "latent point", "asymptotic point", ∞ point were also used. The word "guardian" brings to expression in a more imaginative way the powerful functional quality of these elements.

Archetypal Space. We call "Archetypal Space" the three-dimensional space of projective geometry, which is free of all metrical rigidity; that is to say, that ideal space whose elements are points, lines and planes, the relationships between which are expressed in the archetypal phenomena (axioms) of community (see *"Strahlende Weltgestaltung"* Chapter 3 and L. Locher-Ernst: *Raum und Gegenraum*, 1. Teil).

Breathing and Circling Involutions. These are normally called Hyperbolic and Elliptic Involutions respectively. See also note 26.

Measures. Concerning the three types of measure, Step Measure, Growth Measure and Circling Measure, see Note 26.

19   The formulation is according to Adams in *Die Pflanze in Raum und Gegenraum* (page 41). Locher says the following in *Urphänomene der Geometrie:* The word *axiom* is ... avoided ... Because I do not feel the edifice of mathematical thoughts to be merely a figment of the brain, but the abstract echo of a world of Being whose creation is the physical sense-world, the expression "Archetypal phenomenon", introduced by Goethe in his *Theory of Colour,* impresses me as being more eloquent, more suited to a reality. Its use in mathematics is also understandable, considering that Goethe certainly used the mathematical method in his evaluation of the world of colour phenomena. He used this method in a wider sense, though just as strictly as researchers into the various fields of fact and phenomena generally use the quantitative mathematical method.

20   Figure 33 is taken from *Strahlende Weltgestaltung* (Figure 108), where the formal proof of the theorem of Desargues is also to be found (p. 303). As formal proofs are not included in this book, we mention here a number of suitable textbooks. (For Locher-Ernst's books, see Note 48):

L. Cremona: *Elements of Projective Geometry,* New York 1960.
J. S. L. Hatton: *The Principles of Projective Geometry,* Cambridge 1913.
L. N. G. Filon: *An Introduction to Projective Geometry,* London 1935.
H. F. Baker: *Principles of Geometry,* Cambridge 1943.
H. S. M. Coxeter: *The Real Projective Plane,* Cambridge 1955; *Projective Geometry,* New York 1963.

W. T. Fishbach: *Euclidean and Projective Geometry,* New York 1962.
K. Doehlemann: *Projektive Geometrie in synthetischer Behandlung,* Berlin 1924.
Theodor Reye: *Die Geometrie der Lage,* Leipzig 1909.
The lectures of Felix Klein (Note 11).
Veblen and Young, *Projective Geometry,* Boston 1938.
Clement V. Durell, *Projective Geometry*, London 1962.
The more modern textbooks of projective geometry usually deal with the subject matter analytically and without illustrations.

21 R. G. Boskovič: *Elemente der Kegelschnittlehre,* Venice 1757.
22 J. V. Poncelet: *Traité des propriétés des Figures,* Paris and Metz 1882.
23 Concerning the expressions "Duality" and "Polarity": the expression "Principle of Duality", which stems from the French school at the beginning of last century, is an unfortunate choice. "Dual" gives the impression of two entities of like kind, side by side with one another; "polar" describes two entities which are of opposite nature but precisely because of this are closely related. We are here concerned with a *polarity*, between which there is a harmonising *third* entity. It would indeed be better to use the word *trinity*. Goethe's concept of polarity, so important for natural science, has also a fundamental meaning in relation to projective geometry. See Adams: *Strahlende Weltgestaltung,* Chapter III, Part I and Locher-Ernst: The preface to *Urphänomene der Geometrie* (47).
24 The conic sections are curves of the second order and class; that is to say, a line in the plane of the curve has two and only two points common with it, and a point only two lines.
25 Christian von Staudt: *Geometrie der Lage,* Nuremberg 1847. J. L. Coolidge: *Geometry of the Complex Domain,* Oxford 1924, Chapter VIII, *The von Standt Theory.*
26 It is usual to speak of "parabolic", "hyperbolic" and "elliptic" measures. Adams introduces simpler names, more characteristic of the actual phenomena expressed by these measures. Parabolic measure is called "Step Measure"; hyperbolic measure is called "Growth Measure"; elliptic measure is called "Circling Measure". Locher-Ernst (*Raum und Gegenraum*) speaks of three archetypal measures or scales, which he calls additive, multiplicative and in the wider sense of the word, periodic. These correspond to step measure, growth measure and circling measure.
27 The mathematician distinguishes the inner proportion from the outer by the sign. As the distances AD and DC are in opposite directions, their proportion is expressed by a negative number. Thus: $\frac{AB}{BC} = -\frac{AD}{DC}$ Therefore the anharmonic ratio is also negative: $H(AC, BD) = \frac{AB}{BC} : \frac{AD}{DC} = -1.$
28 See Adams; *Strahlende Weltgestaltung,* pages 18–22 (Goethe's conception of space.)

29 H. Keller von Asten: *Begegnungen mit dem Unendlichen,* Dornach 1969. (Eng. tr: *Encounters with the Infinite,* Dornach 1970). In the ninth chapter, this kind of construction is given in considerable detail. Although this original book is not in all chapters strictly projective, it nevertheless gives excellent opportunity to experience the mobile quality of projective geometry and to practise the principle of duality.

30 Degeneration of curves. We have seen, for example, how a triangle can degenerate into three lines in a point or three points in a line (Desargues). A sudden transformation of this kind takes place with curves also, as for example with the circle (Note 14). Let one axis of an ellipse extend to the infinite, and the ellipse will degenerate into two parallel lines; when one axis becomes infinitely short, the ellipse is changed into a line twice overlaid.

31 Adams insists repeatedly on the significance of the mobile, functional aspect of the mathematical imaginary for the understanding of the formative processes in universal space. See *Die Pflanze in Raum und Gegenraum,* pp. 72ff, and *Strahlende Weltgestaltung,* Chapter IV.

32 See *Strahlende Weltgestaltung,* p. 54.

33 Rudolf Steiner on the three dimensions: *Der Entstehungsmoment der Naturwissenschaft in der Weltgeschichte,* (10 lectures 1922), Dornach 1969. *Menschenfragen und Weltenantworten* (13 lectures 1922,) Dornach 1969 (Eng. tr: *Human Questions and Cosmic Answers,* London 1967). First lecture. *Die Bedeutung der Anthroposophie im Geistesleben der Gegenwart* (6 lectures 1922), Dornach 1957. Answers to questions.

34 See *Strahlende Weltgestaltung,* p. 97.

35 See *Strahlende Weltgestaltung,* p. 114.

36 See Locher-Ernst: *Projektive Geometrie* (Seite 54) about concave point-fields and convex line-realms; *Raum und Gegenraum* (Seite 47ff) about "Hüllen" and "Kerne".

37 *Die Pflanze in Raum und Gegenraum* p. 52.

The following should be added: Just as curves arise in the plane in their two-fold aspect (linewise and pointwise), so also in space there are space-curves whose organism consists of point, line *and* plane. In every point they have a so-called osculating plane and must therefore be considered both pointwise and planewise. On the basis of the Principle of Duality, such space-curves are derivable both from the plane curve and from the cone.

The plastic surfaces which may be created by means of space curves are significant for the understanding of organic forms. In our elementary geometrical considerations, we have only touched shortly the realm of plastic surfaces, when dealing with the line-woven forms (reguli). See Strahlende Weltgestaltung, p. 219.

38 *Die Pflanze in Raum und Gegenraum,* p. 43.

39 *Die Pflanze in Raum und Gegenraum,* pp. 53, 90ff.

40 *Die Pflanze in Raum und Gegenraum,* p. 129.

41 *Strahlende Weltgestaltung,* Chapter V. Reye: *Geometrie der Lage,* parts 2 and 3. Zindler: *Liniengeometrie (analytical).* Klein: *Elementarmathematik* and *Höhere Geometrie.*

42 *Strahlende Weltgestaltung,* Chapter VIII. A. F. Moebius: *Gesammelten Werke,* Leipzig 1885.

43 Werner Boy: *Abbildung der projektiven Ebene auf eine im Endlichen geschlossene singularitätenfreie Fläche,* Leipzig. See also Hilbert and Cohn-Vossen *Geometry and the Imagination,* New York 1952.

44 Rudolf Steiner: *Geisteswissenschaftliche Behandlung sozialer und pädagogischer Fragen* (lecture of 28th September 1919), Dornach 1964.

45 Rudolf Steiner: *Philosophie und Anthroposophie* (Collected essays 1904–1918), Dornach 1965. First essay.
Concerning the Plant, see also pp. 242–244 of the above; Rudolf Steiner and Ita Wegman: *Grundlegendes fur eine Erweiterung der Heilkunst nach geisteswissenschaftlichen Erkenntnissen* (1925), Arlesheim 1953. (Eng. tr: *Fundamentals of Therapy,* London 1967.)

46 Louis Locher-Ernst: *Urphänomene der Geometrie,* Zurich 1937; *Projektive Geometrie,* Zürich 1940; *Raum und Gegenraum,* Dornach 1957; *Zur mathematischen Erfassung des Gegenraumes,* Mathematisch-Astronomische Blätter, Heft 3, Dornach 1941; *Geometrische Metamorphosen,* Dornach 1970.
Louis Locher-Ernst was born on May 7th, 1906. He studied mathematics, astronomy and physics in the University of Zürich. Throughout his life he was much occupied with music and with the theory of knowledge. He met Rudolf Steiner and began to enter into his world of thought while still a student. In 1932 he was asked to become lecturer in mathematics at the Technical College of Winterthur; later he was made Assistant Director and then Director. He continued until his death to build up the good repute of this establishment of learning and to impress upon it the formative qualities of his powerful personality.
A very gifted teacher of mathematics, Locher-Ernst made an unforgettable impression on his audiences, both at the Technicum and at the Goetheanum, where for many years he had the development of the Section for Mathematics and Astronomy at heart. He knew how to enthuse even the less mathematically minded for his subject, through his own clear and dynamic thought and the masterly way he was able to illustrate on the blackboard while speaking. On August 15th, 1962, just as he was about to leave the Technicum in order to devote himself fully to the work at the Goetheanum, he encountered death suddenly, through a fall in the Alps.

47 Georg Unger (Leader of the Section for Mathematics and Astronomy at the Goetheanum, Dornach): *Das Offenbare Geheimnis des Raumes,* Stuttgart 1963; *Vom Bilden physikalischer Begriffe,* Stuttgart 1959–1961; *Physik am Scheideweg,* Stuttgart 1962.

48 Ernst Lehrs: *Man and Matter,* London 1958; *Mensch und Materie,* Frankfurt 1966.

49 Negative Euclidean space is mentioned among the 27 types of space which may be developed from projective space in an essay by Prof. D. M. Y. Sommerville (Proceedings of the Edinburgh Mathematical Society, Vol. 28, 1910. Felix Klein also draws attention to this type of space in his *Vorträgen über Nicheuklidische Geometrie*. For the mathematically trained reader we add the following: Particularly helpful was the way Cayley and Klein showed how Euclidean and non-Euclidean metrics find their place in the wider field of projective geometry, when the Euclidean metrical geometry is seen as the transition between the hyperbolic and the elliptic non-Euclidean geometries. Basic to hyperbolic geometry and taking the place of its Absolute there is a real surface of the second order, a spheroidal type surface, while the Absolute of elliptical geometry is an imaginary surface. The surface is finite in both cases, i.e. not degenerate. If, however, one allows the surfaces to become infinite, they both transform into a plane containing an imaginary conic, thus transforming into one another. In the moment of transition, the Absolute is an infinitely distant plane with its imaginary circle, fulfilling in fact the necessary conditions for a Euclidean-metrical space.

In the process of degeneration, however, there are not merely one but two moments of transition between the real and the imaginary spheroid. The forms degenerate and transform into one another not only as they grow outward into an infinitely distant plane, but also as they tend inward to a point. Only then is the metamorphic cycle completed. While the plane continues to contain an imaginary circle, the point contains an imaginary cone. This it is which gives the idea of a negative space brought forward here. This is the final consequence of the Cayley-Klein mode of thought.

The important fact is shown clearly by hyperbolic geometry. Here, the real absolute plane already determines the membering of space into a positive field and a negative one. The Lobachevski space is only the predominantly pointwise inner space with respect to the Absolute; outside is a space polar to this—a predominantly planewise space, which is normally left out of account. (Whitehead in his *Universal Algebra* calls it "Anti-Space.") This space vanishes into nothing and is, so to speak, pressed against the wall, when the absolute surface becomes infinite in the outward direction, and what is left over is only the predominantly planewise space now become Euclidean. On the other hand, when the degeneration takes place towards the point within, it is the pointwise inner space which vanishes into nothing, and there remains only the predominantly planewise, negative-Euclidean counterspace with its "all relating point" within.

50 Concerning the polarity between the centric, earthly forces and the universal, ethereal forces, see Rudolf Steiner: 2nd Scientific Course (14 lectures 1920) (Publication planned Dornach 1970); *Das*

Verhältnis der verschiedenen naturwissenschaftlichen Gebiete zur Astronomie (18 lectures 1921), Dornach 1926. *Die Brücke zwischen der Weltgeistigkeit und dem physischen des Menschen* (16 lectures 1920), Dornach 1970, 5th and 6th lectures. Guenther Wachsmuth: *Die ätherischen Bildekräfte in Kosmos, Erde und Mensch,* Dornach 1926. English translation: *Etheric Formative Forces in Cosmos, Earth and Man,* London 1932 and New York.
George Adams: *Physical and Ethereal Spaces,* Chapter IV, *Die Pflanze in Raum und Gegenraum,* § 49. *Universalkräfte in der Mechanik,* Mathematisch-Physikalischen Korrespondenz, Dornach 1956–59.
Ernst Lehrs: *Man and Matter,* Chapter 9.
Rudolf Steiner: *Das Wesen der Farben in Licht und Finsternis,* Dornach 1930. *Colour,* London 1971.

51 Rudolf Steiner: 2nd Scientific Course, 1st and 14th lectures; *Das Verhältnis der verschiedenen naturwissenschaftlichen Gebiete zur Astronomie,* lecture 18; *Adam Kadmon. Der Aufbau der Menschenform aus den Konstellationen und Bewegungen der Sterne,* Dornach 1942. (Eng. tr: in *Man's Life on Earth and in the Spiritual World,* London 1941).

52 *Die Pflanze in Raum und Gegenraum; The Living Plant* (1).

53 The Curves of Cassini, one of which is the Lemniscate of Bernoulli, are well-known as curves of constant product. Each of these curves is the locus of a point, whose distance from one focus grows in the same proportion as its distance from the other focus decreases, so that the product of both distances remains constant.

54 Figure 17 is due to Theodor Schwenk, both of whose books are related to the scientific aims of the present work. *Sensibles Chaos, Strömendes Formenschaffen in Wasser und Luft,* 3rd edition Stuttgart 1968; English translation: *Sensitive Chaos,* London 1971; French translation: *Sensible Chaos,* Paris 1964. *Bewegungsformen des Wassers,* Stuttgart 1967.
The photograph in Figure 18 was contributed by Walther Roggenkamp.
The following books are of value for further study: Goethe: *Die Metamorphose der Pflanzen,* Stuttgart 1966.
Goethe's Botanical Writings, translated by Bertha Mueller, Hawaii 1952.
Gerbert Grohmann: *Die Pflanze,* Stuttgart 1959–68.
Fritz von Bothmer: *Gymnastische Erziehung,* Dornach 1959; English translation: London 1959.
Karl König: *Embryologie und Weltentstehung. Studienmaterial zur Medizin,* Freiburg 1967.
Michael Wilson: *What is Colour?* Stourbridge 1949; Michael Wilson and R. W. Brocklebank: *Goethe's Colour Experiments,* Physical Society Yearbook, London 1958; *Colour Experiments,* Palette, Basel 1970.

55 Rudolf Steiner: *Stilformen des Organisch-Lebendigen,* Vortrag 28.12.1921 (Dornach 1933); *Bilder okkulter Siegel und Säulen,* 1907 (Dornach 1957). In connection with Rudolf Steiner's Seal forms and also the

geometrical secrets of the Goetheanum, see Karl Kemper: *Der Bau,* Stuttgart 1966. The Saturn Seal form on page 217 is a drawing by Karl Kemper reproduced from this book.

Rudolf Steiner's indications concerning a true understanding of the sun and the sun-like qualities of the world appear repeatedly in his lectures of the years 1922, 1923 and 1924; for example, see the collection of 12 lectures in *"Das Sonnen Mysterium und das Mysterium von Tod und Auferstehung"*. See also six lectures given in England in 1922: *Man's Life on Earth and in the Spiritual Worlds,* London, 1952.

Additional textbooks for reference on Projective Geometry (See Note 20):
Blattner: *Projective Plane Geometry,* San Francisco, 1968.
Hartshorne: *Foundations of Projective Geometry,* New York, 1967.
Seidenberg: *Lectures in Projective Geometry,* Van Nostrand, 1963.

ACKNOWLEDGEMENTS

Photographs for the reproduction of the Greek Quadriga and for the "Adoration by the Shepherds" by Georges de la Tour were put at our disposal by *Photographie Girandon,* Paris.

The reproduction of photographs of the two works by *Barbara Hepworth* is by kind permission of the artist.

The rights of reproduction of the Architrave by Rudolf Steiner rest with the *Rudolf Steiner-Nachlassverwaltung.*

Index

Absolute 69, 232, 249 ff, 284
Acer campestre 286
Adams 9, 18, 23, 219, 247, 251, 255, 273, 275, 280
Adler 278
Affine geometry 277 f
Agriculture 240
All-embracing plane 248
All-relating point 248, 284
Analytical geometry, concepts, methods 20, 46 f, 70, 156, 219, 239, 262
Angular measure 56, 134, 195 ff
Anharmonic ratio 115 ff, 281
Anti-gravitational forces 243, 258, 284
Anti-space 250, 284
Aquinas, St. Thomas 44, 141
Archetypal phenomenon of community 77, 186, 233, 280
   plane 69, 247
   circle 232, 247, 284
   (primal) polarity of space 10, 186, 218 ff, 232
   (projective) space 118, 248, 250, 280
Arabism 41 f
Archemedes, Spiral of 31 f
Aristotle 38, 44, 277
Art 10, 16, 44
Asymptote 198 f, 208, 212
Atom, atomic 38, 71, 190
Axioms of Community 77, 82, 186, 233, 280

Baker 280
Bamphoid cusp 205
von Baravalle 277
Bartholommeo 72
Bernoulli 178, 285
Biochemistry, biophysics 242
Bohr 238
Bolyai 48, 278
Boskovic 89, 281
von Bothmer 285
Boy 236 f, 283
Brazil Nut (*Bertholletia excelsa*) 268
Breathing Involution 151 ff, 194, 280
Brianchon 21, 89, 127, 138, 148 ff, 156, 185, 194, 231, 278
Brocklebank 285

Caduceus 268
Canna indica 264
Cardioid 207 f
Carnot 21, 89, 278
Cassini curves and surfaces 270, 285
Cave paintings 26
Cayley 18, 19, 49 f, 69, 278, 284
Centre and periphery 28, 125, 179, 190 ff, 200, 201 f, 216, 243, 250, 257
Centrifugal and centripetal 201, 241, 258
Chasles 20, 202, 278
Chemistry 242
Circle, circle-curve 30, 39 ff, 54, 56, 62, 84 ff, 109, 117, 127 ff, 146 ff, 273, 279
Circling involution 153 ff, 194
Circular points 158
Clifford 141
Cohn-Vossen 283

Collineations 140 ff, 183, 186
Cone 39 f, 220 f, 228 ff, 248, 264, 282
   and circle 187, 220, 230, 254, 282, 284
Conic sections 30 f, 39, 49, 52, 84 ff, 127, 280 f
Conjugate elements 155, 167, 192 ff.
Convolvulus arvensis 268
Coolidge 281
Copernicus 46
Correlations 184, 185 ff, 200
Counterspace 241 ff, 247
Cremona 20, 280
Cross, crossing-point 42, 138, 204 ff, 238, 271
Cross ratio 117 f.
Crystal, crystal lattice 27, 69 ff, 234
Cube 52, 69, 181, 223 ff, 234, 255 f, 270
Cubo-octahedron 224
Cusp 203, 205 ff
Cyclic processes, projectivities 87, 114 ff, 153, 157 ff, 160, 234

Dandelion 268
Dead Nettle (*Lamium album*) 266
Degeneration of forms 32, 147 f, 175, 201, 227, 230, 235, 282, 284
Desargues 47 ff, 88 ff, 239
   Theorem of 21, 79 ff, 172 f, 280
Descartes 47, 50, 78, 87, 219, 239
Diagonal (polar, dual) triangle 120 f, 137 f, 154, 194, 232
Diameter 192, 196 ff

Dimensions of space 27, 43, 66 ff, 92, 167 ff, 180 ff, 219 ff, 238 ff, 255, 282
Dimensionless space 243, 268
Directrix 40, 196
Doehlemann 281
Dodecahedron (pentagon, rhombic) 30, 71, 181 ff, 223, 224
Double elements (invariants) 141, 149 ff, 155 ff, 172, 181, 198, 205, 280
Drawing 28 ff, 32, 51, 206, 216
Druid 273
Duality, Principle of 18, 22, 94 f, 111, 131, 184, 186 ff, 202, 281 f
Durell 281
Dürer 18

Earth (see also Sun and Earth) 53, 70, 243, 246, 265
Education 13 ff, 240
Egypt 33 ff
Einstein 238 f
Elation 167 ff
Ellipse 30, 39 f, 87, 108 ff, 127, 147, 159 ff, 189, 196, 210, 282
Elliptical geometry 278, 284
Envelope 30, 39
Ethereal centre, focus 248 ff, 263 ff
   forces 241
   space 243, 247 ff
Euclid, Euclidean geometry 13, 16 f, 30, 36 ff, 47 ff, 52 ff, 140, 144, 251, 278
   space 239, 247
Expansion and contraction 175, 201, 228, 253
Extensive, intensive 184, 231, 279

Fermat 219
Field Cress (*Lepidium campestre*) 265
Figure-of-eight 230 ff, 268
Filon 280
Fishbach 280
Five-pointed star 214
Flower, flowering process 266 f
Focus, Living 248, 263 ff, 267
Focus and directrix 40, 177, 196
Forces, centric, physical, mechanical 241, 258 f
   ethereal, peripheral, planar 241, 258 f

Four-point, four-line 113 ff
Fruiting process 267 f
Functional infinitude 144
Fundamental theorem 99, 102 ff

Gabo 260
Galileo 13, 87
Gauss 48, 89, 273 f, 278
Geometry in a point 74, 94, 187, 218 f, 279
Gmeindl 10
Goethe 135, 139, 232, 246, 275 f, 280 f, 285
Goethean Science Foundation 11, 275
Goetheanum 243, 261, 273, 275 f, 283, 286
Grassmann 278
Gravity, gravitational field 88, 202, 241, 243, 258
Greece, Greek 36 f, 41
Grohmann 285
Growing point of the plant 264 ff
"Guardian" elements 152, 167, 183, 198, 280

Haeckel 271
Hardy 21
Harmonic forms 115, 144 f
   net 31, 58, 177
   web 123
   relationship 115 ff, 144 f, 191 ff
Harebell (*Campanula rotundiflora*) 266
Hatton 21, 277, 280
Heath 48, 278
Hepworth 260
Hexagon, *Hexagrammum Mysticum* 31 f, **85**, **128**
   net 54 f, 125, 134 ff
Hilbert 283
Homology 167 ff, 181
Horizon line, point 54 ff, 96 ff, 116, 231

Hybernia 273
Hyperbola 30, 39 f, 87, 108, 127, 163, 175, 189, 196 f, 210
Hyperboloid 234 f
Hyperbolic geometry 278, 284

Icosahedron 30, 223
Idea (archetypal) 139, 247
Imaginary in geometry 155 ff, 167, 215, 232, 248, 250, 281 f, 284
Inner and outer 61, 189 f
Infinitesimal calculus 44
Infinity, infinitely distant elements 20, 30, 36, 58 ff, 64 ff, 76, 87, 158, 170
"Infinitude within" 125, 230 ff, 248 ff, 251 ff, 267
Inflexion 205 ff
Intensive and extensive 184, 186, 220
Involution 140, 144 ff, 151 ff, 157 ff, 194

Keller-von Asten 10, 281
Kemper 286
Kepler 46, 49, 52, 78
Klein 20, 48, 278, 283 f
König 285

Lambert 89
Legendre 89, 278
Lehrs 248, 283, 285
Lemniscate, lemniscatory surfaces, spaces 236, 261, 268 ff
Leonardo da Vinci 18, 46
Levity 258
Linear transformations 181, 186
Line 62, 74, 140 ff, 186, 279
   geometry, congruence 233 ff
   -woven net 54 ff, 90, 122, 126 f
Lippold 260
Lobachevski 48, 284
Locher-Ernst 23, 244, 247, 255, 276, 280 ff, 283

Logarithmic spirals, spiral cone 32, 116, 180, 231
Loynes 10

Mallow (*Malva Sylvestris*) 266
Matrix 31, 56, 127, 177 ff
Measure, Circling 116, 167, 280 f
   Growth 39, 116, 125, 141, 167, 172, 175 ff, 268, 280 f
   Step 54, 60, 116, 125, 141, 167, 172, 256, 280 f
Median plane 230, 253, 256 f
Medicine 240
Mercury Staff 268
Metamorphosis 16, 69, 135 f, 207, 210, 216, 224 f, 244, 258, 266, 284
Metrical geometry, systems, relationships 18, 39, 69 f, 94, 116 ff, 218, 277 f, 284
Michelangelo 91
Moebius 58, 236, 278, 283
Monge 89, 278
Mueller 285
Mysteries of the East 33, 271
   West 273

Nautilus pompilius 116
Negative space, negative Euclidean space 247, 250, 258, 284
Newton 87 f
Non-Euclidean geometry 16, 47 ff, 239, 278

Octagram 206
Octahedron 233 ff, 234, 256 f, 270
One-dimensional transformations 146, 172, 177
Order and Class of curves 178, 281
Organism, organism of space 94, 282
Osculating plane 282

Paleolithic period 27
Pappos 103 ff, 148, 234
Parabola 30, 39 f, 87, 108, 127, 163, 168, 175, 189, 196 f, 210
Paraboloid 234
Paradoxa of Zeno 38, 277
Parallelism 32, 47, 69, 134, 254 f
Parallel perspective 223 ff
Parallelepipedon 69
Pascal, Theorem of 21, 49, 52, 78, 84 ff, 127 f, 134 ff, 147, 185, 194, 231
Path Curves 158, 178
Pentagram 130 ff, 181 f
Pentagon, pentagon dodecahedron 30, 181 ff, 223
Perspective, perspectivity 52 f, 95 f, 168 ff, 181, 233
Physical and ethereal spaces 247 ff, 270
Physics 16, 238, 242
Planar forces 92, 241
   organs 265
Planck 70
Plant 240, 262, 264 ff, 283
Plastic perspective 180 ff
Plato, platonic forms 30, 36, 39, 181, 222 ff
Plücker 278
Point, line, plane 52, 73 ff, 77, 186, 188, 248, 282
Point, Geometry in a 74, 94, 187, 218, 279
   of contact 134 f, 188, 194
   of inflexion 205 ff, 186 f
Polar conjugates 193
Polarity, Principle of 10, 18, 94, 184, 186 f, 202, 281
Polar relationships (pole and polar) 92, 139, 185 ff, 188 ff, 194 ff, 201 ff, 216, 218 ff, 239
   reciprocation, transformation 184, 158 ff, 200 ff

triangle (see Diagonal triangle)
Poncelet 19 f, 49, 89, 202, 278, 281
Potentizing process 118, 141 ff, 150 ff, 172
Progression, Arithmetical 60, 116
    Geometrical 116
Projection, projectivity 95, 119, 124, 130 ff, 140 ff, 158 ff, 181, 233
Projective concentric circles 127, 174 ff, 279
Proportion 32, 38, 40, 116 f
Pythagoras 30, 33, 36, 277

Quadrangle, quadrilateral 56 ff, 60 ff, 113 ff, 137, 144 f, 154
    net 58 ff, 125
Quartz crystal 68 f, 235

Rainbow 108
Ranunculus repens 268
Reguli 234 ff, 282
Rembrandt 18, 246
Relativity, Theory of 239
Renaissance 17, 44
Reye 20, 281, 283
Rhododendron 264
Rhombic dodecahedron 244 ff
Riemann 48
Right angle 34, 52, 119, 195 ff, 231, 238
    -angled circling property 154 ff, 178
Roggenkamp 7, 285
Rose (*Rosa canina*) 266
Rumex acetosa 268
Russell 48, 21, 278
Rutherford 238

Satellite 271 f
Saturn 53, 217
Schiller 262
Schwenk 276, 285
"Seal" forms 216, 285 f

Seed, seed-vessel 267 f
Singularities 205 ff, 236
Skew lines 77, 233 f
Sommerville 284
Space, Concept of, Experience of 16, 20, 38, 118, 238, 240 ff, 262 ff, 273 f, 284
    Projective transformations in 180 ff, 221 ff, 233 ff
    and Counterspace 239, 243, 251
Spergula arvensis 268
Sphere 187, 221, 248 ff, 257, 268, 270, 278 f
Spiral forms 31 f, 116, 127, 177 ff, 231, 235, 268, 271
"Spiritual Staff" 232
Square 31 f
Star, star-centre, star-point 248 ff, 255 ff, 267, 271
von Staudt 19 f, 49, 114, 156, 218, 278
Steiner, Jacob 20, 89, 133, 278
Steiner, Rudolf 10, 14 ff, 20, 28 f, 39, 46, 70, 232, 239 ff, 247, 257, 258 ff, 261, 262 ff, 275 ff
Strakosch 277
Struik 16, 44, 277
Sun, sunlike space 247, 251, 262 ff, 273 f, 286
Sun and Earth 262 ff
Surfaces 188, 231, 234 ff, 282

Tangent (line, plane, cone) 39, 128, 134 f, 188 ff, 194, 221
Temple 33, 273
Terminology 9, 219, 279
Tetrahedron 223, 234
Textbooks 117, 280 f
Thales 36
Thirteen Configuration 120 ff, 127, 140, 177
Threefold organism 28, 240

Time, time spaces 16, 239, 268
de la Tour 245
Triangle (Three-point and three-line) 32, 54 ff, 66, 78, 167, 193, 229, 255
Trihedron 229, 255
Trinity 94, 281
Tübingen, Hans von 45
Turner 18
Twelve cycle 160
Two-dimensional geometry, forms 27, 167 ff, 172 ff, 177, 187, 252 ff, 264, 268

Unger 244, 276, 283
Universal forces 241, 284
Unit circle 200 ff

Ur-space (Urraum) 118, 220, 248

Vanishing line, scale 54, 123
Veblen and Young 281
Vegetative shoot 263 ff
"Verticon and horizon" 231 ff, 215
Vortex 271 f, 266
Vreede 243, 275

Wachsmuth 275, 285
Wegman, 243, 275
Whitehead 21, 48, 250, 278
Wilson 275, 285

Zeno von Elea 38, 277
Zindler 283